# Contemporary Endocrinology

**Series Editor**

Contemporary Endocrinology offers an array of titles covering clinical as well as bench research topics of interest to practicing endocrinologists and researchers. Topics include obesity management, androgen excess disorders, stem cells in endocrinology, evidence-based endocrinology, diabetes, genomics and endocrinology, as well as others. Series Editor Leonid Poretsky, MD, is Chief of the Division of Endocrinology and Associate Chairman for Research at Lenox Hill Hospital, and Professor of Medicine at Hofstra North Shore-LIJ School of Medicine.

Dimiter Avtanski
Editor

# Adipose Tissue

## Endocrine Functions, Health Implications, and Future Perspectives

Springer

*Editor*
Dimiter Avtanski
Friedman Diabetes Institute
Northwell Health
New York, NY, USA

Feinstein Institutes for Medical Research
Manhasset, NY, USA

Donald and Barbara Zucker School of Medicine at Hofstra/Northwell
Hempstead, NY, USA

ISSN 2523-3785 ISSN 2523-3793 (electronic)
Contemporary Endocrinology
ISBN 978-3-031-72569-2 ISBN 978-3-031-72570-8 (eBook)
https://doi.org/10.1007/978-3-031-72570-8



This Springer imprint is published by the registered company Springer Nature Switzerland AG
The registered company address is: Gewerbestrasse 11, 6330 Cham, Switzerland

# Preface

The understanding of adipose tissue has undergone a remarkable transformation in recent decades. Once viewed simply as a passive storage depot for excess energy, adipose tissue is now recognized as a dynamic endocrine organ with far-reaching effects on metabolism, inflammation, and overall health. This paradigm shift has profound implications for our understanding of obesity, diabetes, and related metabolic disorders, which collectively represent some of the most pressing public health challenges of our time.

In this volume, we aim to provide a comprehensive overview of adipose tissue as an endocrine organ and examine its complex biology and clinical significance. We explore the interplay between adipose tissue and various physiological systems, shedding light on how adipose dysfunction contributes to the pathogenesis of metabolic diseases that affect millions of people worldwide.

Our journey begins with an examination of the fundamental anatomy and physiology of adipose tissue, including the process of adipogenesis and the crucial roles played by both adipocytes and non-adipocytes. We then explore the metabolic functions of adipose tissue from energy storage and release to its involvement in glucose homeostasis and thermogenesis. A significant portion of this book is dedicated to unraveling the endocrine functions of adipose tissue, detailing the actions of various adipokines and their impact on insulin signaling and metabolic health.

Recognizing the critical importance of translating basic science into clinical practice, we present an in-depth analysis of the clinical implications of adipose tissue dysfunction. Through a series of studies, we illustrate the real-world impact of conditions such as obesity, lipodystrophies, and metabolic syndrome, emphasizing the importance of early detection and intervention. We also discuss the current management strategies, including lifestyle modifications and surgical interventions.

As we look to the future, we examine cutting-edge technological advances in adipose tissue research, including single-cell sequencing, multi-omics analyses, and artificial intelligence. These emerging technologies not only enhance our understanding of adipose tissue biology but also offer exciting possibilities for improved diagnostic and therapeutic approaches.

This book is intended for a diverse audience, including researchers, clinicians, and medical students at all levels. Our goal is to bridge the gap between basic science and clinical practice, facilitating connections that inform both future research directions and current patient care strategies.

The editor would like to express his sincere gratitude to the contributing authors for their exceptional work and to the staff at Springer Nature for their support in bringing this volume to fruition. We hope that this book will serve as a valuable resource in the ongoing efforts to understand and address the complex challenges posed by adipose tissue dysfunction in human health.

Salé, Morocco Dimiter Avtanski

# Contents

# Chapter 1
# Significance of Adipose Tissue as an Endocrine Organ

**Dimiter Avtanski and Radoslav Stojchevski**

## Abbreviations

| | |
|---|---|
| ACTH | Adrenocorticotropic hormone |
| A-FABP (*s.* FABP4) | Adipocyte fatty acid-binding protein |
| AGL (*s.* Lawrence syndrome) | Acquired generalized lipodystrophy |
| AGT | Angiotensinogen |
| AI | Artificial intelligence |
| AMPK | AMP-activated protein kinase |
| Ang II | Angiotensin II |
| ANGPTL-4 | Angiopoietin-like protein-4 |
| APL | Acquired partial lipodystrophy |
| ASC | Adipose-derived stem cell |
| AT | Adipose tissue |
| ATGL | Adipose triglyceride lipase |
| ATM | Adipocyte tissue macrophages |
| BAT | Brown adipose tissue |
| CAA | Cancer-associated adipocyte |

D. Avtanski (✉)
Friedman Diabetes Institute, Northwell Health, New York, NY, USA

Feinstein Institutes for Medical Research, Manhasset, NY, USA

Donald and Barbara Zucker School of Medicine at Hofstra/Northwell, Hempstead, NY, USA
e-mail: davtanski@northwell.edu

R. Stojchevski
Friedman Diabetes Institute, Lenox Hill Hospital, New York, NY, USA

Donald and Barbara Zucker School of Medicine at Hofstra/Northwell, Hempstead, NY, USA

Feinstein Institutes for Medical Research, Manhasset, NY, USA
e-mail: rstojchevski@northwell.edu


D. Avtanski (ed.), *Adipose Tissue*, Contemporary Endocrinology,
https://doi.org/10.1007/978-3-031-72570-8_1

| | |
|---|---|
| CAD | Coronary artery disease |
| CGL | Congenital generalized lipodystrophy |
| CTRP | C1q/TNF-related proteins |
| CVD | Cardiovascular diseases |
| DHEA | Dehydroepiandrosterone |
| ECM | Extracellular matrix |
| EGF | Epidermal growth factor |
| ER | Endoplasmic reticulum |
| EV | Extracellular vesicle |
| Fabp4 (*s.* aP2) | Fatty acid-binding protein 4 |
| FetA | Fetuin-A |
| FFA | Free fatty acid |
| FGF | Fibroblast growth factor |
| FPLD | Familial partial lipodystrophy |
| GIP | Glucose-dependent insulinotropic polypeptide |
| GLP-1 | Glucagon-like peptide-1 |
| hADSC | Human adipose-derived stem cell |
| HGF | Hepatocyte growth factor |
| HIF-1α | Hypoxia-inducible factor-1-alpha |
| HPA | Hypothalamic-pituitary-adrenal (axis) |
| HSD | Hydroxysteroid dehydrogenase |
| HSL | Hormone-sensitive lipase |
| IBD | Inflammatory bowel disease |
| IGF-I | Insulin-like growth factor 1 |
| IL | Interleukin |
| IL-1Ra | Interleukin-1 receptor antagonist |
| ILC | Innate lymphoid cell |
| LCN2 | Lipocalin-2 |
| LH | Luteinizing hormone |
| MCP1 (*s.* CCL2) | Monocyte chemoattractant protein 1 |
| Metrnl | Meteorin-like (protein) |
| MetS | Metabolic syndrome |
| ML | Machine learning |
| MSC | Mesenchymal stem cell |
| NAFLD | Non-alcoholic fatty liver disease |
| NAMPT (*s.* visfatin) | Nicotinamide phosphoribosyltransferase |
| NK | Natural killer (cell) |
| PAI-1 | Plasminogen activator inhibitor 1 |
| PCOS | Polycystic ovarian syndrome |
| PDGF | Platelet-derived growth factor |
| PPARα | Peroxisome proliferator-activated receptor-alpha |
| PVAT | Perivascular adipose tissue |
| RAAS | Renin-angiotensin-aldosterone system |
| RBP4 | Retinol-binding protein 4 |
| RIA | Radioimmunoassay |
| SAA | Serum amyloid A |

| | |
|---|---|
| scRNA-seq | Single-cell RNA sequencing |
| SeP | Selenoprotein P |
| SLE | Systemic lupus erythematosus |
| SNS | Sympathetic nervous system |
| SRD5A2 | Steroid 5 alpha-reductase 2 |
| SVF | Stromal vascular fraction |
| sWAT | Subcutaneous white adipose tissue |
| T2DM | Type 2 diabetes mellitus |
| TGFβ | Transforming growth factor-beta |
| TNF | Tumor necrosis factor |
| UCP1 | Uncoupling protein 1 |
| uPA | Urokinase-type plasminogen activator |
| uPAR | Urokinase-type plasminogen activator receptor |
| UPR | Unfolded protein response |
| VEGF | Vascular endothelial growth factor |
| VLDL | Very-low-density lipoprotein |
| vWAT | Visceral white adipose tissue |
| WAT | White adipose tissue |

## Introduction

Adipose tissue (AT) has long been viewed as a mere energy reservoir and depository for storing excess energy as triglycerides. Contrary to this classical view, modern research has revealed that AT is more than an organ of energy storage. It is a dynamic endocrine organ with complex functions that regulates numerous physiological activities. It plays an important role in lipid and glucose metabolism, inflammation, and vascular homeostasis, releasing a variety of bioactive substances called adipokines, through which it communicates with other parts of the body. AT is today recognized for its role in promoting chronic low-grade inflammation, particularly in the context of obesity. The imbalance between pro- and anti-inflammatory factors secreted by AT contributes to the development of metabolic disorders such as insulin resistance, type 2 diabetes mellitus (T2DM), and cardiovascular diseases (CVDs).

This chapter introduces the endocrine function of AT, with particular reference to its diverse range of adipokines and how they contribute to metabolic, inflammatory, and systemic signaling activities. It also discusses how the deregulation of AT can be associated with the development of obesity and metabolic disorders; CVDs; endocrine, neurodegenerative, and immune diseases; and cancer. We attempted to outline the overall complexity of AT, highlighting its importance far beyond fat storage and how this understanding could be beneficial for developing new therapeutic targets and treatment principles for metabolic disorders.

## Changing Perceptions of Adipose Tissue

Throughout the twentieth century, AT was viewed solely as a passive depot for energy storage in the form of triglycerides [1]. However, the remarkable progress in biochemistry, particularly with the invention of electron microscopy and the development of new techniques for cell isolation, uncovered intricate details of the AT anatomy and functions beyond those of passive energy storage. The introduction of radioimmunoassay (RIA) by Rosalyn Sussman Yalow and Solomon Berson in the 1950s [2] allowed for the quantification of hormones (such as insulin) circulating in AT. A milestone in AT research was the discovery of leptin in 1994 by Jeffrey M. Friedman [3], who recognized AT functions beyond passive energy storage and sparked research in the area. AT research has gained even greater momentum with the advancement in molecular biology and genetics, particularly with the advent of single-cell technologies, enabling researchers to analyze the whole genome, transcriptomics, proteomics, and metabolomics at the individual cell level [4]. One notable finding in AT research was the identification of significant amounts of brown AT in adult humans. This discovery opened new perspectives for therapeutic interventions aiming to increase energy expenditure in combating obesity and metabolic disorders.

Today, we comprehend AT as an active endocrine organ that secretes a myriad of factors playing diverse roles in lipid and glucose metabolism, inflammation, and vascular homeostasis. Through a dense network of nerves and vessels, AT is in constant communication with other organs in the control of homeostasis.

## Adipose Tissue-Derived Endocrine Factors

Morphologically, AT can be classified into two main types: white adipose tissue (WAT) and brown adipose tissue (BAT), both of which are spatially distributed as separate depots possessing different secretory profiles and functions [5]. WAT is the predominant type of body fat and is divided into two significant subgroups: subcutaneous WAT (sWAT) stored beneath the skin and visceral WAT (vWAT) surrounding organs in the abdomen [6]. sWAT itself adds up to approximately 80% of the total body fat and serves as the main lipid storage [7]. It also plays a role in heat insulation, protection against infections, and mechanical stress [8]. On the other hand, vWAT is not typically present in large quantities in healthy individuals but is very active metabolically and continuously releases free fatty acids (FFAs) into the bloodstream. However, in a state of obesity, excessive fat accumulates in the vWAT and other ectopic locations, such as around the liver, heart, blood vessels, digestive organs, and kidneys, leading to insulin overproduction and resistance, inflammation, and arterial fat buildup [9, 10].

Although present in minimal amounts in adult humans, BAT is currently appreciated as an important regulator of metabolic health. In adults, BAT is mostly distributed in the upper back, above the clavicles, and around the vertebrae [11]. It derives its color from the high concentration of mitochondria within fat cells. The primary function of BAT is related to thermogenesis and energy expenditure through the conversion of FFAs into ATP via uncoupling protein 1 (UCP1)-facilitated oxidative phosphorylation [12]. Similar to WAT, BAT is endocrine-active and characteristically secretes factors such as fibroblast growth factors (FGFs) and irisin [13, 14].

Besides classical WAT and BAT, a third type of AT (*beige* or *brite*) residing within WAT and comprising cells sharing characteristics of both WAT and BAT adipocytes has been identified. This tissue can successfully adapt to environmental stimuli and represents a promising target for metabolic interventions.

AT is an organ composed of adipocytes and a stromal vascular fraction (SVF) that includes preadipocytes, fibroblasts, and vascular smooth muscle, endothelial, and immune cells [15]. These components produce numerous communication molecules and regulators of metabolism that can be grouped into several major categories: adipokines, inflammatory and immunomodulatory components, growth and differentiation factors, metabolic regulators, and steroid hormones [16, 17]. Some AT-derived endocrine factors are listed in Table 1.1. Adipokines (such as leptin, adiponectin, and resistin) are a heterogeneous group of proteins produced

**Table 1.1** Adipose tissue-derived endocrine factors and their association with obesity, type 2 diabetes mellitus, and cardiovascular diseases

| Name | Obesity-associated changes in levels | Impact on T2DM | Impact on CVD | References |
|---|---|---|---|---|
| Adiponectin | Decreased | Enhances insulin sensitivity, reduces inflammation | Protective | [18–25] |
| Adipsin | Decreased | May promote insulin resistance | Limited data | [26] |
| Ang II | Increased | Contributes to insulin resistance | Promotes vasoconstriction, contributes to hypertension and oxidative stress | [27, 28] |
| Apelin | Increased | Decreases glucose metabolism, enhances glucose uptake, improves insulin sensitivity, reduces inflammation | Decreases blood pressure | [29–34] |
| Chemerin | Increased | Associated with insulin resistance and inflammation | Promotes atherosclerosis, links to MetS | [35–44] |
| CTRP1 | Limited data | Potential benefit | Potential benefit | [45–51] |
| CTRP3 | Decreased | Improves insulin sensitivity, reduces inflammation | Protects against atherosclerosis | [52–59] |
| CTRP5 | Increased | Potential benefit | Potential benefit | [46, 47, 60–62] |
| CTRP9 | Variable | Improves insulin sensitivity, reduces inflammation | Protects against atherosclerosis | [63–69] |

(continued)

**Table 1.1** (continued)

| Name | Obesity-associated changes in levels | Impact on T2DM | Impact on CVD | References |
|---|---|---|---|---|
| IL-1β | Increased | Impairs insulin signaling and β-cell function | Contributes to atherosclerosis and plaque instability | [70, 71] |
| IL-6 | Increased | Contributes to insulin resistance and dyslipidemia | Promotes atherosclerosis and inflammation | [72–77] |
| Leptin | Increased | May increase risk | Increases risk | [78–84] |
| LCN2 | Increased | May worsen insulin resistance and inflammation | Associated with atherosclerosis and heart failure | [85, 86] |
| MCP-1 | Increased | Recruits inflammatory cells, promotes insulin resistance | Contributes to atherosclerosis | [87, 88] |
| Metrnl | Variable | Anti-insulin-resistant effects | Potential benefit | [89–92] |
| Nesfatin-1 | Variable | May be protective, potential role in glucose regulation | Potential role in blood pressure regulation, heart rate, sympathetic nerve activity | [93–95] |
| Omentin | Decreased | Improves insulin sensitivity, reduces inflammation | Protects against atherosclerosis and vascular dysfunction | [96–100] |
| Progranulin | Increased | May increase risk | Complex/mixed | [101–105] |
| PAI-1 | Increased | Contributes to insulin resistance, impairs fibrinolysis | Increases the risk of thrombosis, linked to CVD risk | [106, 107] |
| Resistin | Increased | May increase risk, may promote insulin resistance | May increase risk | [108–117] |
| SAA | Increased | Impairs insulin signaling, promotes inflammation | Marker of CVD risk, may promote atherosclerosis | [96, 118–120] |
| TNF | Increased | Promotes insulin resistance and inflammation | Contributes to endothelial dysfunction, atherosclerosis, and heart failure | [72, 121–128] |
| Vaspin | Variable | Mixed, improves insulin sensitivity | Potential protective role | [129–135] |
| Visfatin | Variable | Mixed | Mixed | [136–142] |

*Ang II* angiotensinogen, *CTRP* C1q/TNF-related protein, *CVD* cardiovascular diseases, *IL* interleukin, *LCN2* lipocalin-2, *MetS* metabolic syndrome, *MCP1* monocyte chemoattractant protein 1, *MetS* metabolic syndrome, *PAI-1* plasminogen activator inhibitor-1, *SAA* serum amyloid A, *T2DM* type 2 diabetes mellitus, *TNF* tumor necrosis factor particularly by the adipocytes [3, 143]. The inflammatory and immune-modulatory factors include various cytokines, chemokines, complement factors, and acute-phase proteins that play roles in the modulation of the inflammatory responses [144, 145]. AT also releases enzymes involved in lipid metabolism and inflammation, as well as lipid-based signaling molecules such as FFAs and triglycerides [146, 147]. The growth factors produced by AT regulate angiogenesis, adipogenesis, and tissue remodeling [148, 149]. Furthermore, AT is a significant source of steroid hormones, including estrogens, androgens, and glucocorticoids [150, 151]. The following sections discuss these categories of AT-derived compounds in more detail.

## *Adipokines*

Adipokines are a large group of loosely classified signaling molecules unified by their production in the AT. These include substances with diverse chemical structures and roles in the control of energy balance, metabolism, and insulin action. Although the major source of adipokines is the adipocytes, other components of SVF, including preadipocytes, endothelial cells, fibroblasts, and immune cells (particularly macrophages), also contribute to adipokine production, albeit to a lesser extent [1]. Classical members of the adipokine group comprise leptin, adiponectin, resistin, and visfatin.

Initially discovered as a regulator of appetite by causing satiety, leptin is today considered a pleiotropic protein associated with energy homeostasis and immunity. Its main roles encompass controlling lipid turnover, energy expenditure, and the sustenance of glucose levels [152, 153]. It also activates the immune system [154, 155], modulates bone metabolism (particularly osteocalcin function) [156], and regulates reproduction [157]. Leptin acts as an adiposity signal, communicating with the brain in a negative feedback loop [156, 158, 159]. In contrast, adiponectin has opposing effects on balancing whole-body glucose and lipid regulation. It improves insulin sensitivity of the liver and skeletal muscles [160, 161] and regulates hepatic fatty acid oxidation [162, 163]. Resistin, gaining its name as a hormone that causes insulin resistance in peripheral tissues [108], is now also seen as a multifunctional hormone with various effects on inflammation, lipid, and vascular function. Resistin levels strongly correlate with the level of adiposity, peaking in obese individuals, and are used as predictive markers for metabolic syndrome (MetS) and T2DM [109]. Visfatin is another adipokine associated with metabolic diseases and is positively correlated with obesity indices [164, 165]. Other significant but less well-studied adipokines include apelin, omentin, vaspin, C1q/TNF-related protein (CTRPs) family members (CTRP3 and CTRP9), nicotinamide phosphoribosyltransferase (NAMPT, also known as visfatin), and retinol-binding protein 4 (RBP4).

The adipokine secretion profile varies significantly among different AT depots. For instance, sWAT produces higher levels of leptin and adiponectin than vWAT. In contrast, vWAT secretes more pro-inflammatory adipokines, such as interleukin (IL)-6 (IL-6), tumor necrosis factor (TNF), and monocyte chemoattractant protein 1

(MCP1, *s.* CCL2). The depot-specific adipokine profile contributes to the distinct metabolic and endocrine effects associated with different body fat distributions. For example, the increased pro-inflammatory adipokine secretion from vWAT is thought to contribute to the higher metabolic risk associated with visceral obesity [1, 166].

## *Inflammatory and Immune-Modulatory Factors*

AT is a significant source of a diverse range of molecules that positively or negatively influence inflammation and immune responses [1]. These molecules can be broadly divided into pro-inflammatory and anti-inflammatory factors [16]. While adipocytes contribute to the production of these inflammatory factors, infiltrating immune cells, particularly macrophages, are major sources of pro-inflammatory cytokines in obese AT [145, 167]. Well-established pro-inflammatory factors secreted by AT include TNF, IL-6, IL-1β, MCP1, and plasminogen activator inhibitor-1 (PAI-1) [144, 168]. Several other molecules with diverse physiological roles have emerged as key regulators of AT inflammation. These include lipocalin-2 (LCN2), chemerin, angiotensin II (Ang II), serum amyloid A (SAA), and adipokines such as adipsin, omentin, vaspin, and CTRP family members [98, 169–175]. The role of these factors in the inflammation of AT and their potential links to obesity-related conditions are currently areas of active investigation [176, 177].

On the other hand, AT also produces anti-inflammatory factors, such as adiponectin, IL-10, and IL-1 receptor antagonist (IL-1Ra), which help to counterbalance pro-inflammatory effects [16]. The balance between anti-inflammatory and pro-inflammatory adipokines is tightly regulated, and its disruption is associated with the pathogenesis of obesity-related complications [178, 179]. The following sections discuss the pro- and anti-inflammatory factors produced by AT in more detail.

### Pro-Inflammatory Factors

Among the pro-inflammatory factors produced by AT, TNF, IL-6, IL-1β, MCP1, PAI-1, LCN2, chemerin, Ang II, SAA, and adipsin play important roles in regulating AT physiology in health and disease [16, 177, 180–184]. AT dysfunction in obesity is associated with hypertrophy (enlarged adipocyte size) and hyperplasia (increased adipocyte numbers). These changes lead to disturbed adipocyte maturation and adipokine production, adipocyte cell death, and infiltration of adipocyte tissue macrophages (ATMs) [185–188]. MCP1 is one of the critical molecules recruiting macrophages to AT, amplifying the inflammatory response [181]. In lean AT, resident macrophages are predominantly of the anti-inflammatory M2 phenotype, while obesity induces a shift toward pro-inflammatory M1 macrophages [189]. The shift toward the production of pro-inflammatory adipokines may cause the progression of chronic low-grade inflammation, which, if persistent, could further lead to dyslipidemia, impaired glucose tolerance of peripheral tissues (such as skeletal muscles and the liver), insulin resistance, T2DM, and CVD [190–195]. Among the pro-inflammatory adipokines, the levels of TNF (produced mainly by ATMs) and IL-6 correlate positively with AT hypertrophy and are important for the initiation of obesity-induced chronic inflammation and impaired insulin signaling [144, 145, 178, 196–198]. TNF and IL-6 can directly interfere with insulin receptor signaling pathways, leading to decreased insulin sensitivity in AT and other metabolic organs [199]. These pro-inflammatory factors act not only locally within AT but also systemically, affecting distant organs, such as the liver, muscle, and vasculature, contributing to whole-body metabolic dysfunction [200].

#### Anti-Inflammatory Factors

In addition to pro-inflammatory factors, AT produces various anti-inflammatory adipokines that are important for maintaining AT homeostasis and preventing the development of obesity-related metabolic disorders [16]. Among these factors, adiponectin, omentin, vaspin, and some CTRP proteins (such as CTRP3 and CTRP9) are well-characterized [201–203]. Adiponectin, the most abundant adipokine, enhances insulin sensitivity, reduces inflammation, and exerts cardioprotective effects [16]. Omentin improves insulin signaling and glucose uptake, whereas vaspin improves insulin sensitivity and reduces food intake [204]. CTRP3 and CTRP9 have emerged as important metabolic regulators with anti-inflammatory properties [205]. These anti-inflammatory adipokines counterbalance the effects of pro-inflammatory factors, helping to resolve inflammation, improve insulin sensitivity, and maintain metabolic homeostasis [206].

The balance between pro- and anti-inflammatory factors in the AT is essential for maintaining metabolic health. In obesity, this balance is often disrupted, with a decrease in the production of anti-inflammatory factors and an increase in the production of factors with pro-inflammatory action [178]. Understanding the roles of anti-inflammatory factors has opened new avenues for potential therapeutic interventions in obesity-related metabolic disorders [206].

### *Growth and Differentiation Factors*

AT is a source of differentiation and growth factors that play a role in maintaining homeostasis, and its dysfunction contributes to the development of obesity-related complications. These factors support AT angiogenesis, growth, and development in both autocrine and paracrine manners while also having systemic effects, particularly in states of obesity [207]. Among these factors, the most prominent are FGFs, vascular endothelial growth factor (VEGF), and transforming growth factor-beta (TGFβ) [208]. FGFs, particularly FGF21, participates in the regulation of glucose and lipid metabolism, VEGF is essential for AT vascularization, which is critical for tissue expansion and metabolic health, while TGFβ is involved in regulating adipocyte differentiation and extracellular matrix (ECM) remodeling [209]. Other important growth factors produced by AT include insulin-like growth factor 1 (IGF-I), which promotes adipocyte differentiation and glucose uptake, and hepatocyte growth factor (HGF), which is involved in adipogenesis and angiogenesis [210].

The growth and differentiation factors are primarily produced by the mesenchymal stem cells (MSCs) within the AT [211, 212], which possess multipotent characteristics that can differentiate into cells of mesodermal origin (such as adipocytes, chondrocytes, and osteoblasts). Additionally, adipose-derived stem cells (ASCs) (a subset of MSCs) can indirectly stimulate angiogenesis and neovascularization through the release of various paracrine factors [213, 214].

The expression and secretion of these growth factors are often dysregulated in obesity. For instance, VEGF levels may be insufficient to support healthy vascularization of the expanding AT, contributing to hypoxia and inflammation [186].

## Angiogenic and Adipogenic Factors

The expansion of AT in obesity requires the formation of new blood vessels (angiogenesis) to transport oxygen and nutrients to the growing tissue [215, 216]. This process is regulated by AT through the production of various pro- and antiangiogenic factors. AT-derived growth factors, including VEGF, FGF21, insulin-like growth factor I (IGF-I), and platelet-derived growth factor (PDGF), promote angiogenesis within AT [148, 186, 207, 217–226] while also playing roles in adipogenesis (differentiation of mature adipocytes from precursor cells) [227]. VEGF is a potent stimulator of endothelial cell proliferation and migration, which is key for forming new blood vessels; FGF21 promotes angiogenesis and regulates glucose and lipid metabolism; and IGF-I stimulates angiogenesis and adipocyte differentiation [209]. In contrast, other factors, such as the serine protease inhibitors serpins (particularly PAI-1) and angiopoietin-like protein-4 (ANGPTL-4), regulate the action of the pro-angiogenic factors, thus contributing to the inhibition of angiogenesis [228, 229]. ANGPTL-4 inhibits angiogenesis by suppressing endothelial cell proliferation and migration, whereas PAI-1 can inhibit angiogenesis by interfering with the proteolytic cascades necessary for blood vessel formation [223]. Besides growth factors, evidence suggests that some adipokines, such as leptin and adiponectin, also regulate various aspects of angiogenesis [230]. Leptin promotes angiogenesis by stimulating endothelial cell tube formation and migration, whereas adiponectin has been shown to have both pro- and antiangiogenic effects depending on the context [148].

In rapidly expanding AT, angiogenesis may not keep pace with tissue growth, leading to areas of hypoxia. This hypoxia can trigger the expression of pro-angiogenic factors such as VEGF but can also contribute to inflammation and insulin resistance if prolonged [186].

**Multifactorial Growth Factors**

Besides the growth factors involved in the regulation of angiogenic processes, AT produces a variety of other growth factors with diverse functions, such as HGF and TGFβ. HGF is involved in insulin signaling, glucose metabolism, β-cell homeostasis, and AT angiogenesis and is a major contributing factor to the pathophysiology of insulin resistance and T2DM [231]. HGF has been shown to induce insulin resistance in adipocytes by activating the PI3K/Akt signaling pathway, which leads to increased glucose uptake and reduced insulin sensitivity. Paradoxically, HGF also exerts protective effects on pancreatic β-cells, highlighting its complex role in metabolic regulation [232]. TGFβ regulates adipose metabolism, inflammation, fibrosis, and angiogenesis. Studies have shown that blocking TGFβ signaling protects against diet-induced obesity and insulin resistance [233–235]. TGFβ expression is negatively associated with AT browning due to its inhibition of the expression of key thermogenic genes, such as UCP1. This suggests that TGFβ signaling may be a potential therapeutic target for increasing energy expenditure in obesity [234]. Epidermal growth factor (EGF) is another multifactorial growth factor with pleiotropic effects on AT. Besides being an angiogenic factor, EGF regulates adipocyte differentiation through its receptor, EGFR. EGF expression is upregulated during adipocyte differentiation, whereas inhibition of EGFR signaling in human adipose-derived stem cells (hADSCs) suppresses adipogenic differentiation [236].

In addition, cross talk exists between these growth factors and adipokines. For example, adiponectin has been shown to modulate HGF expression, while leptin can influence TGFβ signaling, highlighting the complex regulatory network within AT [237, 238].

The complex roles of these growth factors make them potential targets for therapeutic interventions. For instance, inhibition of TGFβ signaling or modulation of HGF activity could potentially improve insulin sensitivity and metabolic health in obesity [239].

## *Metabolic Regulators*

The major function of AT is to maintain metabolism by producing various metabolic regulatory molecules. Besides adipokines discussed earlier, other molecules, such as angiotensinogen (AGT), RBP4, adipose triglyceride lipase (ATGL), hormone-sensitive lipase (HSL), and adipocyte fatty acid-binding protein (A-FABP, also known as FABP4), are essential metabolic regulators exerting a significant impact on adipogenesis and energy homeostasis [240–242]. ATGL initiates lipolysis by hydrolyzing triglycerides to diglycerides, whereas HSL primarily acts on diglycerides to produce monoglycerides. This coordinated action regulates the release of fatty acids from the AT in response to metabolic needs [243].

AGT is a precursor protein of angiotensin II (Ang II), a component of the renin-angiotensin-aldosterone system (RAAS), that regulates fluid balance and blood pressure [244, 245]. AGT also contributes to insulin resistance by interfering with insulin signaling pathways and promoting oxidative stress and inflammation [244, 246, 247]. RBP4 is associated with metabolic diseases and insulin resistance, whereas ATGL and HSL are involved in regulating lipolysis by influencing the release of fatty acids from AT into the bloodstream [248–250]. A-FABP acts as a fatty acid chaperone and hormone that influences lipid homeostasis, inflammation, and thermogenesis [251]. A-FABP not only facilitates fatty acid transport but also interacts with hormone-sensitive lipase to modulate lipolysis. Moreover, it has been implicated in cholesterol trafficking and inflammatory responses [251]. The diverse roles of these metabolic regulators make them potential targets for therapeutic intervention. For example, the inhibition of A-FABP or modulation of RBP4 levels has been explored as a strategy to improve insulin sensitivity and metabolic health in obesity [252].

The metabolic regulatory functions of AT vary depending on its location, with sWAT being more metabolically beneficial than visceral WAT [240]. This difference is evident in obesity, where elevated levels of Ang II associated with increased visceral fat contribute to hypertension and chronic inflammation, further exacerbating metabolic health [242, 246].

## *Steroid Hormones*

AT produces enzymes necessary for de novo steroid biosynthesis from cholesterol and converts steroid precursors into active steroid hormones, including estrogens (estrone, estradiol, and estriol), androgens (androstenedione and testosterone), progestogens (progesterone), and glucocorticoids (cortisol and cortisone) [253–256]. AT also synthesizes dehydroepiandrosterone (DHEA), which can be converted to testosterone, estradiol, or other active sex hormones [257]. The conversion of steroid hormones in AT is catalyzed by locally produced steroidogenic enzymes such as 17β-hydroxysteroid dehydrogenase (HSD17B); cytochrome P450 family members CYP11A1, CYP17A1, and CYP19; hydroxysteroid dehydrogenases HSD3B2 and HSD17B3; and steroid 5 alpha-reductase 2 (SRD5A2) [258–260]. The steroidogenic activity of AT is regulated by various factors, including hormones such as adrenocorticotropic hormone (ACTH) and luteinizing hormone (LH), as well as local factors such as inflammatory cytokines and adipokines. This regulation ensures that steroid production in AT responds to both systemic and local metabolic needs [253].

AT-derived steroid hormones maintain metabolic processes and are important for the normal functioning of adipocytes and gender-related fat depot distribution [254]. The ability of AT to produce and metabolize steroid hormones exemplifies the concept of *intracrinology*, where cells can produce active hormones for their own use or for paracrine effects on neighboring cells without releasing significant amounts into the circulation [261]. The steroid-producing role of AT plays a significant role during the postmenopausal period in women when the main source of estrogen shifts from the ovaries to AT and other tissues [255]. AT-derived estrogens are important for maintaining metabolic health, particularly in postmenopausal women. They also contribute to insulin sensitivity, lipid metabolism, and energy homeostasis. However, excessive estrogen production by AT can lead to adverse metabolic effects [262]. In addition to estrogens, glucocorticoids regulate various aspects of AT function, including adipocyte differentiation, inflammation, oxidative stress, and adipokine expression through the activation of glucocorticoid receptors [263, 264].

Dysregulation of AT-produced steroid hormones has been linked to obesity, insulin resistance, and polycystic ovarian syndrome (PCOS) [265, 266] and contributes to the development and progression of hormone-related cancers, such as breast and endometrial cancers [267, 268]. Understanding steroid hormone production in AT provides opportunities for developing new therapeutic interventions. One such potential option would be modulating the activity of specific steroidogenic enzymes in AT, which could potentially be used to treat hormone-related disorders or metabolic diseases.

## Adipose Tissue and Systemic Signaling

The recognition of AT as an endocrine organ that controls whole-body metabolism and inflammatory responses has led to a reevaluation of the traditional definition of the endocrine system [1, 17, 240, 269, 270]. AT adapts to different physiological states, such as fasting, feeding, exercise, and stress. These adaptations involve changes in the production and secretion of adipokines, which in turn influence the function of other organs [270]. The endocrine function of AI produces hormones and responds to various endocrine signals, creating a complex network of feedback loops [271]. AT is in permanent communication with other organs and tissues with endocrine secretion, such as the hypothalamus, liver, stomach, intestine, and skeletal muscles [178, 180, 272–275]. The liver produces various hormones essential for metabolism and energy homeostasis [276, 277], while the stomach and intestines that secrete hormones, such as ghrelin and incretins, control appetite, energy balance, and glucose homeostasis [278].

The gut microbiome is another important factor that affects AT metabolism, inflammation, and energy balance [279–284]. It influences AT function through multiple mechanisms, including the production of short-chain fatty acids, regulation of bile acid metabolism, and modulation of the intestinal barrier. These interactions can affect AT inflammation, insulin sensitivity, and energy storage capacity [285].

The following sections discuss some of the major organs and tissues with endocrine functions that are closely associated with AT.

### *Hypothalamus*

The hypothalamus is a brain region that regulates appetite, energy balance, and metabolism [286]. The AT and hypothalamus are involved in a complex bidirectional relationship. AT-derived hormones, such as leptin, adiponectin, and resistin, directly modulate hypothalamic function by targeting hypothalamic neurons to regulate feeding behavior and thermogenesis [287–289]. Leptin, often called the *satiety hormone*, acts on the arcuate nucleus of the hypothalamus to suppress appetite and increase energy expenditure [290]. However, in obesity, leptin resistance can develop, leading to a breakdown of this regulatory mechanism [291]. In contrast, during periods of increased energy demand, the hypothalamus can mobilize energy stores by activating the sympathetic nervous system (SNS), which can lead to lipolysis in adipocytes and the release of FFAs in the circulation [292–297]. The sympathetic nervous system not only induces lipolysis but also regulates adipokine secretion and adipocyte differentiation. Chronic sympathetic activation, as observed in obesity, can lead to inflammation and insulin resistance in AT [294].

Disruption of the AT-hypothalamus interaction in obesity and metabolic conditions can lead to impaired hypothalamic function and AT dysfunction. AT dysfunction can affect hypothalamic function through various mechanisms. One such mechanism involves the infiltration of ATMs into the hypothalamus, leading to neuroinflammation and neuronal impairment [298, 299]. Hypothalamic inflammation, which is partly induced by saturated fatty acids and inflammatory cytokines, is now recognized as a key factor in the development of obesity and metabolic syndrome. Inflammation can lead to leptin and insulin resistance in the hypothalamus, further exacerbating metabolic dysfunction [298]. AT-derived extracellular vesicles (EVs) can act on the hypothalamus to modulate energy intake, serving as an additional way of communication between AT and the hypothalamus [300, 301]. The hypothalamic-pituitary-adrenal (HPA) axis is an important pathway through which the hypothalamus regulates AT function [287, 302]. Activation of the HPA axis triggers the release of glucocorticoids, which can influence AT metabolism and distribution [303]. Recent research has shown that the hypothalamus can induce *browning* of WAT through sympathetic activation, leading to increased energy expenditure. This represents another important aspect of the hypothalamus-AT interaction in energy homeostasis [304].

## *Liver*

The liver is a central organ in energy and glucose regulation and serves as a significant site of fat storage [305]. The liver and AT actively exchange signals to maintain their metabolic homeostasis. AT-derived adiponectin, leptin, and resistin target liver function and regulate insulin sensitivity and glucose levels [274]. Decreased serum adiponectin levels in obesity are associated with hepatic adipokine resistance and the development of nonalcoholic fatty liver disease (NAFLD) [306, 307]. Adiponectin improves hepatic insulin sensitivity and shows anti-inflammatory and anti-fibrotic effects in the liver. It acts through AMP-activated protein kinase (AMPK) and peroxisome proliferator-activated receptor-alpha (PPARα) pathways to enhance fatty acid oxidation and reduce gluconeogenesis [308].

In contrast, liver-produced growth factors such as IGF-I and FDF21 influence AT metabolism and insulin sensitivity [309]. FGF21, primarily produced by the liver, acts on AT to enhance glucose uptake, improve insulin sensitivity, and promote the browning of WAT. In turn, AT sensitizes the body to the effects of FGF21, creating a feedback loop between these organs [310]. The liver also produces hepatokines, such as fetuin-A (FetA) and selenoprotein P (SeP), which can influence AT function. For instance, FetA has been shown to impair insulin signaling in adipocytes and promote inflammation in AT [311].

Dysregulation of AT function in obesity state can lead to an increased influx of FFAs to the liver, contributing to the development of NAFLD and other metabolic disorders [312]. The increased influx of FFAs to the liver in obesity overwhelms the liver's capacity for fatty acid oxidation and very-low-density lipoprotein (VLDL) secretion. This leads to hepatic lipid accumulation, which triggers inflammation and oxidative stress, key features of NAFLD progression [312]. Besides its interaction with AT, the liver engages in cross talk with other organs and tissues, including the pancreas, skeletal muscles, and brain, to maintain overall metabolic homeostasis [313].

### *Pancreas*

The pancreas is a vital organ in both endocrine and exocrine systems. The pancreas secretes hormones, such as insulin and glucagon, which are essential for regulating blood glucose levels [314]. Insulin resistance occurs when tissues become less responsive to insulin, leading to decreased glucose uptake in the muscle and AT and impaired suppression of hepatic glucose production. This results in elevated blood glucose levels and compensatory hyperinsulinemia, which can eventually lead to β-cell exhaustion and T2DM [315]. Glucagon, produced by pancreatic α-cells, acts as a counter-regulatory hormone to insulin. It promotes hepatic glucose production and lipolysis in the AT. A balance between insulin and glucagon is essential for maintaining glucose homeostasis [316].

Adipokines secreted by AT can influence pancreatic function, thereby affecting insulin sensitivity [317]. Leptin regulates pancreatic function by promoting cell proliferation, glucose metabolism, and insulin secretion [318, 319]. Resistin affects various aspects of pancreatic function by impairing insulin action and contributing to insulin resistance [108]. On the other hand, adiponectin exerts its effects on the pancreas by promoting β-cell regeneration and improving pancreatic islet lipid metabolism [320]. IL-6 and TNF also impact pancreatic function and contribute to insulin resistance [144, 317, 321].

The incretin hormones glucagon-like peptide-1 (GLP-1) and glucose-dependent insulinotropic polypeptide (GIP), synthesized by the gut in response to food intake, enhance insulin secretion and sensitivity. Interestingly, AT has been shown to express receptors for these incretins, suggesting a direct link between incretin activity and AT function [322].

The expansion of AT and the resulting hypoxia in obesity leads to the release of adipocyte-specific fatty acid-binding protein 4 (Fabp4, *s.* aP2) and other factors, which can stimulate acute insulin secretion from β-cells [323, 324]. In the early stages of obesity, pancreatic β-cells often compensate for insulin resistance by increasing insulin secretion. This compensation is supported by factors released from expanding AT, including Fabp4. However, chronic overstimulation can lead to β-cell dysfunction and eventual failure, contributing to T2DM [325].

## *Skeletal Muscles*

AT and skeletal muscles interact extensively through the secretion of adipokines and myokines (cytokines and other peptides released by skeletal muscle cells that act in an endocrine, paracrine, and autocrine manner) to maintain metabolic homeostasis and regulate insulin sensitivity [326–331]. The cross talk between AT and skeletal muscles represents a complex endocrine loop where adipokines and myokines mutually influence each other's production and function [329]. AT-produced adiponectin enhances insulin sensitivity in skeletal muscles, promotes glucose uptake, and improves metabolic control [332]. Additionally, adiponectin and leptin can activate AMPK, thereby increasing skeletal muscle fatty acid oxidation [333, 334]. Conversely, myokines produced by the skeletal muscles can affect AT function. For example, irisin may increase the thermogenic capacity of WAT by promoting "browning," a process characterized by the expression of UCP1 and increased mitochondrial activity [335, 336]. IL-6 can exert both pro- and anti-inflammatory effects on AT and stimulate AT lipolysis [337–340]. Other important myokines include myostatin, which inhibits muscle growth and influences AT metabolism, and meteorin-like (Metrnl), which promotes AT browning and improves insulin sensitivity [341].

AT dysfunction affects the cross talk between AT and skeletal muscles by increasing the influence of resistin and other pro-inflammatory cytokines on skeletal muscle insulin resistance, potentially leading to the development of metabolic disorders [108, 144, 342]. Intramuscular AT, which increases with obesity and aging, can directly affect muscle function and insulin sensitivity. Ectopic fat deposition can lead to local inflammation and impaired insulin signaling in muscle cells [343]. Exercise significantly alters AT-skeletal muscle cross talk as it increases the production of beneficial myokines, enhances AT insulin sensitivity, and promotes a favorable adipokine profile. Exercise-induced changes contribute to improved metabolic health and highlight the importance of physical activity in maintaining balance between these tissues [344].

## *Immune System*

AT plays a key role in the initiation and modulation of both the innate and adaptive immune responses [345, 346]. AT hosts a diverse repertoire of immune cells, including macrophages, T and B cells, natural killer (NK) cells, innate lymphoid cells (ILCs), and eosinophils, which, alongside adipocytes, produce a variety of pro- and anti-inflammatory cytokines [347]. The balance and phenotype of these cells change with obesity, contributing to the inflammatory state [348]. Adipokines, such as leptin and adiponectin, directly modulate immune cell function. Leptin promotes T-cell activation and inflammatory cytokine production, whereas adiponectin has anti-inflammatory effects, suppressing macrophage activation and promoting the generation of anti-inflammatory M2 macrophages [183].

Under healthy conditions, AT cytokines are maintained in tight balance, supporting normal immune responses against infections. In lean AT, macrophages are predominantly of the M2 anti-inflammatory phenotype; however, obesity induces a shift toward the M1 pro-inflammatory phenotype [167]. AT dysfunction leads to the infiltration and accumulation of immune cells within AT, promoting the development of low-grade chronic inflammation. This altered inflammatory state affects immune system function and triggers insulin resistance [349–351]. Chronic low-grade inflammation associated with obesity is often referred to as *metabolic inflammation* or *metaflammation*. This type of inflammation is unique because it is triggered by metabolic signals and involves many of the same molecular pathways activated during classical inflammation [195]. Conversely, in cases of malnutrition, the shrinking of AT diminishes immune system function, thus increasing susceptibility to infections [352–354].

## *Endothelium*

AT interacts with the endothelium through endocrine and paracrine mechanisms and plays a role in angiogenesis and vascular function [148]. Many adipokines, such as TNF, IL-6, and VEGF, act as pro-angiogenic factors, while others, such as serpins, including vaspin and PAI-1, have antiangiogenic functions [355, 356]. Adiponectin promotes endothelial nitric oxide production, suppresses endothelial cell activation, and inhibits vascular smooth muscle cell proliferation, protecting against atherosclerosis [16].

TNF is a potent inducer of VEGF expression, one of the most powerful drivers of angiogenesis, promoting endothelial cell survival and proliferation [357, 358]. TNF also induces the expression of other pro-angiogenic factors, such as basic fibroblast growth factor (bFGF) and IL-8 [359, 360]. IL-6 also stimulates VEGF production by directly influencing endothelial cell proliferation, migration, and tube formation. It is important to note that TNF and IL-6 may also exert antiangiogenic effects, depending on the specific context [361]. In obesity, rapid AT expansion can outpace angiogenesis, leading to areas of hypoxia. Hypoxia triggers the expression of hypoxia-inducible factor-1-alpha (HIF-1$\alpha$), which in turn upregulates VEGF and other pro-angiogenic factors. However, chronic hypoxia can also contribute to inflammation and insulin resistance in AT [362].

PAI-1 is a primary inhibitor of urokinase-type plasminogen activator (PA), which activates plasmin, an enzyme that degrades the ECM [363]. Decreased ECM breakdown creates a physical barrier that hinders endothelial cell movement, which is crucial for angiogenesis [364]. Conversely, PAI-1 can bind to the uPA receptor (uPAR), triggering a pro-angiogenic response through the modulation of cell adhesion and signaling pathways [365, 366]. Perivascular AT (PVAT), which surrounds most blood vessels, plays a unique role in vascular function. PVAT produces vasoactive substances that directly influence vascular tone and inflammation. In obesity, PVAT dysfunction contributes to vascular insulin resistance and endothelial dysfunction [367].

Serpins play diverse roles in cellular processes, particularly in reducing oxidative stress and improving endothelial cell survival [135]. They also suppress vascular inflammation, a major driver of vascular dysfunction and diseases such as atherosclerosis [368]. Moreover, serpins are involved in the regulation of fibrinolysis, and their overexpression can contribute to atherogenic changes [369].

The interaction between AT and endothelium has significant implications for metabolic health. Dysregulation of this interaction, as observed in obesity, contributes to endothelial dysfunction, a key early step in the development of atherosclerosis and other cardiovascular diseases. Understanding these interactions may reveal new therapeutic targets for obesity-related vascular complications [370].

## Adipose Tissue Dysfunction

### *Adipose Tissue Hypertrophy*

AT expansion occurs through hyperplasia (increase in cell number) or hypertrophy (increase in cell size) [371, 372]. The number of adipocytes in fat depots stabilizes during neonatal development and early childhood. Later in life, hypertrophic expansion typically occurs because of persistent overnutrition [373], as existing adipocytes capture and retain excessive circulating lipids [374]. As a result of an imbalance between energy intake and expenditure, AT initially expands because of subcutaneous hyperplasia. Overnutrition can also activate WAT browning in response to increased energy expenditure; however, this adaptive response is often insufficient to counteract the negative effects of AT hypertrophy [375]. After this process reaches its limit, hypertrophy occurs, which leads to AT dysfunction. Enlarged adipocytes have a reduced capacity for lipid storage. Consequently, excessive fat accumulates in visceral and ectopic sites, including the skeletal muscles, liver, and pancreas [376]. Hyperplasia is limited in conditions of extreme hypertrophic obesity, and further hypertrophic expansion causes lipotoxicity—a MetS that contributes to the onset of obesity-related diseases [377–379]. Adipocyte hypertrophy in obesity is manifested by pathological AT remodeling, which includes deregulated adipo- and angiogenesis, hypoxia, and adipocyte death. Dysregulated adipogenesis affects normal adipocyte secretion and, together with the increased infiltration of ATMs, leads to the development of chronic low-grade inflammation manifested by increased expression of pro-inflammatory factors (such as TNF, IL-6, and leptin) and decreased expression of anti-inflammatory factors (adiponectin and others) [186, 380, 381]. This dramatic change in AT links AT hypertrophy to many pathophysiological conditions, including dyslipidemia, insulin resistance, T2DM, endothelial dysfunction, CVD, and NAFLD [382–384].

#### Insulin Resistance and Diabetes

AT dysfunction, in the context of obesity-induced hypertrophy, is the main driver for the development of insulin resistance and T2DM. As summarized from the previous sections, this is mediated through several mechanisms: (1) *altered adipokine secretion* (hypertrophic adipocytes exhibit a shift in their adipokine profile toward the production of pro-inflammatory cytokines, leading to a chronic low-grade inflammatory state) [16]; (2) *ectopic lipid accumulation* (as adipocytes reach their storage capacity, excess lipids are stored in non-AT such as the liver, skeletal muscle, and pancreas, which impairs insulin signaling in these tissues) [385]; (3) *chronic low-grade inflammation* (as mentioned above, the resulting chronic low-grade inflammation caused by dysfunctional adipocytes leads to activation of stress-response signaling pathways, such as JNK and IKK-β/NF-κB, which interfere with insulin signaling) [195]; (4) *FFA influx* (hypertrophic adipocytes exhibit increased lipolysis, leading to elevated circulating FFAs, which can directly induce insulin resistance in the skeletal muscle and liver, and contribute to pancreatic β-cell dysfunction) [386]; (5) *AT hypoxia* (rapid AT expansion can outpace angiogenesis, resulting in areas of hypoxia, which can trigger the expression of pro-inflammatory factors and contribute to insulin resistance) [362]; and (6) *endoplasmic reticulum (ER) stress* (adipocyte hypertrophy induces ER stress, which can activate the unfolded protein response (UPR); chronic UPR activation can lead to insulin resistance through different mechanisms, including inhibition of insulin receptor signaling) [387].

The progression from insulin resistance to T2DM occurs when pancreatic β-cells can no longer compensate for the reduced insulin sensitivity by increasing insulin production. This β-cell failure is partly because of the lipotoxic and glucotoxic effects of chronic hyperglycemia and hyperlipidemia, as well as the impact of pro-inflammatory cytokines produced by dysfunctional AT [388].

#### Cardiovascular Diseases

AT dysfunction is also associated with the pathogenesis of various CVDs, including hypertension, atherosclerosis, coronary artery disease (CAD), and heart failure [273]. AT dysfunction contributes to CVDs through multiple mechanisms, including systemic inflammation, altered lipid metabolism, insulin resistance, and endothelial dysfunction. Excess visceral adiposity is associated with increased cardiovascular risk [389]. Among the cytokines produced by AT, adiponectin has particularly been shown to exert cardioprotective effects by maintaining endothelial function, reducing inflammation, and modulating lipid metabolism [390]. Adiponectin exerts its cardioprotective effects by activating AMPK and other signaling pathways. This leads to improved endothelial function, reduced vascular inflammation, and enhanced insulin sensitivity [390]. In contrast, high levels of pro-inflammatory adipokines, such as leptin and resistin, are positively associated with the risk of cardiovascular events [391]. Other adipokines also play a role in cardiovascular health. For example, omentin-1 has anti-inflammatory and vasodilatory effects, whereas chemerin and visfatin are associated with increased inflammation and atherosclerosis risk [392]. Recent research has highlighted the existence of an *AT-myocardial axis*, in which dysfunctional AT, particularly epicardial fat, can directly influence cardiac function through paracrine effects [393].

### Endocrine Disorders

Through endocrine secretion, AT also affects the function of other endocrine organs. AT dysfunction can disrupt the hypothalamic-pituitary-gonadal axis, alter steroid hormone metabolism, and influence thyroid hormone function through the secretion of various adipokines and inflammatory mediators [240]. Thus, AT dysfunction may lead to the development of endocrine disorders such as PCOS, Cushing's syndrome, hypogonadism, or thyroid dysfunction [394–397]. In PCOS, AT dysfunction contributes to insulin resistance and hyperandrogenism. Excess visceral adiposity leads to increased production of inflammatory cytokines and altered adipokine secretion, particularly decreased adiponectin and increased leptin levels, exacerbating the metabolic and reproductive features of PCOS [398]. Consequently, these conditions may worsen the chronic inflammatory state, alter adipokine secretion, and further contribute to metabolic disorders. The relationship between AT dysfunction and endocrine disorders is bidirectional. For instance, although obesity can contribute to thyroid dysfunction, thyroid hormones also play a crucial role in regulating AT metabolism and differentiation [399]. The interplay between AT dysfunction and endocrine disorders can create a vicious cycle, potentially leading to long-term health consequences, such as an increased risk of T2DM, CVD, and certain cancers.

## *Lipodystrophy*

Lipodystrophies are a rare group of conditions characterized by abnormal fat distribution, i.e., loss of fat in some body parts and abnormal gain in others. They can be present at birth (congenital) or develop later in life and can affect the whole body (generalized) or only certain areas (partial) [400]. Lipodystrophies result from genetic mutations affecting adipocyte development, function, or survival or can be acquired due to autoimmune processes or medication side effects. Loss of functional AT leads to ectopic fat deposition in other organs, contributing to metabolic complications [401]. These conditions arise from factors such as impaired differentiation of fat cells and disrupted function or premature death of fat cells [402]. Lipodystrophies are often associated with dyslipidemia, insulin resistance, diabetes, hypertriglyceridemia, fatty liver disease, and CVD [400, 403].

Leptin deficiency is a hallmark of generalized lipodystrophy. Leptin replacement therapy has shown promising results in managing metabolic complications in patients with lipodystrophy, demonstrating the role of leptin in metabolic regulation [404]. Some of the known lipodystrophies are congenital generalized lipodystrophy (CGL), also known as Berardinelli-Seip syndrome, familial partial lipodystrophy (FPLD) or Köbberling–Dunnigan syndrome, acquired generalized lipodystrophy (AGL, also known as Lawrence syndrome), and acquired partial lipodystrophy (APL) or Barraquer-Simons syndrome. The extent of metabolic derangements related to these syndromes is positively correlated with the degree of lipoatrophy [405]. HIV-associated lipodystrophy, a common form of acquired partial lipodystrophy, is associated with antiretroviral therapies. It is characterized by peripheral fat loss and central fat accumulation, leading to metabolic complications similar to those observed in other forms of lipodystrophy.

## *Other Pathologies Related to Adipose Tissue Dysfunction*

### Immune Diseases

Given that AT is a source of numerous pro- and anti-inflammatory factors, dysfunction within AT may directly influence immune responses and contribute to immune diseases, such as inflammatory bowel disease (IBD), systemic lupus erythematosus (SLE), and rheumatoid arthritis [406]. Dysfunctional AT, particularly mesenteric fat, can contribute to the pathogenesis of IBD. Hypertrophied mesenteric adipocytes produce increased levels of TNF and IL-6, which can exacerbate intestinal inflammation [407]. In SLE, AT dysfunction may lead to disease progression through increased leptin production, which can promote autoantibody production and T-cell activation. Conversely, chronic inflammation in SLE can induce AT dysfunction, creating a potential feedback loop [408]. In rheumatoid arthritis, the AT surrounding the joints (infrapatellar fat pad) can become a source of pro-inflammatory adipokines and cytokines, contributing to joint inflammation and cartilage degradation. In addition, systemic AT dysfunction can exacerbate the inflammatory state in rheumatoid arthritis [409].

AT dysfunction can alter immune responses through multiple mechanisms, including the dysregulated production of adipokines and pro-inflammatory cytokines, changes in immune cell populations within AT, and alterations in systemic metabolic parameters that influence immune function [406].

### Neurodegenerative Disorders

Adipokines also contribute to the pathogenesis of various neurodegenerative disorders. Leptin may affect neuronal apoptosis and exert neuroprotective effects by modulating inflammation and oxidative stress. Adipokines can influence neurodegenerative processes through multiple pathways, including modulation of neuroinflammation, oxidative stress, and insulin signaling in the brain. They can also affect the blood-brain barrier permeability and neuronal energy metabolism [410]. As a result, AT dysfunction may contribute to the pathogenesis of neurodegenerative disorders, such as Parkinson's disease, multiple sclerosis, and Alzheimer's disease [411]. In Parkinson's disease, leptin has been shown to exert neuroprotective effects on dopaminergic neurons, potentially through the activation of neuroprotective signaling pathways and reduction of oxidative stress [412]. In multiple sclerosis, adipokines may induce disease progression via their effects on immune function and inflammation. For example, leptin can promote T-cell proliferation and pro-inflammatory cytokine production, potentially exacerbating autoimmune responses. Conversely, adiponectin may have anti-inflammatory effects that could be beneficial [413]. Altered levels of adipokines have been observed in Alzheimer's disease. For example, low levels of adiponectin are associated with an increased risk of cognitive decline, possibly owing to its anti-inflammatory and insulin-sensitizing effects. Conversely, elevated leptin levels may have neuroprotective effects, although leptin resistance in obesity may negate these benefits [414].

Modulating adipokine signaling or targeting their downstream effects could potentially offer novel neuroprotective strategies for these conditions [415].

### Cancer

AT dysfunction may also contribute to the development and progression of different types of cancer. The mechanisms underlying this association include chronic inflammation, altered hormone signaling (especially in hormone-dependent cancers), dysregulated adipokine production, and changes in the tumor microenvironment [416]. Obesity is linked to the increased incidence and worsening outcomes of multiple cancers, including prostate, colon, endometrial, and premenopausal breast cancers [417]. Studies have demonstrated that AT-derived pro-inflammatory cytokines, such as leptin or resistin, positively influence cancer cell proliferation, migration, and invasion potential [418–420]. Leptin has been shown to promote cancer cell proliferation, migration, and invasion of various cancer types, often through the activation of JAK/STAT, PI3K/AKT, and MAPK. Conversely, adiponectin has anti-tumor effects, inhibits cell proliferation, and promotes apoptosis [421]. Recent research has highlighted the role of cancer-associated adipocytes (CAAs), which can be reprogrammed by cancer cells to support tumor growth. CAAs provide energy-rich fatty acids to cancer cells, secrete pro-inflammatory factors, and contribute to chemoresistance [422]. Further, obesity-induced inflammation and insulin resistance create a favorable microenvironment for tumor growth and metastasis [423].

## Future Research Directions and Clinical Implications

AT research has seen a dramatic shift in recent years, revealing a complex role beyond fat storage. However, many questions regarding the endocrine activity of AT remain unanswered, necessitating further investigation. An immediate imperative is detailed characterization of the AT secretome and identification of new adipokines. Simultaneously, a comprehensive analysis of known cytokines and mechanistic studies are needed to identify the signaling pathways involved in their actions. Defying the precise molecular mechanisms of action is important to determining the role of these factors in the pathogenesis of obesity-related metabolic disorders [424]. Future research should also explore the role of AT dysfunction in age-related metabolic decline and age-related diseases [425].

Another avenue that requires further exploration is the interaction between AT and other endocrine organs. As discussed above, AT has in a dynamic relationship with other endocrine organs, including the hypothalamus, liver, skeletal muscle, and pancreas. Further exploration is needed to define the precise mechanisms that mediate this dynamic communication. Recent studies have demonstrated the importance of AT-derived EVs in regulating systemic insulin sensitivity and glucose homeostasis [426, 427]. Further studies are needed to clarify the endocrine role of EVs in AT.

Obesity is a growing global issue that will continue to challenge humanity in the future. Research on the pathogenesis of obesity-related comorbidities will likely expand the understanding of the molecular mechanisms underlying these associations. These findings will eventually lead to the identification of new targets and strategies for the prevention and treatment of obesity and coexisting conditions. Research into methods to induce "browning" of WAT, increasing energy expenditure, represents a promising aspect of obesity treatment. Future studies should focus on identifying safe and effective ways to promote this process [375]. The interaction between the gut microbiome and AT function is an emerging area of research, and studies on how modulation of the gut microbiome influences adipokine production and AT metabolism would offer new therapeutic approaches [428]. It is also important to investigate the effects of environmental factors, such as increasing amounts of waste, climate change, and higher concentrations of pollutants and endocrine-disrupting chemicals on AT, especially its endocrine function [429].

Novel techniques that provide an unprecedented view of AT at single-cell and molecular levels are being developed, enabling better understanding of AT physiology and potentially offering new individualized therapies [4]. Obesity can be treated with novel pharmacological interventions, including adipokine receptor agonists and antagonists, which can modulate adipokine signaling and improve metabolic health [430]. Ultimately, exploring the therapeutic potential of ATMs and their secretome could lead to the development of cell-based therapies for the treatment of metabolic disorders and other comorbidities associated with obesity [431]. The application of artificial intelligence (AI) and machine learning (ML) to analyze large-scale AT data could potentially reveal new patterns and relationships affecting AT in health and disease [432].

## Conclusion

In this chapter, we briefly introduce the main molecular factors produced by AT, which will be further discussed in more detail in subsequent chapters. We also explored the interaction of AT with other organs and tissues, as well as the clinical implications of our improved understanding of AT physiology for developing novel treatments and individualized therapies for metabolic and cardiovascular diseases. Rapid advancements in technologies such as single-cell RNA sequencing (scRNA-seq), metabolomics, and advanced imaging techniques can uncover new therapeutic options to address the increasing prevalence of metabolic disorders that currently affect a significant portion of the world population. However, perhaps the most crucial principle is to look back to our evolutionary roots and appreciate a lifestyle that maintains a balance with our environment, emphasizing the importance of diet, physical activity, and stress management in maintaining AT health and overall metabolic well-being. As our understanding of AT as an endocrine organ continues to evolve, it promises to revolutionize our approach for treating and preventing obesity-related disorders. The implications of this research extend beyond individual health to global public health strategies, emphasizing the need for interdisciplinary approaches to tackle the worldwide challenges of metabolic disorders.

## References


1. Kershaw EE, Flier JS. Adipose tissue as an endocrine organ. J Clin Endocrinol Metab. 2004;89(6):2548–56. https://doi.org/10.1210/jc.2004-0395.
2. Yalow RS, Berson SA. Immunoassay of endogenous plasma insulin in man. J Clin Invest. 1960;39(7):1157–75. https://doi.org/10.1172/JCI104130.
3. Zhang Y, Proenca R, Maffei M, Barone M, Leopold L, Friedman JM. Positional cloning of the mouse obese gene and its human homologue. Nature. 1994;372(6505):425–32. https://doi.org/10.1038/372425a0.
4. Avtanski D, Hadzi-Petrushev N, Josifovska S, Mladenov M, Reddy V. Emerging technologies in adipose tissue research. Adipocyte. 2023;12(1):2248673. https://doi.org/10.1080/21623945.2023.2248673.
5. Szablewski L, editor. Adipose tissue—an update. London: IntechOpen; 2019. https://doi.org/10.5772/intechopen.80248.
6. Reyes-Farias M, Fos-Domenech J, Serra D, Herrero L, Sánchez-Infantes D. White adipose tissue dysfunction in obesity and aging. Biochem Pharmacol. 2021;192:114723. https://doi.org/10.1016/j.bcp.2021.114723.
7. Reddy P, Lent-Schochet D, Ramakrishnan N, McLaughlin M, Jialal I. Metabolic syndrome is an inflammatory disorder: A conspiracy between adipose tissue and phagocytes. Clin Chim Acta. 2019;496:35–44. https://doi.org/10.1016/j.cca.2019.06.019.
8. Kwok KHM, Lam KSL, Xu A. Heterogeneity of white adipose tissue: molecular basis and clinical implications. Exp Mol Med. 2016;48(3):e215. https://doi.org/10.1038/emm.2016.5.
9. Björntorp P. 'Portal' adipose tissue as a generator of risk factors for cardiovascular disease and diabetes. Arteriosclerosis. 1990;10(4):493–6. https://doi.org/10.1161/01.ATV.10.4.493.
10. Chait A, den Hartigh LJ. Adipose tissue distribution, inflammation and its metabolic consequences, including diabetes and cardiovascular disease. Front Cardiovasc Med. 2020;7:22. https://doi.org/10.3389/fcvm.2020.00022.
11. Jung SM, Sanchez-Gurmaches J, Guertin DA. Brown adipose tissue development and metabolism. In: Handb Exp Pharmacol, vol. 251; 2018. p. 3–36. https://doi.org/10.1007/164_2018_168.
12. Cannon B, Nedergaard J. Brown adipose tissue: function and physiological significance. Physiol Rev. 2004;84(1):277–359. https://doi.org/10.1152/physrev.00015.2003.
13. Townsend K, Tseng Y-H. Brown adipose tissue. Adipocyte. 2012;1(1):13–24. https://doi.org/10.4161/adip.18951.

14. Poretsky L, et al. The effects of irisin and leptin on steroidogenic enzyme gene expression in human granulosa cells: in vitro studies. Metab Open. 2023;17:100230. https://doi.org/10.1016/j.metop.2023.100230.
15. De Francesco F, Ricci G, D'Andrea F, Nicoletti GF, Ferraro GA. Human adipose stem cells: from bench to bedside. Tissue Eng Part B Rev. 2015;21(6):572–84. https://doi.org/10.1089/ten.teb.2014.0608.
16. Ouchi N, Parker JL, Lugus JJ, Walsh K. Adipokines in inflammation and metabolic disease. Nat Rev Immunol. 2011;11(2):85–97. https://doi.org/10.1038/nri2921.
17. Galic S, Oakhill JS, Steinberg GR. Adipose tissue as an endocrine organ. Mol Cell Endocrinol. 2010;316(2):129–39. https://doi.org/10.1016/j.mce.2009.08.018.
18. Achari A, Jain S. Adiponectin, a therapeutic target for obesity, diabetes, and endothelial dysfunction. Int J Mol Sci. 2017;18(6):1321. https://doi.org/10.3390/ijms18061321.
19. Wang Y, Lam KSL, Yau M, Xu A. Post-translational modifications of adiponectin: mechanisms and functional implications. Biochem J. 2008;409(3):623–33. https://doi.org/10.1042/BJ20071492.
20. Ruan H, Dong LQ. Adiponectin signaling and function in insulin target tissues. J Mol Cell Biol. 2016;8(2):101–9. https://doi.org/10.1093/jmcb/mjw014.
21. Woodward L, Akoumianakis I, Antoniades C. Unravelling the adiponectin paradox: novel roles of adiponectin in the regulation of cardiovascular disease. Br J Pharmacol. 2017;174(22):4007–20. https://doi.org/10.1111/bph.13619.
22. Naot D, Musson DS, Cornish J. The activity of adiponectin in bone. Calcif Tissue Int. 2017;100(5):486–99. https://doi.org/10.1007/s00223-016-0216-5.
23. Elsalem L, et al. Genetic variation in the ADIPOQ gene and serum adiponectin increase the risk of bladder cancer. J Appl Biomed. 2022;20(3):106–13. https://doi.org/10.32725/jab.2022.012.
24. Liu Y, et al. Metabolomic profiling in liver of adiponectin-knockout mice uncovers lysophospholipid metabolism as an important target of adiponectin action. Biochem J. 2015;469(1):71–82. https://doi.org/10.1042/BJ20141455.
25. Krumm CS, et al. Fibroblast growth factor-21 (FGF21) administration to early-lactating dairy cows. I. Effects on signaling and indices of insulin action. J Dairy Sci. 2019;102(12):11586–96. https://doi.org/10.3168/jds.2019-16695.
26. Schiffmann R, Banwell B. Brain MRI and motor function in leukodystrophies. Neurology. 2016;87(8):748–9. https://doi.org/10.1212/WNL.0000000000003017.
27. Khataibeh M, Abu-Alruz K, Al-Widyan O, Abu-Samak M, Al-Qudah J. Combined supplementation of soy and garlic modulate biochemical parameters of 7,12-dimethylbenz[alpha]anthracene induced mammary cancer in female albino rats. Pak J Biol Sci. 2007;10(14):2308–13. https://doi.org/10.3923/pjbs.2007.2308.2313.
28. Ibáñez-Escribano A, Nogal-Ruiz JJ, Gómez-Barrio A, Arán VJ, Escario JA. In vitro trichomonacidal activity and preliminary in silico chemometric studies of 5-nitroindazolin-3-one and 3-alkoxy-5-nitroindazole derivatives. Parasitology. 2016;143(1):34–40. https://doi.org/10.1017/S0031182015001419.
29. Mughal A, Sun C, O'Rourke ST. Activation of large conductance, calcium-activated potassium channels by nitric oxide mediates apelin-induced relaxation of isolated rat coronary arteries. J Pharmacol Exp Ther. 2018;366(2):265–73. https://doi.org/10.1124/jpet.118.248682.
30. Masri B, Morin N, Pedebernade L, Knibiehler B, Audigier Y. The apelin receptor is coupled to Gi1 or Gi2 protein and is differentially desensitized by apelin fragments. J Biol Chem. 2006;281(27):18317–26. https://doi.org/10.1074/jbc.M600606200.
31. Zhang X, Hu W, Feng F, Xu J, Wu F. Apelin-13 protects against myocardial infarction-induced myocardial fibrosis. Mol Med Rep. 2016;13(6):5262–8. https://doi.org/10.3892/mmr.2016.5163.
32. Pitkin SL, Maguire JJ, Kuc RE, Davenport AP. Modulation of the apelin/APJ system in heart failure and atherosclerosis in man. Br J Pharmacol. 2010;160(7):1785–95. https://doi.org/10.1111/j.1476-5381.2010.00821.x.

33. Habchi M, et al. Circulating <scp>A</scp> pelin is increased in patients with type 1 or type 2 diabetes and is associated with better glycaemic control. Clin Endocrinol (Oxf). 2014;81(5):696–701. https://doi.org/10.1111/cen.12404.
34. Dray C, et al. Apelin and APJ regulation in adipose tissue and skeletal muscle of type 2 diabetic mice and humans. Am J Physiol Metab. 2010;298(6):E1161–9. https://doi.org/10.1152/ajpendo.00598.2009.
35. Schuster M, Rosanowski F, Schwarz R, Eysholdt U, Lohscheller J. Quantitative detection of substitute voice generator during phonation in patients undergoing laryngectomy. Arch Otolaryngol Head Neck Surg. 2005;131(11):945–52. https://doi.org/10.1001/archotol.131.11.945.
36. Bergstrom HC, McDonald CG, Dey S, Fernandez GM, Johnson LR. Neurons activated during fear memory consolidation and reconsolidation are mapped to a common and new topography in the lateral amygdala. Brain Topogr. 2013;26(3):468–78. https://doi.org/10.1007/s10548-012-0266-6.
37. Buechler C, Feder S, Haberl EM, Aslanidis C. Chemerin isoforms and activity in obesity. Int J Mol Sci. 2019;20(5):1128. https://doi.org/10.3390/ijms20051128.
38. Xiong W, et al. Chemerin stimulates vascular smooth muscle cell proliferation and carotid neointimal hyperplasia by activating mitogen-activated protein kinase signaling. PLoS One. 2016;11(10):e0165305. https://doi.org/10.1371/journal.pone.0165305.
39. Dupont J, Reverchon M, Cloix L, Froment P, Ramé C. Involvement of adipokines, AMPK, PI3K and the PPAR signaling pathways in ovarian follicle development and cancer. Int J Dev Biol. 2012;56:959–67. https://doi.org/10.1387/ijdb.120134jd.
40. Peng L, et al. Chemerin regulates the proliferation and migration of pulmonary arterial smooth muscle cells via the ERK1/2 signaling pathway. Front Pharmacol. 2022;13:767705. https://doi.org/10.3389/fphar.2022.767705.
41. Shang J, et al. Chemerin/ChemR23 axis promotes inflammation of glomerular endothelial cells in diabetic nephropathy. J Cell Mol Med. 2019;23(5):3417–28. https://doi.org/10.1111/jcmm.14237.
42. Ye Y, et al. Chemerin/ChemR23 regulates cementoblast function and tooth resorption in mice via inflammatory factors. J Periodontol. 2021;92(10):1470–82. https://doi.org/10.1002/JPER.20-0675.
43. Yu M, et al. Targeting the chemerin/CMKLR1 axis by small molecule antagonist α-NETA mitigates endometriosis progression. Front Pharmacol. 2022;13:985618. https://doi.org/10.3389/fphar.2022.985618.
44. Jiang Y, Liu P, Jiao W, Meng J, Feng J. Gax suppresses chemerin/CMKLR1-induced preadipocyte biofunctions through the inhibition of Akt/mTOR and ERK signaling pathways. J Cell Physiol. 2018;233(1):572–86. https://doi.org/10.1002/jcp.25918.
45. Ren M, Pan J, Yu X, Chang K, Yuan X, Zhang C. CTRP1 prevents high fat diet-induced obesity and improves glucose homeostasis in obese and STZ-induced diabetic mice. J Transl Med. 2022;20(1):449. https://doi.org/10.1186/s12967-022-03672-5.
46. Xie Y, et al. C1q complement/tumor necrosis factor-associated proteins in cardiovascular disease and COVID-19. Proteomes. 2021;9(1):12. https://doi.org/10.3390/proteomes9010012.
47. Zhang Z-Z, Wang G, Yin S-H, Yu X-H. C1q tumor necrosis factor–related protein 1: a promising therapeutic target for atherosclerosis. J Cardiovasc Pharmacol. 2022;79(3):273–80. https://doi.org/10.1097/FJC.0000000000001186.
48. Rodriguez S, Lei X, Petersen PS, Tan SY, Little HC, Wong GW. Loss of CTRP1 disrupts glucose and lipid homeostasis. Am J Physiol Metab. 2016;311(4):E678–97. https://doi.org/10.1152/ajpendo.00087.2016.
49. Yuasa D, et al. C1q/TNF-related protein-1 functions to protect against acute ischemic injury in the heart. FASEB J. 2016;30(3):1065–75. https://doi.org/10.1096/fj.15-279885.
50. Shen L, Wang S, Ling Y, Liang W. Association of C1q/TNF-related protein-1 (CTRP1) serum levels with coronary artery disease. J Int Med Res. 2019;47(6):2571–9. https://doi.org/10.1177/0300060519847372.

51. Feng S, Su Y, Luo L, Jing F, Yi Q. Serum levels of C1q/tumor necrosis factor-related protein-1 in children with Kawasaki disease. Pediatr Res. 2018;83(5):999–1003. https://doi.org/10.1038/pr.2018.16.
52. Öner Rİ, Karadağ AS. Evaluation of choroidal perfusion changes in obese patients: ocular effects of insulin resistance. Arq Bras Oftalmol. 2018;81(6):461–5. https://doi.org/10.5935/0004-2749.20180088.
53. Peterson JM, Wei Z, Wong GW. C1q/TNF-related Protein-3 (CTRP3), a novel adipokine that regulates hepatic glucose output. J Biol Chem. 2010;285(51):39691–701. https://doi.org/10.1074/jbc.M110.180695.
54. Deng W, et al. Serum C1q/TNF-related protein-3 (CTRP3) levels are decreased in obesity and hypertension and are negatively correlated with parameters of insulin resistance. Diabetol Metab Syndr. 2015;7(1):33. https://doi.org/10.1186/s13098-015-0029-0.
55. Peterson JM, Seldin MM, Wei Z, Aja S, Wong GW. CTRP3 attenuates diet-induced hepatic steatosis by regulating triglyceride metabolism. Am J Physiol Liver Physiol. 2013;305(3):G214–24. https://doi.org/10.1152/ajpgi.00102.2013.
56. Trogen G, Alamian A, Peterson JM. High molecular weight, but not total, CTRP3 levels are associated with serum triglyceride levels. Physiol Rep. 2019;7(23):e14306. https://doi.org/10.14814/phy2.14306.
57. Cai Z, Li S, Yu T, Deng J, Li X, Jin J. Non-coding RNA regulatory network in ischemic stroke. Front Neurol. 2022;13:820858. https://doi.org/10.3389/fneur.2022.820858.
58. Li X, Jiang L, Yang M, Wu Y, Sun J, Sun S. Białko CTRP3 zwiększa wrażliwość na insulinę adipocytów 3T3-L1 przez hamowanie procesu zapalnego i poprawę przekazywania sygnału insulinowego. Endokrynol Pol. 2014;65(4):252–8. https://doi.org/10.5603/EP.2014.0034.
59. Wang Y-J, Zhao J-L, Lau WB, Liu J, Guo R, Ma X-L. Adipose tissue-derived cytokines, CTRPs as biomarkers and therapeutic targets in metabolism and the cardiovascular system. Vessel Plus. 2017;1:202. https://doi.org/10.20517/2574-1209.2017.29.
60. Si Y, Fan W, Sun L. A review of the relationship between CTRP family and coronary artery disease. Curr Atheroscler Rep. 2020;22(6):22. https://doi.org/10.1007/s11883-020-00840-0.
61. Lei X, et al. Loss of CTRP5 improves insulin action and hepatic steatosis. Am J Physiol Metab. 2016;310(11):E1036–52. https://doi.org/10.1152/ajpendo.00010.2016.
62. Zhang C, et al. Circulating complement-1q tumor necrosis factor-α-related protein isoform 5 levels are low in type 2 diabetes patients and reduced by dapagliflozin. J Diabetes Investig. 2020;11(1):88–95. https://doi.org/10.1111/jdi.13069.
63. Sivapalan ST, et al. Plasmonic enhancement of the two photon absorption cross section of an organic chromophore using polyelectrolyte-coated gold nanorods. Langmuir. 2012;28(24):9147–54. https://doi.org/10.1021/la300762k.
64. Yamaguchi S, et al. C1q/TNF-related protein 9 promotes revascularization in response to ischemia via an eNOS-dependent manner. Front Pharmacol. 2020;11:1313. https://doi.org/10.3389/fphar.2020.01313.
65. Zhang L, Ding H, Shi Y, Zhang D, Yang X. CTRP9 decreases high glucose-induced trophoblast cell damage by reducing endoplasmic reticulum stress. Mol Med Rep. 2022;25(5):185. https://doi.org/10.3892/mmr.2022.12701.
66. Han F, et al. C1q/TNF-related protein 9 improves the anti-contractile effects of perivascular adipose tissue via the AMPK—eNOS pathway in diet-induced obese mice. Clin Exp Pharmacol Physiol. 2018;45(1):50–7. https://doi.org/10.1111/1440-1681.12851.
67. Zhang L, et al. C1q/TNF-related protein 9 inhibits THP-1 macrophage foam cell formation by enhancing autophagy. J Cardiovasc Pharmacol. 2018;72(4):167–75. https://doi.org/10.1097/FJC.0000000000000612.
68. Zhou Q, et al. C1q/TNF-related protein-9 is elevated in hypertension and associated with the occurrence of hypertension-related atherogenesis. Cell Biol Int. 2021;45(5):989–1000. https://doi.org/10.1002/cbin.11542.

69. Yang MM, et al. Circulating CTRP9 is associated with severity of systemic sclerosis–associated interstitial lung disease. Arthritis Care Res. 2023;75(1):152–7. https://doi.org/10.1002/acr.24749.
70. Costa MMS, Lima WG, Figueiredo MM, Michalick MSM, Tafuri WL, Tafuri WL. Cervical, mandibular, and parotid lymph nodes of dogs naturally infected with Leishmania infantum: a histopathologic and immunohistochemistry study and its correlation with facial skin lesions. Vet Pathol. 2008;45(5):613–6. https://doi.org/10.1354/vp.45-5-613.
71. Gohel A, Oda M, Katkar AS, Sakai O. Multidetector row computed tomography in maxillofacial imaging. Dent Clin N Am. 2018;62(3):453–65. https://doi.org/10.1016/j.cden.2018.03.005.
72. Bastard J-P, et al. Recent advances in the relationship between obesity, inflammation, and insulin resistance. Eur Cytokine Netw. 2006;17(1):4–12. http://www.ncbi.nlm.nih.gov/pubmed/16613757.
73. Acharya S, Saha S, Pradhan P. Novel symmetry-based gene-gene dissimilarity measures utilizing Gene Ontology: application in gene clustering. Gene. 2018;679:341–51. https://doi.org/10.1016/j.gene.2018.08.062.
74. Schmidt-Arras D, Rose-John S. IL-6 pathway in the liver: from physiopathology to therapy. J Hepatol. 2016;64(6):1403–15. https://doi.org/10.1016/j.jhep.2016.02.004.
75. Cho SH, et al. Age-related increased prevalence of asthma and nasal polyps in chronic rhinosinusitis and its association with altered IL-6 trans-signaling. Am J Respir Cell Mol Biol. 2015;53(5):601–6. https://doi.org/10.1165/rcmb.2015-0207RC.
76. Zhang N, et al. The EGFR pathway is involved in the regulation of PD-L1 expression via the IL-6/JAK/STAT3 signaling pathway in EGFR-mutated non-small cell lung cancer. Int J Oncol. 2016;49(4):1360–8. https://doi.org/10.3892/ijo.2016.3632.
77. Peters AT, et al. Evidence for altered activity of the IL-6 pathway in chronic rhinosinusitis with nasal polyps. J Allergy Clin Immunol. 2010;125(2):397–403.e10. https://doi.org/10.1016/j.jaci.2009.10.072.
78. Cummings BP. Leptin therapy in type 2 diabetes. Diabetes Obes Metab. 2013;15(7):607–12. https://doi.org/10.1111/dom.12048.
79. Mohammadzadeh G, Zarghami N. Serum leptin level is reduced in non-obese subjects with type 2 diabetes. Int J Endocrinol Metab. 2013;11:3–10. https://doi.org/10.5812/ijem.6535.
80. Scarpace PJ, Matheny M, Kirichenko N, Gao YX, Tümer N, Zhang Y. Leptin overexpression in VTA trans-activates the hypothalamus whereas prolonged leptin action in either region cross-desensitizes. Neuropharmacology. 2013;65:90–100. https://doi.org/10.1016/j.neuropharm.2012.09.005.
81. Wauters M, Considine RV, Yudkin JS, Peiffer F, De Leeuw I, Van Gaal LF. Leptin levels in type 2 diabetes: associations with measures of insulin resistance and insulin secretion. Horm Metab Res. 2003;35(2):92–6. https://doi.org/10.1055/s-2003-39054.
82. Frühbeck G. A heliocentric view of leptin. Proc Nutr Soc. 2001;60(3):301–18. https://doi.org/10.1079/PNS200196.
83. Ge H, Huang L, Pourbahrami T, Li C. Generation of soluble leptin receptor by ectodomain shedding of membrane-spanning receptors in vitro and in vivo. J Biol Chem. 2002;277(48):45898–903. https://doi.org/10.1074/jbc.M205825200.
84. Müller G, Ertl J, Gerl M, Preibisch G. Leptin impairs metabolic actions of insulin in isolated rat adipocytes. J Biol Chem. 1997;272(16):10585–93. https://doi.org/10.1074/jbc.272.16.10585.
85. Romei C, Di Prisco S, Raiteri M, Raiteri L. Glycine release provoked by disturbed $Na^+$, $Na^+$ and $Ca^{2+}$ homeostasis in cerebellar nerve endings: roles of $Ca^{2+}$ channels, $Na^+/Ca^{2+}$ exchangers and GlyT2 transporter reversal. J Neurochem. 2011;119(1):50–63. https://doi.org/10.1111/j.1471-4159.2011.07401.x.
86. Kim MK, et al. Associations of dietary calcium intake with metabolic syndrome and bone mineral density among the Korean population: KNHANES 2008-2011. Osteoporos Int. 2017;28(1):299–308. https://doi.org/10.1007/s00198-016-3717-1.

87. Qu W, Hashimoto S-I, Morishita S. Efficient frequency-based de novo short-read clustering for error trimming in next-generation sequencing. Genome Res. 2009;19(7):1309–15. https://doi.org/10.1101/gr.089151.108.
88. Cao S, Du R, Zhou Y. Development of a denitrification system using primary sludge as solid carbon source—potential to couple with anammox process. Sci Total Environ. 2020;737:140315. https://doi.org/10.1016/j.scitotenv.2020.140315.
89. Miao C-Y. Metrnl: a novel secreted protein against insulin resistance. Proc Annu Meet Jpn Pharmacol Soc. 2018;WCP2018:PO1-2-70. https://doi.org/10.1254/jpssuppl.WCP2018.0_PO1-2-70.
90. Jung TW, et al. METRNL attenuates lipid-induced inflammation and insulin resistance via AMPK or PPARδ-dependent pathways in skeletal muscle of mice. Exp Mol Med. 2018;50(9):1–11. https://doi.org/10.1038/s12276-018-0147-5.
91. Miao Z, Hu W, Li Z, Miao C. Involvement of the secreted protein Metrnl in human diseases. Acta Pharmacol Sin. 2020;41(12):1525–30. https://doi.org/10.1038/s41401-020-00529-9.
92. Hu C, et al. Meteorin-like protein attenuates doxorubicin-induced cardiotoxicity via activating cAMP/PKA/SIRT1 pathway. Redox Biol. 2020;37:101747. https://doi.org/10.1016/j.redox.2020.101747.
93. Caroleo M, et al. Does NUCB2/Nesfatin-1 influence eating behaviors in obese patients with binge eating disorder? Toward a neurobiological pathway. Nutrients. 2023;15(2):348. https://doi.org/10.3390/nu15020348.
94. Stengel A, Mori M, Taché Y. The role of nesfatin-1 in the regulation of food intake and body weight: recent developments and future endeavors. Obes Rev. 2013;14(11):859–70. https://doi.org/10.1111/obr.12063.
95. Psilopanagioti A, Nikou S, Papadaki H. Nucleobindin-2/Nesfatin-1 in the human hypothalamus is reduced in obese subjects and colocalizes with oxytocin, vasopressin, melanin-concentrating hormone, and cocaine- and amphetamine-regulated transcript. Neuroendocrinology. 2019;108(3):190–200. https://doi.org/10.1159/000496731.
96. Wronkowitz N, Romacho T, Sell H, Eckel J. Adipose tissue dysfunction and inflammation in cardiovascular disease. Front Horm Res. 2014;43:79–92. https://doi.org/10.1159/000360560.
97. Rajaei SM, Mood MA, Ghaffari MS, Williams DL. Measurement of tear production using phenol red thread and standardized endodontic absorbent paper points in European pond turtles (*Emys orbicularis*). J Zoo Wildl Med. 2014;45(4):825–9. https://doi.org/10.1638/2013-0306.1.
98. Yang R-Z, et al. Identification of omentin as a novel depot-specific adipokine in human adipose tissue: possible role in modulating insulin action. Am J Physiol Endocrinol Metab. 2006;290(6):E1253–61. https://doi.org/10.1152/ajpendo.00572.2004.
99. Tan BK, et al. Omentin-1, a novel adipokine, is decreased in overweight insulin-resistant women with polycystic ovary syndrome. Diabetes. 2008;57(4):801–8. https://doi.org/10.2337/db07-0990.
100. de Souza Batista CM, et al. Omentin plasma levels and gene expression are decreased in obesity. Diabetes. 2007;56(6):1655–61. https://doi.org/10.2337/db06-1506.
101. Matsubara T, et al. PGRN is a key adipokine mediating high fat diet-induced insulin resistance and obesity through IL-6 in adipose tissue. Cell Metab. 2012;15(1):38–50. https://doi.org/10.1016/j.cmet.2011.12.002.
102. Shafaei A, Marjani A, Khoshnia M. Serum progranulin levels in type 2 diabetic patients with metabolic syndrome. Rom J Intern Med. 2016;54(4):211–6. https://doi.org/10.1515/rjim-2016-0034.
103. Saeedi-Boroujeni A, Purrahman D, Shojaeian A, Poniatowski ŁA, Rafiee F, Mahmoudian-Sani M-R. Progranulin (PGRN) as a regulator of inflammation and a critical factor in the immunopathogenesis of cardiovascular diseases. J Inflamm. 2023;20(1):1. https://doi.org/10.1186/s12950-023-00327-0.

104. Nakamura K, Sasaki T, Cheng XW, Iguchi A, Sato K, Kuzuya M. Statin prevents plaque disruption in apoE-knockout mouse model through pleiotropic effect on acute inflammation. Atherosclerosis. 2009;206(2):355–61. https://doi.org/10.1016/j.atherosclerosis.2009.02.014.
105. McElhanon KE, Huff TC, Hirenallur-Shanthappa D, Miller RA, Christoforou N. Increased circulating progranulin is not sufficient to induce cardiac dysfunction or supraventricular arrhythmia. Sci Rep. 2023;13(1):21541. https://doi.org/10.1038/s41598-023-47311-5.
106. Pawłowski T, Kiejna A. Pathways to psychiatric care and reform of the public health care system in Poland. Eur Psychiatry. 2004;19(3):168–71. https://doi.org/10.1016/j.eurpsy.2003.09.009.
107. Skiba G, Raj S, Sobol M, Czauderna M, Kowalczyk P, Grela ER. Effects of fish oil and dietary antioxidant supplementation on bone health of growing lambs. Animals. 2021;11(1):230. https://doi.org/10.3390/ani11010230.
108. Steppan CM, et al. The hormone resistin links obesity to diabetes. Nature. 2001;409(6818):307–12. https://doi.org/10.1038/35053000.
109. McTernan CL, McTernan PG, Harte AL, Levick PL, Barnett AH, Kumar S. Resistin, central obesity, and type 2 diabetes. Lancet. 2002;359(9300):46–7. https://doi.org/10.1016/s0140-6736(02)07281-1.
110. Degawa-Yamauchi M, et al. Serum resistin (FIZZ3) protein is increased in obese humans. J Clin Endocrinol Metab. 2003;88(11):5452–5. https://doi.org/10.1210/jc.2002-021808.
111. Savage DB, et al. Resistin/Fizz3 expression in relation to obesity and peroxisome proliferator–activated receptor-γ action in humans. Diabetes. 2001;50(10):2199–202. https://doi.org/10.2337/diabetes.50.10.2199.
112. Lazar MA. How obesity causes diabetes: not a tall tale. Science (80-). 2005;307(5708):373–5. https://doi.org/10.1126/science.1104342.
113. Reilly MP, Lehrke M, Wolfe ML, Rohatgi A, Lazar MA, Rader DJ. Resistin is an inflammatory marker of atherosclerosis in humans. Circulation. 2005;111(7):932–9. https://doi.org/10.1161/01.CIR.0000155620.10387.43.
114. Pischon T, et al. Association of plasma resistin levels with coronary heart disease in women. Obes Res. 2005;13(10):1764–71. https://doi.org/10.1038/oby.2005.215.
115. Patel SD, Rajala MW, Rossetti L, Scherer PE, Shapiro L. Disulfide-dependent multimeric assembly of resistin family hormones. Science. 2004;304(5674):1154–8. https://doi.org/10.1126/science.1093466.
116. Banerjee RR, et al. Regulation of fasted blood glucose by resistin. Science. 2004;303(5661):1195–8. https://doi.org/10.1126/science.1092341.
117. Ghantous CM, Azrak Z, Hanache S, Abou-Kheir W, Zeidan A. Differential role of leptin and adiponectin in cardiovascular system. Int J Endocrinol. 2015;2015:534320. https://doi.org/10.1155/2015/534320.
118. Sidorov RA, Zhukov AV, Pchelkin VP, Vereshchagin AG, Tsydendambaev VD. Dynamics of fatty-acid composition of neutral acylglycerols in maturing euonymus fruits. Chem Biodivers. 2014;11(4):581–92. https://doi.org/10.1002/cbdv.201300305.
119. Sack GH. Serum amyloid A—a review. Mol Med. 2018;24(1):46. https://doi.org/10.1186/s10020-018-0047-0.
120. Carty CL, et al. Association of genetic variation in serum amyloid-A with cardiovascular disease and interactions with IL6, IL1RN, IL1beta and TNF genes in the Cardiovascular Health Study. J Atheroscler Thromb. 2009;16(4):419–30. https://doi.org/10.5551/jat.no968.
121. Coppack SW. Pro-inflammatory cytokines and adipose tissue. Proc Nutr Soc. 2001;60(3):349–56. https://doi.org/10.1079/pns2001110.
122. Petersen AMW, Pedersen BK. The anti-inflammatory effect of exercise. J Appl Physiol. 2005;98(4):1154–62. https://doi.org/10.1152/japplphysiol.00164.2004.
123. Redondo M, et al. Serum adiposity-induced biomarkers in obese and lean children with recently diagnosed autoimmune type 1 diabetes. Pediatr Diabetes. 2014;15(8):543–9. https://doi.org/10.1111/pedi.12159.

124. Raskin Erusan R, Nalini D, Manohar G, Malathi R. Correlation between obesity and inflammation in cardiovascular diseases—evaluation of leptin and inflammatory cytokines. Open J Endocr Metab Dis. 2012;02(02):7–15. https://doi.org/10.4236/ojemd.2012.22002.
125. Hohensinner PJ, Niessner A, Huber K, Weyand CM, Wojta J. Inflammation and cardiac outcome. Curr Opin Infect Dis. 2011;24(3):259–64. https://doi.org/10.1097/QCO.0b013e328344f50f.
126. Channual J, Wu JJ, Dann FJ. Effects of tumor necrosis factor-α blockade on metabolic syndrome components in psoriasis and psoriatic arthritis and additional lessons learned from rheumatoid arthritis. Dermatol Ther. 2009;22(1):61–73. https://doi.org/10.1111/j.1529-8019.2008.01217.x.
127. Roubille C, Martel-Pelletier J, Haraoui B, Tardif J-C, Pelletier J-P. Biologics and the cardiovascular system: a double-edged sword. Antiinflamm Antiallergy Agents Med Chem. 2013;12(1):68–82. https://doi.org/10.2174/1871523011312010009.
128. Nurmohamed M, Bao Y, Signorovitch J, Trahey A, Mulani P, Furst DE. Longer durations of antitumour necrosis factor treatment are associated with reduced risk of cardiovascular events in patients with rheumatoid arthritis. RMD Open. 2015;1(1):e000080. https://doi.org/10.1136/rmdopen-2015-000080.
129. D'Ercole L, et al. Patient dose during carotid artery stenting with embolic-protection devices: evaluation with radiochromic films and related diagnostic reference levels according to factors influencing the procedure. Cardiovasc Intervent Radiol. 2013;36(2):320–9. https://doi.org/10.1007/s00270-012-0392-2.
130. Youn B-S, et al. Serum vaspin concentrations in human obesity and type 2 diabetes. Diabetes. 2008;57(2):372–7. https://doi.org/10.2337/db07-1045.
131. Heiker JT. Vaspin (serpinA12) in obesity, insulin resistance, and inflammation. J Pept Sci. 2014;20(5):299–306. https://doi.org/10.1002/psc.2621.
132. Pilarski Ł, Pelczyńska M, Koperska A, Seraszek-Jaros A, Szulińska M, Bogdański P. Association of serum vaspin concentration with metabolic disorders in obese individuals. Biomolecules. 2023;13(3):508. https://doi.org/10.3390/biom13030508.
133. Tan BK, et al. Metformin decreases the adipokine vaspin in overweight women with polycystic ovary syndrome concomitant with improvement in insulin sensitivity and a decrease in insulin resistance. Diabetes. 2008;57(6):1501–7. https://doi.org/10.2337/db08-0127.
134. Al-kuraishy H, Al-Gareeb A, Al-Buhadilly A. Rosuvastatin improves vaspin serum levels in obese patients with acute coronary syndrome. Diseases. 2018;6(1):9. https://doi.org/10.3390/diseases6010009.
135. Nakatsuka A, et al. Vaspin is an adipokine ameliorating ER stress in obesity as a ligand for cell-surface GRP78/MTJ-1 complex. Diabetes. 2012;61(11):2823–32. https://doi.org/10.2337/db12-0232.
136. Berndt J, et al. Plasma visfatin concentrations and fat depot–specific mRNA expression in humans. Diabetes. 2005;54(10):2911–6. https://doi.org/10.2337/diabetes.54.10.2911.
137. Fukuhara A, et al. Visfatin: a protein secreted by visceral fat that mimics the effects of insulin. Science (80-). 2005;307(5708):426–30. https://doi.org/10.1126/science.1097243.
138. Chang Y-H, Chang D-M, Lin K-C, Shin S-J, Lee Y-J. Visfatin in overweight/obesity, type 2 diabetes mellitus, insulin resistance, metabolic syndrome and cardiovascular diseases: a meta-analysis and systemic review. Diabetes Metab Res Rev. 2011;27(6):515–27. https://doi.org/10.1002/dmrr.1201.
139. Revollo JR, et al. Nampt/PBEF/Visfatin regulates insulin secretion in β cells as a systemic NAD biosynthetic enzyme. Cell Metab. 2007;6(5):363–75. https://doi.org/10.1016/j.cmet.2007.09.003.
140. Abdalla MMI. Role of visfatin in obesity-induced insulin resistance. World J Clin Cases. 2022;10(30):10840–51. https://doi.org/10.12998/wjcc.v10.i30.10840.
141. Koch A, et al. Visfatin serum levels predict mortality in critically ill patients. Dis Markers. 2018;2018:1–8. https://doi.org/10.1155/2018/7315356.

142. Hosseinzadeh-Attar MJ, Golpaie A, Janani L, Derakhshanian H. Effect of weight reduction following bariatric surgery on serum visfatin and adiponectin levels in morbidly obese subjects. Obes Facts. 2013;6(2):193–202. https://doi.org/10.1159/000351162.
143. Scherer PE, Williams S, Fogliano M, Baldini G, Lodish HF. A novel serum protein similar to C1q, produced exclusively in adipocytes. J Biol Chem. 1995;270(45):26746–9. https://doi.org/10.1074/jbc.270.45.26746.
144. Hotamisligil GS, Shargill NS, Spiegelman BM. Adipose expression of tumor necrosis factor-alpha: direct role in obesity-linked insulin resistance. Science. 1993;259(5091):87–91. https://doi.org/10.1126/science.7678183.
145. Weisberg SP, McCann D, Desai M, Rosenbaum M, Leibel RL, Ferrante AW. Obesity is associated with macrophage accumulation in adipose tissue. J Clin Invest. 2003;112(12):1796–808. https://doi.org/10.1172/JCI19246.
146. Jaworski K, Sarkadi-Nagy E, Duncan RE, Ahmadian M, Sul HS. Regulation of Triglyceride metabolism. IV. Hormonal regulation of lipolysis in adipose tissue. Am J Physiol Liver Physiol. 2007;293(1):G1–4. https://doi.org/10.1152/ajpgi.00554.2006.
147. Frühbeck G, Méndez-Giménez L, Fernández-Formoso J-A, Fernández S, Rodríguez A. Regulation of adipocyte lipolysis. Nutr Res Rev. 2014;27(1):63–93. https://doi.org/10.1017/S095442241400002X.
148. Cao Y. Angiogenesis modulates adipogenesis and obesity. J Clin Invest. 2007;117(9):2362–8. https://doi.org/10.1172/JCI32239.
149. Fukumura D, et al. Paracrine regulation of angiogenesis and adipocyte differentiation during in vivo adipogenesis. Circ Res. 2003;93(9):e88. https://doi.org/10.1161/01.RES.0000099243.20096.FA.
150. Simpson E. Sources of estrogen and their importance. J Steroid Biochem Mol Biol. 2003;86(3–5):225–30. https://doi.org/10.1016/S0960-0760(03)00360-1.
151. Bélanger C, Luu-The V, Dupont P, Tchernof A. Adipose tissue intracrinology: potential importance of local androgen/estrogen metabolism in the regulation of adiposity. Horm Metab Res. 2002;34(11/12):737–45. https://doi.org/10.1055/s-2002-38265.
152. McMillen IC, Muhlhausler BS, Duffield JA, Yuen BSJ. Prenatal programming of postnatal obesity: fetal nutrition and the regulation of leptin synthesis and secretion before birth. Proc Nutr Soc. 2004;63(3):405–12. https://doi.org/10.1079/pns2004370.
153. Wu L, et al. Dissimilar regulation of glucose and lipid metabolism by leptin in two strains of gibel carp (Carassius gibelio). Br J Nutr. 2021;125(11):1215–29. https://doi.org/10.1017/S0007114520003608.
154. Fernández-Riejos P, et al. Role of leptin in the activation of immune cells. Mediators Inflamm. 2010;2010:568343. https://doi.org/10.1155/2010/568343.
155. Takeda M, et al. Obesity and eosinophilic inflammation: does leptin play a role. Int Arch Allergy Immunol. 2012;158(Suppl):87–91. https://doi.org/10.1159/000337799.
156. Upadhyay J, Farr OM, Mantzoros CS. The role of leptin in regulating bone metabolism. Metabolism. 2015;64(1):105–13. https://doi.org/10.1016/j.metabol.2014.10.021.
157. Feuermann Y, Mabjeesh SJ, Niv-Spector L, Levin D, Shamay A. Prolactin affects leptin action in the bovine mammary gland via the mammary fat pad. J Endocrinol. 2006;191(2):407–13. https://doi.org/10.1677/joe.1.06913.
158. Huang L, Li C. Leptin: a multifunctional hormone. Cell Res. 2000;10(2):81–92. https://doi.org/10.1038/sj.cr.7290038.
159. Park H-K, Ahima RS. Physiology of leptin: energy homeostasis, neuroendocrine function and metabolism. Metabolism. 2015;64(1):24–34. https://doi.org/10.1016/j.metabol.2014.08.004.
160. Lee B, Shao J. Adiponectin and energy homeostasis. Rev Endocr Metab Disord. 2014;15(2):149–56. https://doi.org/10.1007/s11154-013-9283-3.
161. Kim AY, et al. Obesity-induced DNA hypermethylation of the adiponectin gene mediates insulin resistance. Nat Commun. 2015;6:7585. https://doi.org/10.1038/ncomms8585.

162. Liu Q, Yuan B, Lo KA, Patterson HC, Sun Y, Lodish HF. Adiponectin regulates expression of hepatic genes critical for glucose and lipid metabolism. Proc Natl Acad Sci U S A. 2012;109(36):14568–73. https://doi.org/10.1073/pnas.1211611109.
163. Christou GA, Kiortsis DN. Adiponectin and lipoprotein metabolism. Obes Rev. 2013;14(12):939–49. https://doi.org/10.1111/obr.12064.
164. Hug C, Lodish HF. Medicine. Visfatin: a new adipokine. Science. 2005;307(5708):366–7. https://doi.org/10.1126/science.1106933.
165. Romacho T, Sánchez-Ferrer CF, Peiró C. Visfatin/Nampt: an adipokine with cardiovascular impact. Mediators Inflamm. 2013;2013:946427. https://doi.org/10.1155/2013/946427.
166. Ibrahim MM. Subcutaneous and visceral adipose tissue: structural and functional differences. Obes Rev. 2010;11(1):11–8. https://doi.org/10.1111/j.1467-789X.2009.00623.x.
167. Castoldi A, Naffah de Souza C, Câmara NOS, Moraes-Vieira PM. The macrophage switch in obesity development. Front Immunol. 2015;6:637. https://doi.org/10.3389/fimmu.2015.00637.
168. Fried SK, Bunkin DA, Greenberg AS. Omental and subcutaneous adipose tissues of obese subjects release interleukin-6: depot difference and regulation by glucocorticoid. J Clin Endocrinol Metab. 1998;83(3):847–50. https://doi.org/10.1210/jcem.83.3.4660.
169. Wang Y, et al. Lipocalin-2 is an inflammatory marker closely associated with obesity, insulin resistance, and hyperglycemia in humans. Clin Chem. 2007;53(1):34–41. https://doi.org/10.1373/clinchem.2006.075614.
170. Bozaoglu K, et al. Chemerin is a novel adipokine associated with obesity and metabolic syndrome. Endocrinology. 2007;148(10):4687–94. https://doi.org/10.1210/en.2007-0175.
171. Yvan-Charvet L, Quignard-Boulangé A. Role of adipose tissue renin-angiotensin system in metabolic and inflammatory diseases associated with obesity. Kidney Int. 2011;79(2):162–8. https://doi.org/10.1038/ki.2010.391.
172. Yang R-Z, et al. Acute-phase serum amyloid A: an inflammatory adipokine and potential link between obesity and its metabolic complications. PLoS Med. 2006;3(6):e287. https://doi.org/10.1371/journal.pmed.0030287.
173. Cianflone K, Xia Z, Chen LY. Critical review of acylation-stimulating protein physiology in humans and rodents. Biochim Biophys Acta. 2003;1609(2):127–43. https://doi.org/10.1016/s0005-2736(02)00686-7.
174. Hida K, et al. Visceral adipose tissue-derived serine protease inhibitor: a unique insulin-sensitizing adipocytokine in obesity. Proc Natl Acad Sci U S A. 2005;102(30):10610–5. https://doi.org/10.1073/pnas.0504703102.
175. Seldin MM, Tan SY, Wong GW. Metabolic function of the CTRP family of hormones. Rev Endocr Metab Disord. 2014;15(2):111–23. https://doi.org/10.1007/s11154-013-9255-7.
176. Wellen KE, Hotamisligil GS. Obesity-induced inflammatory changes in adipose tissue. J Clin Invest. 2003;112(12):1785–8. https://doi.org/10.1172/JCI20514.
177. Xu H, et al. Chronic inflammation in fat plays a crucial role in the development of obesity-related insulin resistance. J Clin Invest. 2003;112(12):1821–30. https://doi.org/10.1172/JCI19451.
178. Makki K, Froguel P, Wolowczuk I. Adipose tissue in obesity-related inflammation and insulin resistance: cells, cytokines, and chemokines. ISRN Inflamm. 2013;2013:139239. https://doi.org/10.1155/2013/139239.
179. Atawia RT, Bunch KL, Toque HA, Caldwell RB, Caldwell RW. Mechanisms of obesity-induced metabolic and vascular dysfunctions. Front Biosci. 2019;24(5):890–934. https://doi.org/10.2741/4758.
180. Trayhurn P, Wood IS. Adipokines: inflammation and the pleiotropic role of white adipose tissue. Br J Nutr. 2004;92(3):347–55. https://doi.org/10.1079/bjn20041213.
181. Kanda H, et al. MCP-1 contributes to macrophage infiltration into adipose tissue, insulin resistance, and hepatic steatosis in obesity. J Clin Invest. 2006;116(6):1494–505. https://doi.org/10.1172/JCI26498.

182. Alessi M-C, Juhan-Vague I. PAI-1 and the metabolic syndrome: links, causes, and consequences. Arterioscler Thromb Vasc Biol. 2006;26(10):2200–7. https://doi.org/10.1161/01.ATV.0000242905.41404.68.
183. Tilg H, Moschen AR. Adipocytokines: mediators linking adipose tissue, inflammation and immunity. Nat Rev Immunol. 2006;6(10):772–83. https://doi.org/10.1038/nri1937.
184. Rodríguez-Hernández H, Simental-Mendía LE, Rodríguez-Ramírez G, Reyes-Romero MA. Obesity and inflammation: epidemiology, risk factors, and markers of inflammation. Int J Endocrinol. 2013;2013:678159. https://doi.org/10.1155/2013/678159.
185. Cinti S, et al. Adipocyte death defines macrophage localization and function in adipose tissue of obese mice and humans. J Lipid Res. 2005;46(11):2347–55. https://doi.org/10.1194/jlr.M500294-JLR200.
186. Sun K, Kusminski CM, Scherer PE. Adipose tissue remodeling and obesity. J Clin Invest. 2011;121(6):2094–101. https://doi.org/10.1172/JCI45887.
187. Goossens GH, Blaak EE. Adipose tissue dysfunction and impaired metabolic health in human obesity: a matter of oxygen? Front Endocrinol. 2015;6:55. https://doi.org/10.3389/fendo.2015.00055.
188. Lee CH, Lam KS. Obesity-induced insulin resistance and macrophage infiltration of the adipose tissue: a vicious cycle. J Diabetes Investig. 2019;10(1):29–31. https://doi.org/10.1111/jdi.12918.
189. Lumeng CN, Bodzin JL, Saltiel AR. Obesity induces a phenotypic switch in adipose tissue macrophage polarization. J Clin Invest. 2007;117(1):175–84. https://doi.org/10.1172/JCI29881.
190. Shoelson SE, Lee J, Goldfine AB. Inflammation and insulin resistance. J Clin Invest. 2006;116(7):1793–801. https://doi.org/10.1172/JCI29069.
191. Van Gaal LF, Mertens IL, De Block CE. Mechanisms linking obesity with cardiovascular disease. Nature. 2006;444(7121):875–80. https://doi.org/10.1038/nature05487.
192. Donath MY, Shoelson SE. Type 2 diabetes as an inflammatory disease. Nat Rev Immunol. 2011;11(2):98–107. https://doi.org/10.1038/nri2925.
193. Esser N, Legrand-Poels S, Piette J, Scheen AJ, Paquot N. Inflammation as a link between obesity, metabolic syndrome and type 2 diabetes. Diabetes Res Clin Pract. 2014;105(2):141–50. https://doi.org/10.1016/j.diabres.2014.04.006.
194. Nosalski R, Guzik TJ. Perivascular adipose tissue inflammation in vascular disease. Br J Pharmacol. 2017;174(20):3496–513. https://doi.org/10.1111/bph.13705.
195. Hotamisligil GS. Inflammation, metaflammation and immunometabolic disorders. Nature. 2017;542(7640):177–85. https://doi.org/10.1038/nature21363.
196. Rotter V, Nagaev I, Smith U. Interleukin-6 (IL-6) induces insulin resistance in 3T3-L1 adipocytes and is, like IL-8 and tumor necrosis factor-alpha, overexpressed in human fat cells from insulin-resistant subjects. J Biol Chem. 2003;278(46):45777–84. https://doi.org/10.1074/jbc.M301977200.
197. Sjöstrand M, Eriksson JW. Neuroendocrine mechanisms in insulin resistance. Mol Cell Endocrinol. 2009;297(1–2):104–11. https://doi.org/10.1016/j.mce.2008.05.010.
198. Tateya S, Kim F, Tamori Y. Recent advances in obesity-induced inflammation and insulin resistance. Front Endocrinol. 2013;4:93. https://doi.org/10.3389/fendo.2013.00093.
199. Hotamisligil GS, Peraldi P, Budavari A, Ellis R, White MF, Spiegelman BM. IRS-1-mediated inhibition of insulin receptor tyrosine kinase activity in TNF-alpha- and obesity-induced insulin resistance. Science. 1996;271(5249):665–8. https://doi.org/10.1126/science.271.5249.665.
200. Saltiel AR, Olefsky JM. Inflammatory mechanisms linking obesity and metabolic disease. J Clin Invest. 2017;127(1):1–4. https://doi.org/10.1172/JCI92035.
201. Tan BK, Adya R, Randeva HS. Omentin: a novel link between inflammation, diabesity, and cardiovascular disease. Trends Cardiovasc Med. 2010;20(5):143–8. https://doi.org/10.1016/j.tcm.2010.12.002.

202. Wada J. Vaspin: a novel serpin with insulin-sensitizing effects. Expert Opin Investig Drugs. 2008;17(3):327–33. https://doi.org/10.1517/13543784.17.3.327.
203. Wong GW, Krawczyk SA, Kitidis-Mitrokostas C, Revett T, Gimeno R, Lodish HF. Molecular, biochemical and functional characterizations of C1q/TNF family members: adipose-tissue-selective expression patterns, regulation by PPAR-gamma agonist, cysteine-mediated oligomerizations, combinatorial associations and metabolic functions. Biochem J. 2008;416(2):161–77. https://doi.org/10.1042/BJ20081240.
204. Blüher M. Vaspin in obesity and diabetes: pathophysiological and clinical significance. Endocrine. 2012;41(2):176–82. https://doi.org/10.1007/s12020-011-9572-0.
205. Schäffler A, Buechler C. CTRP family: linking immunity to metabolism. Trends Endocrinol Metab. 2012;23(4):194–204. https://doi.org/10.1016/j.tem.2011.12.003.
206. Cao H. Adipocytokines in obesity and metabolic disease. J Endocrinol. 2014;220(2):T47–59. https://doi.org/10.1530/JOE-13-0339.
207. Christiaens V, Lijnen HR. Angiogenesis and development of adipose tissue. Mol Cell Endocrinol. 2010;318(1–2):2–9. https://doi.org/10.1016/j.mce.2009.08.006.
208. Saiki A, Watanabe F, Murano T, Miyashita Y, Shirai K. Hepatocyte growth factor secreted by cultured adipocytes promotes tube formation of vascular endothelial cells in vitro. Int J Obes (Lond). 2006;30(11):1676–84. https://doi.org/10.1038/sj.ijo.0803316.
209. Cao Y. Angiogenesis and vascular functions in modulation of obesity, adipose metabolism, and insulin sensitivity. Cell Metab. 2013;18(4):478–89. https://doi.org/10.1016/j.cmet.2013.08.008.
210. Hausman GJ, Hausman DB. Search for the preadipocyte progenitor cell. J Clin Invest. 2006;116(12):3103–6. https://doi.org/10.1172/JCI30666.
211. Suzuki E, Fujita D, Takahashi M, Oba S, Nishimatsu H. Adipose tissue-derived stem cells as a therapeutic tool for cardiovascular disease. World J Cardiol. 2015;7(8):454–65. https://doi.org/10.4330/wjc.v7.i8.454.
212. Hutchings G, et al. The proliferation and differentiation of adipose-derived stem cells in neovascularization and angiogenesis. Int J Mol Sci. 2020;21(11):3790. https://doi.org/10.3390/ijms21113790.
213. Ramakrishnan VM, Boyd NL. The adipose stromal vascular fraction as a complex cellular source for tissue engineering applications. Tissue Eng Part B Rev. 2018;24(4):289–99. https://doi.org/10.1089/ten.TEB.2017.0061.
214. Ghatreh K, Eliyasi M, Alaei S, Saki G. Differentiation potential of adipose tissue-derived mesenchymal stem cells into germ cells with and without growth factors. Andrologia. 2021;53(1):e13892. https://doi.org/10.1111/and.13892.
215. Corvera S, Gealekman O. Adipose tissue angiogenesis: impact on obesity and type-2 diabetes. Biochim Biophys Acta. 2014;1842(3):463–72. https://doi.org/10.1016/j.bbadis.2013.06.003.
216. Corvera S, Solivan-Rivera J, Yang Loureiro Z. Angiogenesis in adipose tissue and obesity. Angiogenesis. 2022;25(4):439–53. https://doi.org/10.1007/s10456-022-09848-3.
217. Jonker JW, et al. A PPARγ-FGF1 axis is required for adaptive adipose remodelling and metabolic homeostasis. Nature. 2012;485(7398):391–4. https://doi.org/10.1038/nature10998.
218. Lin S, et al. IGF-1 promotes angiogenesis in endothelial cells/adipose-derived stem cells co-culture system with activation of PI3K/Akt signal pathway. Cell Prolif. 2017;50(6):e12390. https://doi.org/10.1111/cpr.12390.
219. Laschke MW, Kontaxi E, Scheuer C, Heß A, Karschnia P, Menger MD. Insulin-like growth factor 1 stimulates the angiogenic activity of adipose tissue-derived microvascular fragments. J Tissue Eng. 2019;10:2041731419879837. https://doi.org/10.1177/2041731419879837.
220. Ferrara N, Gerber H-P, LeCouter J. The biology of VEGF and its receptors. Nat Med. 2003;9(6):669–76. https://doi.org/10.1038/nm0603-669.
221. Kharitonenkov A, et al. FGF-21 as a novel metabolic regulator. J Clin Invest. 2005;115(6):1627–35. https://doi.org/10.1172/JCI23606.

222. Nishimura S, et al. Adipogenesis in obesity requires close interplay between differentiating adipocytes, stromal cells, and blood vessels. Diabetes. 2007;56(6):1517–26. https://doi.org/10.2337/db06-1749.
223. Lijnen HR. Angiogenesis and obesity. Cardiovasc Res. 2008;78(2):286–93. https://doi.org/10.1093/cvr/cvm007.
224. Andrae J, Gallini R, Betsholtz C. Role of platelet-derived growth factors in physiology and medicine. Genes Dev. 2008;22(10):1276–312. https://doi.org/10.1101/gad.1653708.
225. Cao Y. Adipose tissue angiogenesis as a therapeutic target for obesity and metabolic diseases. Nat Rev Drug Discov. 2010;9(2):107–15. https://doi.org/10.1038/nrd3055.
226. Sun K, et al. Dichotomous effects of VEGF-A on adipose tissue dysfunction. Proc Natl Acad Sci U S A. 2012;109(15):5874–9. https://doi.org/10.1073/pnas.1200447109.
227. Tang QQ, Lane MD. Adipogenesis: from stem cell to adipocyte. Annu Rev Biochem. 2012;81:715–36. https://doi.org/10.1146/annurev-biochem-052110-115718.
228. Fernández-Hernando C, Suárez Y. ANGPTL4: a multifunctional protein involved in metabolism and vascular homeostasis. Curr Opin Hematol. 2020;27(3):206–13. https://doi.org/10.1097/MOH.0000000000000580.
229. Ito Y, et al. Inhibition of angiogenesis and vascular leakiness by angiopoietin-related protein 4. Cancer Res. 2003;63(20):6651–7. http://www.ncbi.nlm.nih.gov/pubmed/14583458.
230. Adya R, Tan BK, Randeva HS. Differential effects of leptin and adiponectin in endothelial angiogenesis. J Diabetes Res. 2015;2015:648239. https://doi.org/10.1155/2015/648239.
231. Oliveira AG, Araújo TG, de Carvalho B, Rocha GZ, Santos A, Saad MJA. The role of hepatocyte growth factor (HGF) in insulin resistance and diabetes. Front Endocrinol. 2018;9:503. https://doi.org/10.3389/fendo.2018.00503.
232. Fain JN, Tichansky DS, Madan AK. Transforming growth factor beta1 release by human adipose tissue is enhanced in obesity. Metabolism. 2005;54(11):1546–51. https://doi.org/10.1016/j.metabol.2005.05.024.
233. Wankhade UD, et al. TGF-β receptor 1 regulates progenitors that promote browning of white fat. Mol Metab. 2018;16:160–71. https://doi.org/10.1016/j.molmet.2018.07.008.
234. Yadav H, et al. Protection from obesity and diabetes by blockade of TGF-β/Smad3 signaling. Cell Metab. 2011;14(1):67–79. https://doi.org/10.1016/j.cmet.2011.04.013.
235. Koncarevic A, et al. A novel therapeutic approach to treating obesity through modulation of TGFβ signaling. Endocrinology. 2012;153(7):3133–46. https://doi.org/10.1210/en.2012-1016.
236. Latorre J, et al. The relevance of EGFR, ErbB receptors and neuregulins in human adipocytes and adipose tissue in obesity. Biomed Pharmacother. 2022;156:113972. https://doi.org/10.1016/j.biopha.2022.113972.
237. Kusunoki H, Taniyama Y, Otsu R, Rakugi H, Morishita R. Anti-inflammatory effects of hepatocyte growth factor on the vicious cycle of macrophages and adipocytes. Hypertens Res. 2014;37(6):500–6. https://doi.org/10.1038/hr.2014.41.
238. Leung JCK, Chan LYY, Tang SCW, Chu KM, Lai KN. Leptin induces TGF-beta synthesis through functional leptin receptor expressed by human peritoneal mesothelial cell. Kidney Int. 2006;69(11):2078–86. https://doi.org/10.1038/sj.ki.5000409.
239. Guo T, et al. Adipocyte ALK7 links nutrient overload to catecholamine resistance in obesity. Elife. 2014;3:e03245. https://doi.org/10.7554/eLife.03245.
240. Coelho M, Oliveira T, Fernandes R. Biochemistry of adipose tissue: an endocrine organ. Arch Med Sci. 2013;9(2):191–200. https://doi.org/10.5114/aoms.2013.33181.
241. Rutkowski JM, Stern JH, Scherer PE. The cell biology of fat expansion. J Cell Biol. 2015;208(5):501–12. https://doi.org/10.1083/jcb.201409063.
242. Pahlavani M, Kalupahana NS, Ramalingam L, Moustaid-Moussa N. Regulation and functions of the renin-angiotensin system in white and brown adipose tissue. In: Comprehensive physiology. Hoboken, NJ: Wiley; 2017. p. 1137–50. https://doi.org/10.1002/cphy.c160031.
243. Zechner R, et al. FAT SIGNALS—lipases and lipolysis in lipid metabolism and signaling. Cell Metab. 2012;15(3):279–91. https://doi.org/10.1016/j.cmet.2011.12.018.

244. Frigolet ME, Torres N, Tovar AR. The renin–angiotensin system in adipose tissue and its metabolic consequences during obesity. J Nutr Biochem. 2013;24(12):2003–15. https://doi.org/10.1016/j.jnutbio.2013.07.002.
245. Kalupahana NS, Moustaid-Moussa N. The adipose tissue renin-angiotensin system and metabolic disorders: a review of molecular mechanisms. Crit Rev Biochem Mol Biol. 2012;47(4):379–90. https://doi.org/10.3109/10409238.2012.694843.
246. Ramalingam L, et al. The renin angiotensin system, oxidative stress and mitochondrial function in obesity and insulin resistance. Biochim Biophys Acta Mol Basis Dis. 2017;1863(5):1106–14. https://doi.org/10.1016/j.bbadis.2016.07.019.
247. Okada S, et al. Adipose tissue–specific dysregulation of angiotensinogen by oxidative stress in obesity. Metabolism. 2010;59(9):1241–51. https://doi.org/10.1016/j.metabol.2009.11.016.
248. Yang Q, et al. Serum retinol binding protein 4 contributes to insulin resistance in obesity and type 2 diabetes. Nature. 2005;436(7049):356–62. https://doi.org/10.1038/nature03711.
249. Cerk IK, Wechselberger L, Oberer M. Adipose triglyceride lipase regulation: an overview. Curr Protein Pept Sci. 2017;19(2):221–33. https://doi.org/10.2174/1389203718666170918160110
.
250. Althaher AR. An overview of hormone-sensitive lipase (HSL). Sci World J. 2022;2022:1–9. https://doi.org/10.1155/2022/1964684.
251. Furuhashi M, Hotamisligil GS. Fatty acid-binding proteins: role in metabolic diseases and potential as drug targets. Nat Rev Drug Discov. 2008;7(6):489–503. https://doi.org/10.1038/nrd2589.
252. Kralisch S, Fasshauer M. Adipocyte fatty acid binding protein: a novel adipokine involved in the pathogenesis of metabolic and vascular disease? Diabetologia. 2013;56(1):10–21. https://doi.org/10.1007/s00125-012-2737-4.
253. Li J, Papadopoulos V, Vihma V. Steroid biosynthesis in adipose tissue. Steroids. 2015;103:89–104. https://doi.org/10.1016/j.steroids.2015.03.016.
254. Wawrzkiewicz-Jałowiecka A, Lalik A, Soveral G. Recent update on the molecular mechanisms of gonadal steroids action in adipose tissue. Int J Mol Sci. 2021;22(10):5226. https://doi.org/10.3390/ijms22105226.
255. Laforest S, et al. Estrogens and glucocorticoids in mammary adipose tissue: relationships with body mass index and breast cancer features. J Clin Endocrinol Metab. 2020;105(4):e1504–16. https://doi.org/10.1210/clinem/dgz268.
256. Wagner IV, et al. De novo and depot-specific androgen production in human adipose tissue: a source of hyperandrogenism in women with obesity. Obes Facts. 2022;15(2):281–91. https://doi.org/10.1159/000521571.
257. Liu L, Kang J, Ding X, Chen D, Zhou Y, Ma H. Dehydroepiandrosterone-regulated testosterone biosynthesis via activation of the ERK1/2 signaling pathway in primary rat Leydig cells. Cell Physiol Biochem. 2015;36(5):1778–92. https://doi.org/10.1159/000430150.
258. Paatela H, et al. Steroid sulfatase activity in subcutaneous and visceral adipose tissue: a comparison between pre- and postmenopausal women. Eur J Endocrinol. 2016;174(2):167–75. https://doi.org/10.1530/EJE-15-0831.
259. Grino PB, Griffin JE, Wilson JD. Testosterone at high concentrations interacts with the human androgen receptor similarly to dihydrotestosterone. Endocrinology. 1990;126(2):1165–72. https://doi.org/10.1210/endo-126-2-1165.
260. Blouin K, Veilleux A, Luu-The V, Tchernof A. Androgen metabolism in adipose tissue: recent advances. Mol Cell Endocrinol. 2009;301(1–2):97–103. https://doi.org/10.1016/j.mce.2008.10.035.
261. Labrie F. All sex steroids are made intracellularly in peripheral tissues by the mechanisms of intracrinology after menopause. J Steroid Biochem Mol Biol. 2015;145:133–8. https://doi.org/10.1016/j.jsbmb.2014.06.001.
262. Kuryłowicz A. Estrogens in adipose tissue physiology and obesity-related dysfunction. Biomedicine. 2023;11(3):690. https://doi.org/10.3390/biomedicines11030690.

263. Hill JL, Solomon MB, Nguyen ET, Caldwell JL, Wei Y, Foster MT. Glucocorticoids regulate adipose tissue protein concentration in a depot- and sex-specific manner. Stress. 2020;23(2):243–7. https://doi.org/10.1080/10253890.2019.1658736.
264. Akalestou E, Genser L, Rutter GA. Glucocorticoid metabolism in obesity and following weight loss. Front Endocrinol. 2020;11:59. https://doi.org/10.3389/fendo.2020.00059.
265. Legeza B, et al. Fructose, glucocorticoids and adipose tissue: implications for the metabolic syndrome. Nutrients. 2017;9(5):426. https://doi.org/10.3390/nu9050426.
266. Spritzer PM, Lecke SB, Satler F, Morsch DM. Adipose tissue dysfunction, adipokines, and low-grade chronic inflammation in polycystic ovary syndrome. Reproduction. 2015;149(5):R219–27. https://doi.org/10.1530/REP-14-0435.
267. Simpson ER, Brown KA. Obesity and breast cancer: role of inflammation and aromatase. J Mol Endocrinol. 2013;51(3):T51–9. https://doi.org/10.1530/JME-13-0217.
268. Gjorgoska M, Rizner TL. Integration of androgen hormones in endometrial cancer biology. Trends Endocrinol Metab. 2022;33(9):639–51. https://doi.org/10.1016/j.tem.2022.06.001.
269. Fasshauer M, Blüher M. Adipokines in health and disease. Trends Pharmacol Sci. 2015;36(7):461–70. https://doi.org/10.1016/j.tips.2015.04.014.
270. Luo L, Liu M. Adipose tissue in control of metabolism. J Endocrinol. 2016;231(3):R77–99. https://doi.org/10.1530/JOE-16-0211.
271. Scherer PE. Adipose tissue: from lipid storage compartment to endocrine organ. Diabetes. 2006;55(6):1537–45. https://doi.org/10.2337/db06-0263.
272. Ahima RS, Flier JS. Adipose tissue as an endocrine organ. Trends Endocrinol Metab. 2000;11(8):327–32. https://doi.org/10.1016/s1043-2760(00)00301-5.
273. Nakamura K, Fuster JJ, Walsh K. Adipokines: a link between obesity and cardiovascular disease. J Cardiol. 2014;63(4):250–9. https://doi.org/10.1016/j.jjcc.2013.11.006.
274. Stern JH, Rutkowski JM, Scherer PE. Adiponectin, leptin, and fatty acids in the maintenance of metabolic homeostasis through adipose tissue crosstalk. Cell Metab. 2016;23(5):770–84. https://doi.org/10.1016/j.cmet.2016.04.011.
275. Francisco V, et al. Obesity, fat mass and immune system: role for leptin. Front Physiol. 2018;9:640. https://doi.org/10.3389/fphys.2018.00640.
276. Meex RCR, Watt MJ. Hepatokines: linking nonalcoholic fatty liver disease and insulin resistance. Nat Rev Endocrinol. 2017;13(9):509–20. https://doi.org/10.1038/nrendo.2017.56.
277. Afraz S, Kapila N. Endocrinology for the hepatologist. Clin Liver Dis. 2023;22(1):1–6. https://doi.org/10.1097/CLD.0000000000000037.
278. Rosendo-Silva D, Matafome P. Gut-adipose tissue crosstalk: a bridge to novel therapeutic targets in metabolic syndrome? Obes Rev. 2021;22(2):e13130. https://doi.org/10.1111/obr.13130.
279. Musso G, Gambino R, Cassader M. Interactions between gut microbiota and host metabolism predisposing to obesity and diabetes. Annu Rev Med. 2011;62:361–80. https://doi.org/10.1146/annurev-med-012510-175505.
280. Tilg H, Kaser A. Gut microbiome, obesity, and metabolic dysfunction. J Clin Invest. 2011;121(6):2126–32. https://doi.org/10.1172/JCI58109.
281. Tremaroli V, Bäckhed F. Functional interactions between the gut microbiota and host metabolism. Nature. 2012;489(7415):242–9. https://doi.org/10.1038/nature11552.
282. Saad MJA, Santos A, Prada PO. Linking gut microbiota and inflammation to obesity and insulin resistance. Physiology. 2016;31(4):283–93. https://doi.org/10.1152/physiol.00041.2015.
283. Cani PD. Human gut microbiome: hopes, threats and promises. Gut. 2018;67(9):1716–25. https://doi.org/10.1136/gutjnl-2018-316723.
284. Avtanski D, Reddy V, Stojchevski R, Hadzi-Petrushev N, Mladenov M. The microbiome in the obesity-breast cancer axis: diagnostic and therapeutic potential. Pathogens. 2023;12(12):1402. https://doi.org/10.3390/pathogens12121402.
285. Rastelli M, Knauf C, Cani PD. Gut microbes and health: a focus on the mechanisms linking microbes, obesity, and related disorders. Obesity. 2018;26(5):792–800. https://doi.org/10.1002/oby.22175.

286. Timper K, Brüning JC. Hypothalamic circuits regulating appetite and energy homeostasis: pathways to obesity. Dis Model Mech. 2017;10(6):679–89. https://doi.org/10.1242/dmm.026609.
287. Scherer T, Buettner C. Yin and Yang of hypothalamic insulin and leptin signaling in regulating white adipose tissue metabolism. Rev Endocr Metab Disord. 2011;12(3):235–43. https://doi.org/10.1007/s11154-011-9190-4.
288. Kwon O, Kim KW, Kim M-S. Leptin signalling pathways in hypothalamic neurons. Cell Mol Life Sci. 2016;73(7):1457–77. https://doi.org/10.1007/s00018-016-2133-1.
289. Jiang Y, Rezai-Zadeh K, Desmoulins LD, Muenzberg H, Derbenev AV, Zsombok A. GABAergic leptin receptor-expressing neurons in the dorsomedial hypothalamus project to brown adipose tissue-related neurons in the paraventricular nucleus of mice. Auton Neurosci. 2023;245:103058. https://doi.org/10.1016/j.autneu.2022.103058.
290. Stojchevski R, Velichkovikj S, Arsov T. Genetic and epigenetic basis of obesity-induced inflammation and diabetes. In: Obesity, diabetes and inflammation; 2023. p. 129–46. https://doi.org/10.1007/978-3-031-39721-9_6.
291. Myers MG, Leibel RL, Seeley RJ, Schwartz MW. Obesity and leptin resistance: distinguishing cause from effect. Trends Endocrinol Metab. 2010;21(11):643–51. https://doi.org/10.1016/j.tem.2010.08.002.
292. Cone RD. Anatomy and regulation of the central melanocortin system. Nat Neurosci. 2005;8(5):571–8. https://doi.org/10.1038/nn1455.
293. Duncan RE, Ahmadian M, Jaworski K, Sarkadi-Nagy E, Sul HS. Regulation of lipolysis in adipocytes. Annu Rev Nutr. 2007;27:79–101. https://doi.org/10.1146/annurev.nutr.27.061406.093734.
294. Bartness TJ, Liu Y, Shrestha YB, Ryu V. Neural innervation of white adipose tissue and the control of lipolysis. Front Neuroendocrinol. 2014;35(4):473–93. https://doi.org/10.1016/j.yfrne.2014.04.001.
295. Herman JP, et al. Regulation of the hypothalamic-pituitary-adrenocortical stress response. Compr Physiol. 2016;6(2):603–21. https://doi.org/10.1002/cphy.c150015.
296. Sato K, Idelevich A, Nagano K, Rowe GC, Gori F, Baron R. Hypothalamic ΔFosB prevents age-related metabolic decline and functions via SNS. Aging. 2017;9(2):353–69. https://doi.org/10.18632/aging.101157.
297. Guilherme A, Henriques F, Bedard AH, Czech MP. Molecular pathways linking adipose innervation to insulin action in obesity and diabetes mellitus. Nat Rev Endocrinol. 2019;15(4):207–25. https://doi.org/10.1038/s41574-019-0165-y.
298. Valdearcos M, Xu AW, Koliwad SK. Hypothalamic inflammation in the control of metabolic function. Annu Rev Physiol. 2015;77:131–60. https://doi.org/10.1146/annurev-physiol-021014-071656.
299. Chen K-HE, Lainez NM, Nair MG, Coss D. Visceral adipose tissue imparts peripheral macrophage influx into the hypothalamus. J Neuroinflammation. 2021;18(1):140. https://doi.org/10.1186/s12974-021-02183-2.
300. Gao Y, et al. Hormones and diet, but not body weight, control hypothalamic microglial activity. Glia. 2014;62(1):17–25. https://doi.org/10.1002/glia.22580.
301. Gao J, et al. Adipocyte-derived extracellular vesicles modulate appetite and weight through mTOR signalling in the hypothalamus. Acta Physiol. 2020;228(2):e13339. https://doi.org/10.1111/apha.13339.
302. Ahima RS, Qi Y, Singhal NS. Adipokines that link obesity and diabetes to the hypothalamus. Prog Brain Res. 2006;153:155–74. https://doi.org/10.1016/S0079-6123(06)53009-2.
303. Tsigos C, Chrousos GP. Hypothalamic-pituitary-adrenal axis, neuroendocrine factors and stress. J Psychosom Res. 2002;53(4):865–71. https://doi.org/10.1016/s0022-3999(02)00429-4.
304. Contreras C, et al. The brain and brown fat. Ann Med. 2015;47(2):150–68. https://doi.org/10.3109/07853890.2014.919727.

305. Rui L. Energy metabolism in the liver. Compr Physiol. 2014;4(1):177–97. https://doi.org/10.1002/cphy.c130024.
306. Jones JG. Hepatic glucose and lipid metabolism. Diabetologia. 2016;59(6):1098–103. https://doi.org/10.1007/s00125-016-3940-5.
307. Gamberi T, Magherini F, Modesti A, Fiaschi T. Adiponectin signaling pathways in liver diseases. Biomedicine. 2018;6(2):52. https://doi.org/10.3390/biomedicines6020052.
308. Polyzos SA, Kountouras J, Zavos C, Tsiaousi E. The role of adiponectin in the pathogenesis and treatment of non-alcoholic fatty liver disease. Diabetes Obes Metab. 2010;12(5):365–83. https://doi.org/10.1111/j.1463-1326.2009.01176.x.
309. Woo YC, Xu A, Wang Y, Lam KSL. Fibroblast growth factor 21 as an emerging metabolic regulator: clinical perspectives. Clin Endocrinol (Oxf). 2013;78(4):489–96. https://doi.org/10.1111/cen.12095.
310. BonDurant LD, Potthoff MJ. Fibroblast growth factor 21: a versatile regulator of metabolic homeostasis. Annu Rev Nutr. 2018;38:173–96. https://doi.org/10.1146/annurev-nutr-071816-064800.
311. Stefan N, Häring H-U. The role of hepatokines in metabolism. Nat Rev Endocrinol. 2013;9(3):144–52. https://doi.org/10.1038/nrendo.2012.258.
312. Buzzetti E, Pinzani M, Tsochatzis EA. The multiple-hit pathogenesis of non-alcoholic fatty liver disease (NAFLD). Metabolism. 2016;65(8):1038–48. https://doi.org/10.1016/j.metabol.2015.12.012.
313. Katagiri H, Yamada T, Oka Y. Adiposity and cardiovascular disorders: disturbance of the regulatory system consisting of humoral and neuronal signals. Circ Res. 2007;101(1):27–39. https://doi.org/10.1161/CIRCRESAHA.107.151621.
314. Röder PV, Wu B, Liu Y, Han W. Pancreatic regulation of glucose homeostasis. Exp Mol Med. 2016;48(3):e219. https://doi.org/10.1038/emm.2016.6.
315. Kahn SE, Hull RL, Utzschneider KM. Mechanisms linking obesity to insulin resistance and type 2 diabetes. Nature. 2006;444(7121):840–6. https://doi.org/10.1038/nature05482.
316. Quesada I, Tudurí E, Ripoll C, Nadal A. Physiology of the pancreatic alpha-cell and glucagon secretion: role in glucose homeostasis and diabetes. J Endocrinol. 2008;199(1):5–19. https://doi.org/10.1677/JOE-08-0290.
317. Scheja L, Heeren J. The endocrine function of adipose tissues in health and cardiometabolic disease. Nat Rev Endocrinol. 2019;15(9):507–24. https://doi.org/10.1038/s41574-019-0230-6.
318. Marroquí L, et al. Role of leptin in the pancreatic β-cell: effects and signaling pathways. J Mol Endocrinol. 2012;49(1):R9–17. https://doi.org/10.1530/JME-12-0025.
319. Xu Y, et al. Leptin receptor mediates the proliferation and glucose metabolism of pancreatic cancer cells via AKT pathway activation. Mol Med Rep. 2020;21(2):945–52. https://doi.org/10.3892/mmr.2019.10855.
320. Ye R, Wang M, Wang QA, Scherer PE. Adiponectin-mediated antilipotoxic effects in regenerating pancreatic islets. Endocrinology. 2015;156(6):2019–28. https://doi.org/10.1210/en.2015-1066.
321. Teper Y, Eibl G. Pancreatic macrophages: critical players in obesity-promoted pancreatic cancer. Cancer. 2020;12(7):1946. https://doi.org/10.3390/cancers12071946.
322. Mezza T, Kulkarni RN. The regulation of pre- and post-maturational plasticity of mammalian islet cell mass. Diabetologia. 2014;57(7):1291–303. https://doi.org/10.1007/s00125-014-3251-7.
323. Ye R, Scherer PE. Fabp4, a new player in the adipo-pancreatic axis. Mol Metab. 2014;3(4):347–8. https://doi.org/10.1016/j.molmet.2014.03.009.
324. Wu LE, et al. Identification of fatty acid binding protein 4 as an adipokine that regulates insulin secretion during obesity. Mol Metab. 2014;3(4):465–73. https://doi.org/10.1016/j.molmet.2014.02.005.
325. Prentki M, Nolan CJ. Islet beta cell failure in type 2 diabetes. J Clin Invest. 2006;116(7):1802–12. https://doi.org/10.1172/JCI29103.

326. Rodríguez A, Becerril S, Ezquerro S, Méndez-Giménez L, Frühbeck G. Crosstalk between adipokines and myokines in fat browning. Acta Physiol. 2017;219(2):362–81. https://doi.org/10.1111/apha.12686.
327. Li F, Li Y, Duan Y, Hu C-AA, Tang Y, Yin Y. Myokines and adipokines: Involvement in the crosstalk between skeletal muscle and adipose tissue. Cytokine Growth Factor Rev. 2017;33:73–82. https://doi.org/10.1016/j.cytogfr.2016.10.003.
328. Eckel J. Myokines in metabolic homeostasis and diabetes. Diabetologia. 2019;62(9):1523–8. https://doi.org/10.1007/s00125-019-4927-9.
329. Pedersen BK, Febbraio MA. Muscles, exercise and obesity: skeletal muscle as a secretory organ. Nat Rev Endocrinol. 2012;8(8):457–65. https://doi.org/10.1038/nrendo.2012.49.
330. Garneau L, Aguer C. Role of myokines in the development of skeletal muscle insulin resistance and related metabolic defects in type 2 diabetes. Diabetes Metab. 2019;45(6):505–16. https://doi.org/10.1016/j.diabet.2019.02.006.
331. Sheptulina AF, Antyukh KY, Kiselev AR, Mitkovskaya NP, Drapkina OM. Possible mechanisms linking obesity, steroidogenesis, and skeletal muscle dysfunction. Life. 2023;13(6):1415. https://doi.org/10.3390/life13061415.
332. Yamauchi T, et al. Adiponectin stimulates glucose utilization and fatty-acid oxidation by activating AMP-activated protein kinase. Nat Med. 2002;8(11):1288–95. https://doi.org/10.1038/nm788.
333. Tomas E, et al. Metabolic and hormonal interactions between muscle and adipose tissue. Proc Nutr Soc. 2004;63(2):381–5. https://doi.org/10.1079/PNS2004356.
334. Kahn BB, Alquier T, Carling D, Hardie DG. AMP-activated protein kinase: ancient energy gauge provides clues to modern understanding of metabolism. Cell Metab. 2005;1(1):15–25. https://doi.org/10.1016/j.cmet.2004.12.003.
335. Boström P, et al. A PGC1-α-dependent myokine that drives brown-fat-like development of white fat and thermogenesis. Nature. 2012;481(7382):463–8. https://doi.org/10.1038/nature10777.
336. Elsen M, Raschke S, Eckel J. Browning of white fat: does irisin play a role in humans? J Endocrinol. 2014;222(1):R25–38. https://doi.org/10.1530/JOE-14-0189.
337. Pedersen BK, Febbraio MA. Muscle as an endocrine organ: focus on muscle-derived interleukin-6. Physiol Rev. 2008;88(4):1379–406. https://doi.org/10.1152/physrev.90100.2007.
338. Wueest S, Konrad D. The role of adipocyte-specific IL-6-type cytokine signaling in FFA and leptin release. Adipocyte. 2018;7(3):226–8. https://doi.org/10.1080/21623945.2018.1493901.
339. Wedell-Neergaard A-S, et al. Exercise-induced changes in visceral adipose tissue mass are regulated by IL-6 signaling: a randomized controlled trial. Cell Metab. 2019;29(4):844–855.e3. https://doi.org/10.1016/j.cmet.2018.12.007.
340. Ackermann J, et al. IL-6 signaling drives self-renewal and alternative activation of adipose tissue macrophages. Front Immunol. 2024;15:1201439. https://doi.org/10.3389/fimmu.2024.1201439.
341. Severinsen MCK, Pedersen BK. Muscle-organ crosstalk: the emerging roles of myokines. Endocr Rev. 2020;41(4):594–609. https://doi.org/10.1210/endrev/bnaa016.
342. Plomgaard P, Bouzakri K, Krogh-Madsen R, Mittendorfer B, Zierath JR, Pedersen BK. Tumor necrosis factor-alpha induces skeletal muscle insulin resistance in healthy human subjects via inhibition of Akt substrate 160 phosphorylation. Diabetes. 2005;54(10):2939–45. https://doi.org/10.2337/diabetes.54.10.2939.
343. Addison O, Marcus RL, Lastayo PC, Ryan AS. Intermuscular fat: a review of the consequences and causes. Int J Endocrinol. 2014;2014:309570. https://doi.org/10.1155/2014/309570.
344. Stanford KI, Goodyear LJ. Exercise regulation of adipose tissue. Adipocyte. 2016;5(2):153–62. https://doi.org/10.1080/21623945.2016.1191307.
345. Poulos SP, Hausman DB, Hausman GJ. The development and endocrine functions of adipose tissue. Mol Cell Endocrinol. 2010;323(1):20–34. https://doi.org/10.1016/j.mce.2009.12.011.
346. Vadde R, Gupta MK, Nagaraju GP. Is adipose tissue an immunological organ? Crit Rev Immunol. 2019;39(6):481–90. https://doi.org/10.1615/CritRevImmunol.2020033457.

347. Trim W, Turner JE, Thompson D. Parallels in immunometabolic adipose tissue dysfunction with ageing and obesity. Front Immunol. 2018;9:169. https://doi.org/10.3389/fimmu.2018.00169.
348. Russo L, Lumeng CN. Properties and functions of adipose tissue macrophages in obesity. Immunology. 2018;155(4):407–17. https://doi.org/10.1111/imm.13002.
349. Ferrante AW. Obesity-induced inflammation: a metabolic dialogue in the language of inflammation. J Intern Med. 2007;262(4):408–14. https://doi.org/10.1111/j.1365-2796.2007.01852.x.
350. Exley MA, Hand L, O'Shea D, Lynch L. Interplay between the immune system and adipose tissue in obesity. J Endocrinol. 2014;223(2):R41–8. https://doi.org/10.1530/JOE-13-0516.
351. Fischer-Posovszky P, Möller P. The immune system of adipose tissue: obesity-associated inflammation. Pathologe. 2020;41(3):224–9. https://doi.org/10.1007/s00292-020-00782-z.
352. Chandra RK. Nutrition and the immune system: an introduction. Am J Clin Nutr. 1997;66(2):460S–3S. https://doi.org/10.1093/ajcn/66.2.460S.
353. Rytter MJH, Kolte L, Briend A, Friis H, Christensen VB. The immune system in children with malnutrition—a systematic review. PLoS One. 2014;9(8):e105017. https://doi.org/10.1371/journal.pone.0105017.
354. Gerriets VA, MacIver NJ. Role of T cells in malnutrition and obesity. Front Immunol. 2014;5:379. https://doi.org/10.3389/fimmu.2014.00379.
355. Li FYL, Cheng KKY, Lam KSL, Vanhoutte PM, Xu A. Cross-talk between adipose tissue and vasculature: role of adiponectin. Acta Physiol. 2011;203(1):167–80. https://doi.org/10.1111/j.1748-1716.2010.02216.x.
356. Fitzgibbons TP, Czech MP. Epicardial and perivascular adipose tissues and their influence on cardiovascular disease: basic mechanisms and clinical associations. J Am Heart Assoc. 2014;3(2):e000582. https://doi.org/10.1161/JAHA.113.000582.
357. Ryuto M, et al. Induction of vascular endothelial growth factor by tumor necrosis factor alpha in human glioma cells. Possible roles of SP-1. J Biol Chem. 1996;271(45):28220–8. https://doi.org/10.1074/jbc.271.45.28220.
358. Giraudo E, et al. Tumor necrosis factor-alpha regulates expression of vascular endothelial growth factor receptor-2 and of its co-receptor neuropilin-1 in human vascular endothelial cells. J Biol Chem. 1998;273(34):22128–35. https://doi.org/10.1074/jbc.273.34.22128.
359. Yoshida S, et al. Involvement of interleukin-8, vascular endothelial growth factor, and basic fibroblast growth factor in tumor necrosis factor alpha-dependent angiogenesis. Mol Cell Biol. 1997;17(7):4015–23. https://doi.org/10.1128/MCB.17.7.4015.
360. Namba S, Nakano R, Kitanaka T, Kitanaka N, Nakayama T, Sugiya H. ERK2 and JNK1 contribute to TNF-α-induced IL-8 expression in synovial fibroblasts. PLoS One. 2017;12(8):e0182923. https://doi.org/10.1371/journal.pone.0182923.
361. Zegeye MM, Andersson B, Sirsjö A, Ljungberg LU. IL-6 trans-signaling impairs sprouting angiogenesis by inhibiting migration, proliferation and tube formation of human endothelial cells. Cells. 2020;9(6):1414. https://doi.org/10.3390/cells9061414.
362. Trayhurn P. Hypoxia and adipose tissue function and dysfunction in obesity. Physiol Rev. 2013;93(1):1–21. https://doi.org/10.1152/physrev.00017.2012.
363. Gils A, Declerck PJ. Plasminogen activator inhibitor-1. Curr Med Chem. 2004;11(17):2323–34. https://doi.org/10.2174/0929867043364595.
364. Smith HW, Marshall CJ. Regulation of cell signalling by uPAR. Nat Rev Mol Cell Biol. 2010;11(1):23–36. https://doi.org/10.1038/nrm2821.
365. Nykjaer A, et al. Recycling of the urokinase receptor upon internalization of the uPA:serpin complexes. EMBO J. 1997;16(10):2610–20. https://doi.org/10.1093/emboj/16.10.2610.
366. Blasi F, Carmeliet P. uPAR: a versatile signalling orchestrator. Nat Rev Mol Cell Biol. 2002;3(12):932–43. https://doi.org/10.1038/nrm977.
367. Britton KA, Fox CS. Perivascular adipose tissue and vascular disease. Clin Lipidol. 2011;6(1):79–91. https://doi.org/10.2217/clp.10.89.

368. Sánchez-Navarro A, González-Soria I, Caldiño-Bohn R, Bobadilla NA. An integrative view of serpins in health and disease: the contribution of SerpinA3. Am J Physiol Cell Physiol. 2021;320(1):C106–18. https://doi.org/10.1152/ajpcell.00366.2020.
369. Iwaki T, Urano T, Umemura K. PAI-1, progress in understanding the clinical problem and its aetiology. Br J Haematol. 2012;157(3):291–8. https://doi.org/10.1111/j.1365-2141.2012.09074.x.
370. Lovren F, Teoh H, Verma S. Obesity and atherosclerosis: mechanistic insights. Can J Cardiol. 2015;31(2):177–83. https://doi.org/10.1016/j.cjca.2014.11.031.
371. Arner E, et al. Adipocyte turnover: relevance to human adipose tissue morphology. Diabetes. 2010;59(1):105–9. https://doi.org/10.2337/db09-0942.
372. Kiess W, et al. Adipocytes and adipose tissue. Best Pract Res Clin Endocrinol Metab. 2008;22(1):135–53. https://doi.org/10.1016/j.beem.2007.10.002.
373. Spalding KL, et al. Dynamics of fat cell turnover in humans. Nature. 2008;453(7196):783–7. https://doi.org/10.1038/nature06902.
374. Serra MC, Ryan AS, Sorkin JD, Favor KH, Goldberg AP. High adipose LPL activity and adipocyte hypertrophy reduce visceral fat and metabolic risk in obese, older women. Obesity. 2015;23(3):602–7. https://doi.org/10.1002/oby.20998.
375. Bartelt A, Heeren J. Adipose tissue browning and metabolic health. Nat Rev Endocrinol. 2014;10(1):24–36. https://doi.org/10.1038/nrendo.2013.204.
376. Hammarstedt A, Gogg S, Hedjazifar S, Nerstedt A, Smith U. Impaired adipogenesis and dysfunctional adipose tissue in human hypertrophic obesity. Physiol Rev. 2018;98(4):1911–41. https://doi.org/10.1152/physrev.00034.2017.
377. Jo J, et al. Hypertrophy and/or hyperplasia: dynamics of adipose tissue growth. PLoS Comput Biol. 2009;5(3):e1000324. https://doi.org/10.1371/journal.pcbi.1000324.
378. Symons JD, Abel ED. Lipotoxicity contributes to endothelial dysfunction: a focus on the contribution from ceramide. Rev Endocr Metab Disord. 2013;14(1):59–68. https://doi.org/10.1007/s11154-012-9235-3.
379. van Herpen NA, Schrauwen-Hinderling VB. Lipid accumulation in non-adipose tissue and lipotoxicity. Physiol Behav. 2008;94(2):231–41. https://doi.org/10.1016/j.physbeh.2007.11.049.
380. Choe SS, Huh JY, Hwang IJ, Kim JI, Kim JB. Adipose tissue remodeling: its role in energy metabolism and metabolic disorders. Front Endocrinol. 2016;7:30. https://doi.org/10.3389/fendo.2016.00030.
381. Gustafson B, Gogg S, Hedjazifar S, Jenndahl L, Hammarstedt A, Smith U. Inflammation and impaired adipogenesis in hypertrophic obesity in man. Am J Physiol Metab. 2009;297(5):E999–E1003. https://doi.org/10.1152/ajpendo.00377.2009.
382. Reilly SM, Saltiel AR. Adapting to obesity with adipose tissue inflammation. Nat Rev Endocrinol. 2017;13(11):633–43. https://doi.org/10.1038/nrendo.2017.90.
383. Blüher M. Obesity: global epidemiology and pathogenesis. Nat Rev Endocrinol. 2019;15(5):288–98. https://doi.org/10.1038/s41574-019-0176-8.
384. Longo M, et al. Adipose tissue dysfunction as determinant of obesity-associated metabolic complications. Int J Mol Sci. 2019;20(9):2358. https://doi.org/10.3390/ijms20092358.
385. Samuel VT, Shulman GI. Mechanisms for insulin resistance: common threads and missing links. Cell. 2012;148(5):852–71. https://doi.org/10.1016/j.cell.2012.02.017.
386. Boden G. Obesity, insulin resistance and free fatty acids. Curr Opin Endocrinol Diabetes Obes. 2011;18(2):139–43. https://doi.org/10.1097/MED.0b013e3283444b09.
387. Cnop M, Foufelle F, Velloso LA. Endoplasmic reticulum stress, obesity and diabetes. Trends Mol Med. 2012;18(1):59–68. https://doi.org/10.1016/j.molmed.2011.07.010.
388. Kahn SE, Cooper ME, Del Prato S. Pathophysiology and treatment of type 2 diabetes: perspectives on the past, present, and future. Lancet. 2014;383(9922):1068–83. https://doi.org/10.1016/S0140-6736(13)62154-6.
389. Oikonomou EK, Antoniades C. The role of adipose tissue in cardiovascular health and disease. Nat Rev Cardiol. 2019;16(2):83–99. https://doi.org/10.1038/s41569-018-0097-6.

390. Hui X, Lam KSL, Vanhoutte PM, Xu A. Adiponectin and cardiovascular health: an update. Br J Pharmacol. 2012;165(3):574–90. https://doi.org/10.1111/j.1476-5381.2011.01395.x.
391. Katsiki N, Mikhailidis DP, Banach M. Leptin, cardiovascular diseases and type 2 diabetes mellitus. Acta Pharmacol Sin. 2018;39(7):1176–88. https://doi.org/10.1038/aps.2018.40.
392. Northcott JM, Yeganeh A, Taylor CG, Zahradka P, Wigle JT. Adipokines and the cardiovascular system: mechanisms mediating health and disease. Can J Physiol Pharmacol. 2012;90(8):1029–59. https://doi.org/10.1139/y2012-053.
393. Packer M. Epicardial adipose tissue may mediate deleterious effects of obesity and inflammation on the myocardium. J Am Coll Cardiol. 2018;71(20):2360–72. https://doi.org/10.1016/j.jacc.2018.03.509.
394. Villa J, Pratley RE. Adipose tissue dysfunction in polycystic ovary syndrome. Curr Diab Rep. 2011;11(3):179–84. https://doi.org/10.1007/s11892-011-0189-8.
395. Lee IT, et al. Active Cushing disease is characterized by increased adipose tissue macrophage presence. J Clin Endocrinol Metab. 2019;104(6):2453–61. https://doi.org/10.1210/jc.2018-02552.
396. Genchi VA, et al. Adipose tissue dysfunction and obesity-related male hypogonadism. Int J Mol Sci. 2022;23(15):8194. https://doi.org/10.3390/ijms23158194.
397. Chen Y, et al. Changes in profile of lipids and adipokines in patients with newly diagnosed hypothyroidism and hyperthyroidism. Sci Rep. 2016;6(1):26174. https://doi.org/10.1038/srep26174.
398. Rosenfield RL, Ehrmann DA. The pathogenesis of polycystic ovary syndrome (PCOS): the hypothesis of PCOS as functional ovarian hyperandrogenism revisited. Endocr Rev. 2016;37(5):467–520. https://doi.org/10.1210/er.2015-1104.
399. Biondi B. Thyroid and obesity: an intriguing relationship. J Clin Endocrinol Metab. 2010;95(8):3614–7. https://doi.org/10.1210/jc.2010-1245.
400. Hussain I, Garg A. Lipodystrophy syndromes. Endocrinol Metab Clin North Am. 2016;45(4):783–97. https://doi.org/10.1016/j.ecl.2016.06.012.
401. Garg A. Clinical review#: Lipodystrophies: genetic and acquired body fat disorders. J Clin Endocrinol Metab. 2011;96(11):3313–25. https://doi.org/10.1210/jc.2011-1159.
402. Knebel B, Müller-Wieland D, Kotzka J. Lipodystrophies—disorders of the fatty tissue. Int J Mol Sci. 2020;21(22):8778. https://doi.org/10.3390/ijms21228778.
403. Garg A. Acquired and inherited lipodystrophies. N Engl J Med. 2004;350(12):1220–34. https://doi.org/10.1056/NEJMra025261.
404. Brown RJ, et al. Long-term effectiveness and safety of metreleptin in the treatment of patients with generalized lipodystrophy. Endocrine. 2018;60(3):479–89. https://doi.org/10.1007/s12020-018-1589-1.
405. Araújo-Vilar D, Santini F. Diagnosis and treatment of lipodystrophy: a step-by-step approach. J Endocrinol Invest. 2019;42(1):61–73. https://doi.org/10.1007/s40618-018-0887-z.
406. Versini M, Jeandel P-Y, Rosenthal E, Shoenfeld Y. Obesity in autoimmune diseases: not a passive bystander. Autoimmun Rev. 2014;13(9):981–1000. https://doi.org/10.1016/j.autrev.2014.07.001.
407. Peyrin-Biroulet L, et al. Mesenteric fat in Crohn's disease: a pathogenetic hallmark or an innocent bystander? Gut. 2007;56(4):577–83. https://doi.org/10.1136/gut.2005.082925.
408. Lourenço EV, Liu A, Matarese G, La Cava A. Leptin promotes systemic lupus erythematosus by increasing autoantibody production and inhibiting immune regulation. Proc Natl Acad Sci U S A. 2016;113(38):10637–42. https://doi.org/10.1073/pnas.1607101113.
409. Šenolt L. Adipokines: role in local and systemic inflammation of rheumatic diseases. Expert Rev Clin Immunol. 2017;13(1):1–3. https://doi.org/10.1080/1744666X.2017.1249850.
410. Forny-Germano L, De Felice FG, Vieira N. The role of leptin and adiponectin in obesity-associated cognitive decline and Alzheimer's disease. Front Neurosci. 2018;12:1027. https://doi.org/10.3389/fnins.2018.01027.

411. Arnoldussen IAC, Kiliaan AJ, Gustafson DR. Obesity and dementia: adipokines interact with the brain. Eur Neuropsychopharmacol. 2014;24(12):1982–99. https://doi.org/10.1016/j.euroneuro.2014.03.002.
412. Folch J, et al. Neuroprotective and anti-ageing role of leptin. J Mol Endocrinol. 2012;49(3):R149–56. https://doi.org/10.1530/JME-12-0151.
413. Signoriello E, et al. Adiponectin profile at baseline is correlated to progression and severity of multiple sclerosis. Eur J Neurol. 2019;26(2):348–55. https://doi.org/10.1111/ene.13822.
414. Letra L, Santana I, Seiça R. Obesity as a risk factor for Alzheimer's disease: the role of adipocytokines. Metab Brain Dis. 2014;29(3):563–8. https://doi.org/10.1007/s11011-014-9501-z.
415. Gao Y, et al. Lipoprotein lipase maintains microglial innate immunity in obesity. Cell Rep. 2017;20(13):3034–42. https://doi.org/10.1016/j.celrep.2017.09.008.
416. Avgerinos KI, Spyrou N, Mantzoros CS, Dalamaga M. Obesity and cancer risk: emerging biological mechanisms and perspectives. Metabolism. 2019;92:121–35. https://doi.org/10.1016/j.metabol.2018.11.001.
417. Lengyel E, Makowski L, DiGiovanni J, Kolonin MG. Cancer as a matter of fat: the crosstalk between adipose tissue and tumors. Trends Cancer. 2018;4(5):374–84. https://doi.org/10.1016/j.trecan.2018.03.004.
418. Avtanski DB, et al. Honokiol abrogates leptin-induced tumor progression by inhibiting Wnt1-MTA1-β-catenin signaling axis in a microRNA-34a dependent manner. Oncotarget. 2015;6(18):16396–410. https://doi.org/10.18632/oncotarget.3844.
419. Yan D, Avtanski D, Saxena NK, Sharma D. Leptin-induced epithelial-mesenchymal transition in breast cancer cells requires β-catenin activation via Akt/GSK3- and MTA1/Wnt1 protein-dependent pathways. J Biol Chem. 2012;287(11):8598–612. https://doi.org/10.1074/jbc.M111.322800.
420. Avtanski D, et al. Resistin induces breast cancer cells epithelial to mesenchymal transition (EMT) and stemness through both adenylyl cyclase-associated protein 1 (CAP1)-dependent and CAP1-independent mechanisms. Cytokine. 2019;120:155–64. https://doi.org/10.1016/j.cyto.2019.04.016.
421. Deng T, Lyon CJ, Bergin S, Caligiuri MA, Hsueh WA. Obesity, inflammation, and cancer. Annu Rev Pathol. 2016;11:421–49. https://doi.org/10.1146/annurev-pathol-012615-044359.
422. Dirat B, et al. Cancer-associated adipocytes exhibit an activated phenotype and contribute to breast cancer invasion. Cancer Res. 2011;71(7):2455–65. https://doi.org/10.1158/0008-5472.CAN-10-3323.
423. Iyengar NM, Gucalp A, Dannenberg AJ, Hudis CA. Obesity and cancer mechanisms: tumor microenvironment and inflammation. J Clin Oncol. 2016;34(35):4270–6. https://doi.org/10.1200/JCO.2016.67.4283.
424. Lehr S, Hartwig S, Sell H. Adipokines: a treasure trove for the discovery of biomarkers for metabolic disorders. Proteomics Clin Appl. 2012;6(1–2):91–101. https://doi.org/10.1002/prca.201100052.
425. Stout MB, Justice JN, Nicklas BJ, Kirkland JL. Physiological aging: links among adipose tissue dysfunction, diabetes, and frailty. Physiology (Bethesda). 2017;32(1):9–19. https://doi.org/10.1152/physiol.00012.2016.
426. Kulaj K, et al. Adipocyte-derived extracellular vesicles increase insulin secretion through transport of insulinotropic protein cargo. Nat Commun. 2023;14(1):709. https://doi.org/10.1038/s41467-023-36148-1.
427. Thomou T, et al. Adipose-derived circulating miRNAs regulate gene expression in other tissues. Nature. 2017;542(7642):450–5. https://doi.org/10.1038/nature21365.
428. Cani PD, Jordan BF. Gut microbiota-mediated inflammation in obesity: a link with gastrointestinal cancer. Nat Rev Gastroenterol Hepatol. 2018;15(11):671–82. https://doi.org/10.1038/s41575-018-0025-6.
429. Heindel JJ, et al. Metabolism disrupting chemicals and metabolic disorders. Reprod Toxicol. 2017;68:3–33. https://doi.org/10.1016/j.reprotox.2016.10.001.

430. Rodríguez A, Ezquerro S, Méndez-Giménez L, Becerril S, Frühbeck G. Revisiting the adipocyte: a model for integration of cytokine signaling in the regulation of energy metabolism. Am J Physiol Endocrinol Metab. 2015;309(8):E691–714. https://doi.org/10.1152/ajpendo.00297.2015.
431. Mikłosz A, Nikitiuk BE, Chabowski A. Using adipose-derived mesenchymal stem cells to fight the metabolic complications of obesity: where do we stand? Obes Rev. 2022;23(5):e13413. https://doi.org/10.1111/obr.13413.
432. Rajkomar A, Dean J, Kohane I. Machine learning in medicine. N Engl J Med. 2019;380(14):1347–58. https://doi.org/10.1056/NEJMra1814259.

# Chapter 2
# Anatomy and Physiology of Adipose Tissue

**Yijie Huang, Yunwen Hu, Haiping Chen, Zhaohua Cai, and Ben He**

## Abbreviations

| | |
|---|---|
| 12,13-diHOME | 12,13-Dihydroxy-9Z-octadecenoic acid |
| 18F-FDG | Fluorine-18 fluorodeoxyglucose |
| 9-PAHSA | 9-Hydroxy-stearic acid |
| $\beta_3$-AR | $\beta_3$-Adrenergic receptor |
| Acetyl-CoA | Acetyl-coenzyme A |
| ADicerKO | Adipose Dicer knockout |
| AdipoR1 and AdipoR2 | Adiponectin receptors 1 and 2 |
| ADSC | Adipose-derived mesenchymal stem cell |
| AMPK$\alpha$ | AMP-activated protein kinase alpha subunit |
| aPVAT | Abdominal periaortic perivascular adipose tissue |
| ASC | Adipose stem cell |
| ATM | Adipose tissue macrophage |
| A-tPVAT | Anterior thoracic periaortic perivascular adipose tissue |
| BAT | Brown adipose tissue |
| BMAT | Bone marrow adipose tissue |
| cAMP | Cyclic adenosine monophosphate |

Yijie Huang, Yunwen Hu and Haiping Chen have contributed equally to this work.

Y. Huang · Y. Hu · H. Chen · Z. Cai (✉) · B. He
Department of Cardiology, Shanghai Chest Hospital, Shanghai Jiao Tong University School of Medicine, Shanghai, China
e-mail: 540290682@sjtu.edu.cn; hu-long@sjtu.edu.cn; hpchan@sjtu.edu.cn; caizhaohua@shchest.org; heben@shchest.org


D. Avtanski (ed.), *Adipose Tissue*, Contemporary Endocrinology, https://doi.org/10.1007/978-3-031-72570-8_2

| | |
|---|---|
| cBMAT | Constitutive bone marrow adipose tissue |
| CLSTN3β | Calsyntenin 3β |
| Col | Collagen |
| CT | Computed tomography |
| CXCL | C-X-C motif chemokine ligand |
| dAT | Dermal adipose tissue |
| DNL | De novo lipogenesis |
| EAT | Epicardial adipose tissue |
| ECM | Extracellular matrix |
| Elf5 | E74-like factor 5 |
| EPDR1 | Ependymin-related protein 1 |
| EV | Extracellular vesicle |
| eWAT | Epididymal white adipose tissue |
| FAHFA | Fatty acid esters of hydroxy fatty acid |
| FFA | Free fatty acid |
| FGF21 | Fibroblast growth factor 21 |
| GIP | Glucose-dependent insulinotropic polypeptide |
| GLUT | Glucose transporter |
| gWAT | Gonadal white adipose tissue |
| HFD | High-fat diet |
| hMSC | Human mesenchymal stem cell |
| HSC | Hematopoietic stem cell |
| HSL | Hormone-sensitive lipase |
| iBAT | Interscapular brown adipose tissue |
| IGF | Insulin-like growth factor |
| IL | Interleukin |
| ILCs | Innate lymphoid cells |
| iNKT cells | Invariant natural killer T cells |
| iWAT | Inguinal white adipose tissue |
| LepR | Leptin receptor |
| lPRAT | Lateral perirenal adipose tissue |
| L-tPVAT | Lateral thoracic periaortic perivascular adipose tissue |
| MAPK | Mitogen-activated protein kinase |
| MAT | Mesenteric adipose tissue |
| MCP-1 | Monocyte chemoattractant protein 1 |
| mEAT | Myocardial epicardial adipose tissue |
| MI | Myocardial ischemia |
| MI/R | Myocardial ischemia and reperfusion |
| miRNA | MicroRNA |
| mPRAT | Medial perirenal adipose tissue |
| MRI | Magnetic resonance imaging |
| MSC | Mesenchymal stem cell |
| MSCA1 | Mesenchymal stem cell antigen-1 |
| Msln | Mesothelin |
| MVB | Multivesicular compartment |

| | |
|---|---|
| Myh11 | Myosin heavy chain 11 |
| NE | Norepinephrine |
| NO | Nitric oxide |
| NOS1 | Nitric oxide synthase 1 |
| PA | Palmitic acid |
| PAME | Palmitic acid methyl ester |
| PAO | Palmitoleic acid (C16:1n7-palmitoleate) |
| PDGFRα | Platelet-derived growth factor receptor alpha |
| PET/CT | Positron emission tomography/computed tomography |
| PKA | Protein kinase A |
| PLC | Phospholipase C |
| PPARγ | Peroxisome proliferator-activated receptor gamma |
| PRAT | Perirenal adipose tissue |
| PVAT | Perivascular adipose tissue |
| PVCF | Perivascular adipose tissue-derived contractile factor |
| PVRF | Perivascular adipose tissue-derived relaxing factor |
| sAT | Subcutaneous adipose tissue |
| scRNAseq | Single-cell RNA sequencing |
| SHRSP | Stroke-prone spontaneously hypertensive rats |
| SM22α | Smooth muscle protein 22-alpha |
| SMC | Smooth muscle cell |
| SVF | Stromal vascular fraction |
| sWAT | Subcutaneous white adipose tissue |
| TNFα | Tumor necrosis factor-alpha |
| tPVAT | Thoracic periaortic perivascular adipose tissue |
| TZD | Thiazolidinedione |
| UCP1 | Uncoupling protein 1 |
| VSMC | Vascular smooth muscle cells |
| vWAT | Visceral white adipose tissue |
| WAT | White adipose tissue |
| Wt1 | Wilms tumor 1 |

## Overview of Adipose Tissue

### *Physiological Functions of Adipocytes*

Adipose tissue can generally be classified into white adipose tissue (WAT) and brown adipose tissue (BAT). This binary classification was originally based on macroscopic and histological observations. Appearing yellow and homogeneous, WAT is primarily composed of uniform spherical adipocytes. These cells, also known as white adipocytes, exhibit a characteristic "ringlike" morphology, in which a single large lipid droplet occupies most of the space, thereby displacing organelles such as the mitochondria and nuclei to one side. Adipocytes in BAT are typically polygonal and smaller than those in the WAT. Their most prominent feature is the presence of multiple small lipid droplets and abundant mitochondria, which impart a brown appearance to the tissue; hence, the origin of the name BAT [1, 2].

Adipose tissue is widely distributed throughout the body (Table 2.1). In rodents and humans, WAT is primarily distributed subcutaneously, as well as around internal organs such as the omentum and mesentery [3, 4] (Table 2.1). During the period from birth to adolescence, human WAT expands through hyperplasia (an increase in number) and hypertrophy (an increase in cell size) [5, 6]. The number of adipocytes remains relatively stable in adulthood in lean and obese individuals, with adipocytes turning over at a rate of 10% per year in human [6].

Unlike widely distributed WAT, BAT is relatively sparse. In mice, it is primarily distributed in the cervical, supraclavicular, axillary, perirenal-adrenal, and paravertebral regions, as well as around major arteries [7–9] (Table 2.1). The BAT distribution in humans differs slightly from that in rodents (Table 2.1). In infants and young children, BAT is primarily found in the same supraclavicular region [10]. However, as individuals grow older, most BAT is replaced with WAT [11]. The presence of metabolically active BAT in adults was first demonstrated in vivo using fluorine-18 fluorodeoxyglucose ($^{18}$F-FDG) positron emission tomography/computed tomography (PET/CT) [12, 13]. Levy et al. [14] suggested that adult humans have three main BAT depots: the supraclavicular section located in the neck and above the

**Table 2.1** Key characteristics of white, brown, and beige adipose tissue

| | White | Brown | Beige |
|---|---|---|---|
| Macroscopic morphology | White color, less vascularization | Brown color, high vascularization | N/A |
| Microscopic morphology | Low dense mitochondrial with a big lipid droplet | High dense mitochondrial with several small lipid droplets | High dense mitochondrial with many intermediate lipid droplets |
| UCP1 expression | – | +++ | + (Baseline)<br>++ (Activated) |
| Distribution in mice | Omental, mesenteric, perirenal, gonadal (epididymal), aPVAT, interscapular, inguinal | Interscapular, cervical, axillary, perirenal, tPVAT | Inguinal (widely known), epididymal (cachexia) |
| Distribution in human | Omental, mesenteric, perirenal, gonadal, epicardial depot sAT (represented by thigh, abdominal depot) | Infant:<br>Interscapular depot<br>Adult:<br>Suprascapular, cervical, axillary, paravertebral, mediastinal, perirenal depot | Uncertain |
| Thermogenic capacity | Low | High | Inducible |
| Function | Energy storage | Thermogenesis, metabolically healthy | Thermogenesis, metabolically healthy |

*UCP1* uncoupling protein 1, *aPVAT* abdominal periaortic adipose tissue, *tPVAT* thoracic periaortic adipose tissue, *sAT* subcutaneous adipose tissue

clavicle, the mediastinal zone situated around the spine and heart, and the infradiaphragmatic portion surrounding the kidneys/adrenal glands, pancreas, liver, and spleen. However, because of the intrinsic limits of PET/CT for measuring the uptake of $^{18}$F-FDG tracer in tissues, research focusing on adult BAT often occurs under stimulation conditions, such as cold exposure. Consequently, these findings may not accurately reflect the physiological state of BAT.

Currently, it is widely accepted that in addition to its physical functions such as mechanical protection and heat insulation, adipose tissue also has metabolic and energy properties. White adipocytes take in fatty acids carried by circulating lipoproteins and convert them into triglycerides for storage in large lipid droplets within cells [15]. Serving as an energy reservoir, WAT depots initiate lipolysis during fasting and other catabolic states, releasing free fatty acids (FFAs) to supply other organs [15]. With abundant mitochondria and high expression of uncoupling protein 1 (UCP1), BAT can dissipate energy through the production of heat and is associated with maintaining the core temperature during cold exposure through a process referred to as nonshivering thermogenesis [16].

Furthermore, a series of specific cell markers have been identified that differentiate between classical brown and white adipocytes. For example, in rodents, brown adipocytes are characterized by high expression of *Ucp1*, *Prdm16*, and *Ppargc1a* [17–19], whereas white adipocytes are specifically identified by gene expression of *Hoxc8*, *Hoxc9*, and *Tcf21* [20, 21]. Along with the distinct morphological and physiological characteristics of BAT and WAT, the classification of "white" and "brown" has emerged as a standard paradigm for studying newly determined adipose tissue.

Of note, not all adipocytes remain metabolically static throughout life; rather, they exhibit dynamic changes. In response to certain stimuli, such as cold exposure [22] and exercise [23], inguinal adipocytes in mice exhibit a "browning" characteristic supported by a multilocular appearance, upregulation of BAT-specific marker genes such as *Ucp1*, *Prdm16*, and *Ppargc1a*, together with increased mitochondrial activity and respiration rates [24, 25]. After 5 weeks of warm adaptation, the cold-induced brown-like inguinal adipocytes in mice converted into white adipocytes with a dramatic decrement in *Ucp1* to an undetectable level, but with the acquisition of typical white features [26]. Combining the features of white and brown fat cells, the specific adipocytes specialized in inducible and reversible conversion are named beige/brite (brown-in-white) adipocytes in a humorous way, which is an example of the learning paradigm of adipose tissue. Specific markers of different adipocytes have been reviewed by some researchers with subtle disparities, which certainly need duplication and further demonstration [21, 27–29].

Questions remain to be answered about whether all adipocytes present the reversible pattern, whether all beige fat cells come from the conversion of white adipocytes or share the same origins as WAT, and which stage/factor in the developmental course serves as the determinant of the classification. Moreover, there remains a gap in our understanding of the detailed regulatory mechanisms of the conversion process. It is worth noting that a study combining PET/CT and histology proposed that the adipose tissue used to be considered BAT in the supraclavicular region of adult humans is actually composed of beige adipocytes [24], further verified by a gene expression profile of human BAT samples isolated from many anatomical locations with a large array of beige cell-selective genes [30]. Therefore, caution must be exercised when applying the learning pattern and extrapolating the findings from experimental studies in mice to clinical applications in humans.

## *Endocrine Function of Adipocytes*

The complex and highly active role of adipose tissue as an endocrine organ is widely recognized via the secretion of adipokines [31]. To date, the definition and constitution of adipokines have remained unclear and controversial. Some researchers have identified adipokines as peptides produced by adipocytes and/or other situated cells within adipose tissue [32–34]. To avoid ambiguity in the sources of the factors discussed below, adipokines are defined as the bioactive substances secreted by adipocytes (not equal to adipocyte-specific) to display an integral metabolic profile, performing in an autocrine, paracrine, and endocrine manner. Based on their ingredients, adipokines can be classified into three groups: inorganic molecules such as nitric oxide (NO) [35]; organic compounds such as peptides, lipokines, and microRNAs (miRNAs); and extracellular vehicles (EVs). Those produced by BAT are referred to as batokines habitually, while actually belonging to adipokines. Certain well-studied batokines that are preferentially secreted by BAT are also discussed below.

### Main Polypeptides

Polypeptides derived from adipocytes, particularly leptin and adiponectin, are among the most extensively studied bioactive substances with endocrine function. Leptin, a 16 kDa protein hormone that is essential for metabolism and energy balance expressed by adipocytes, was discovered in 1994 [36]. Predominantly produced by subcutaneous WAT (sWAT) rather than visceral WAT (vWAT), leptin primarily exerts its effects by binding to the leptin receptor (LepR), which is widely distributed in the central nervous system, adipose tissue, heart, muscle, lungs, small intestine, and liver tissues [37, 38]. Not only exerts its appetite-suppressing effects by crossing the blood-brain barrier and acting on the hypothalamus, but leptin also interacts with hormones involved in regulating the body's substance and energy metabolism [39]. Leptin indirectly mediates the effects of insulin, glucagon, insulin-like growth factor (IGF), growth hormone, and glucocorticoids [40]. Loss of leptin or LepR in both rodents and humans not only leads to extreme obesity but also manifests elevated blood sugar levels and severe insulin resistance [41]. Moreover, hyperglycemia and insulin resistance also occur in individuals with lipodystrophy, a condition characterized by very low levels of leptin [41]. Although the regulatory mechanisms remain to be determined, these findings support the involvement of leptin in maintaining metabolic homeostasis.

Adiponectin, initially discovered in 1995, was first identified as a 30 kDa adipocyte complement-related protein by Scherer et al., who isolated it from 3 T3-L1 adipocytes [42]. Adiponectin is primarily secreted by adipose tissue and expressed in minimal amounts in other cell types [43–45]. It interacts with the conserved receptors AdipoR1 and AdipoR2, which are present in various tissues in both humans and rodents [46]. Adiponectin is ultimately degraded in the liver, following its circulation and action throughout the body [47]. Adiponectin activity is crucial for maintaining whole-body insulin sensitivity, as evidenced by observations of insulin resistance in mice lacking adiponectin [48] and, conversely, retained insulin sensitivity in obese mice with elevated adiponectin expression [49]. The ceramidase activity of adiponectin receptors regulates key metabolic pathways, including PI3K/AKT, AMPK, mitogen-activated protein kinase (MAPK), phospholipase C (PLC), and peroxisome proliferator-activated receptor gamma (PPARγ), explaining the insulin-sensitizing effects of adiponectin through the modulation of ceramide levels that critically influence glucose and lipid metabolism [47]. Additionally, adiponectin exhibits potent anti-inflammatory and antifibrotic actions across multiple organs by engaging macrophages and fibroblasts [50–52], contributing to protection against diseases such as atherosclerosis [53], gestational diabetes [54], and nonalcoholic fatty liver disease [55]. This collective evidence highlights adiponectin as a critical endocrine hormone secreted by fat cells, with well-established roles in maintaining metabolic balance, especially under conditions of prolonged metabolic strain.

### Main Lipokines

Lipokines are specific species of fatty acids and derivatives produced by adipocytes and have well-defined physiological roles in an endocrine manner [56]. In 2008, through genetic intervention in de novo lipogenesis (DNL) of WAT in obese mice, it was the first time to postulate a lipokine known as palmitoleic acid or C16:1n7-palmitoleate (PAO), a monounsaturated fatty acid [57]. PAO is capable of exerting an effect on many organs to increase systemic insulin sensitivity, thereby contributing to metabolic health. Rat muscle cells were found to increase glucose uptake by PAO by increasing the abundance of glucose transporters (GLUTs) GLUT1 and GLUT4 in the plasma membrane [58]. PAO also acts in an endocrine manner to decrease hepatic steatosis and promote the viability and mitogenesis of the insulin-producing pancreatic β-cells [59, 60].

Fatty acid esters of hydroxy fatty acids (FAHFAs) were first discovered in 2014 from lipidomic analysis of mice, derived from BAT, WAT, and the liver [58]. FAHFAs represent a new class of lipids with more than 24 members distinguished by the branched ester sites on hydroxy fatty acids that can be synthesized in vivo [58, 61, 62]. Most studies on the bioactive roles of FAHFAs have been in a pathological context. A systemic reduction in FAHFA levels is associated with the deterioration of the metabolic profile of insulin sensitivity in both mice and human [58]. Conversely, oral administration of the family member 9-hydroxy-stearic acid (9-PAHSA) can stimulate glucose uptake by adipocytes and promote insulin secretion under hyperglycemic conditions in insulin-resistant mice [58]. The metabolic benefits of FAHFAs are further clarified to attenuate the inflammation of pancreatic tissue and promote β-cell survival in nonobese diabetic mice, showing the therapeutic potential of type1 diabetes [63].

## Main Extracellular Vesicles and MicroRNAs

Despite the long-term debate of the nomenclature and classification [64], membrane-enclosed functional vesicles derived from a variety of cell types, including adipocytes, are referred to as extracellular vesicles (EVs)/exosomes in this section. EVs can be released by budding off the cell surface, or by fusion of internal multivesicular compartments (MVBs) with the plasma membrane [64, 65]. As a mixture of biochemical compounds, such as proteins, lipids, and miRNAs, EVs serve as channels for cell communication of metabolic signals [64]. The circulating levels of miRNAs and adipocyte-derived EVs are considered novel forms of adipokines.

Adipose tissues are a primary source of circulating exosomal miRNAs. This is supported by a dramatic reduction in circulating exosomal miRNAs observed in mice with adipose tissue-specific knockdown of the miRNA-processing enzyme Dicer (ADicerKO) and in patients with lipodystrophy [66]. The levels of circulating exosomal miRNAs in ADicerKO mice can be restored by transplantation of normal inguinal WAT (iWAT), epididymal WAT (eWAT), and BAT. However, impaired glucose tolerance tests markedly improved only in mice transplanted with BAT [66], indicating that BAT-derived miRNAs may contribute to the regulation of glucose metabolism. Throughout the circulation, exosomal miRNAs derived from adipocytes have a distant endocrine effect in other organs. Circulating exosomal miR-27a in exosomes released by adipose tissue can be taken up by the skeletal muscle, leading to obesity-related insulin resistance state in mice [67]. In addition, these fat-derived circulating miRNAs can act on other tissues such as the liver and muscles, to regulate mRNA translation and stability [66]. Exosomal miRNAs derived from adipose tissues can have different regional effects on the regulation of adipogenesis, adipocyte hypertrophy, and lipid accumulation [68]. In line with this, ADicerKO mice exhibit an abnormal fat distribution pattern characterized by the loss of intra-abdominal and subcutaneous white fat and an enlargement and "whitening" of interscapular brown fat [69]. Exosomal miRNAs can also be produced by adipose-derived mesenchymal stem cells (ADSCs). ADSC-derived exosomal miR-342-5p has been proposed as a potential mechanism for protecting endothelial cells against atherosclerosis [70]. The precise regulatory mechanisms of the cross talk between adipose tissue and adipose tissue-derived exosomal miRNAs and depot-specific disparity remain to be elucidated.

### Main Batokines

Batokines are adipokines derived from BAT that play pivotal roles in promoting adipogenesis and angiogenesis as well as in conveying the advantageous metabolic effects associated with activated BAT [71]. A thorough classification and explanation of the multifarious functions of batokines have been comprehensively summarized by Villarroya et al. [72]. Some important batokines are mentioned below.

Fibroblast growth factor 21 (FGF21), a 23 kDa protein composed of 209 amino acids, which is abundantly secreted by BAT during thermogenesis, stimulates the uptake of FFAs and accelerates lipid metabolism in both WAT and BAT [73–75]. Its efficacy in managing obesity is well documented, demonstrating effects such as facilitating fat reduction and alleviating conditions such as hyperglycemia, insulin resistance, and dyslipidemia [76]. A recently discovered human batokine, ependymin-related protein 1 (EPDR1), has been found to improve the function of pancreatic β-cells and boost overall metabolic efficiency [77, 78]. 12,13-Dihydroxy-9Z-octadecenoic acid (12,13-diHOME) is a converting product of linoleic acid and is increasingly produced by BAT in response to exercise and cold exposure [79, 80]. 12,13-diHOME enhances the uptake of fatty acids by heart muscle cells and promotes their respiratory capability [81]. It also facilitates improved contraction and respiration in these cells through a mechanism dependent on nitric oxide synthase 1 (NOS1) [81]. Long-term administration of 12,13-diHOME counteracts the detrimental effects of a high-fat diet (HFD) on heart function and structural remodeling in mice [81]. EVs secreted by BAT facilitate interorgan communication and exert cardioprotective effects. A study further demonstrated that intramyocardial injection of BAT-derived EVs attenuates the injury caused by MI/R [82]. BAT miRNAs (miR-125b-5p, miR-128-3p, and miR-30d-5p) have been identified as cardioprotective factors in MI/R mice by suppressing the proapoptotic MAPK pathway, an effect that can be reversed when co-using BAT miRNA inhibitors [82].

In summary, adipokines play a vital role in metabolic homeostasis by releasing biochemical substances including polypeptides, lipokines, EVs, and exosomal miRNAs. However, until now, most studies have focused on systematic metabolic effects and circulating levels of certain adipokines, especially in the pathological context. There is still a large gap in our understanding of the cross talk between adipokines and their local physiological functions in adipose tissue.

## *The Integral Role of Other Cells and Extracellular Matrix (ECM) in Adipose Tissue*

Adipose depots are composed of lobules crowded with adipocytes, supported by arterioles, regulated by neuronal processes, and separated by fibrous septa. Adipose tissue also comprises immune cells, fibroblasts, neuronal processes, etc. [83].

## Immune Cells Within Adipose Tissue

Adipose tissue harbors a diverse population of innate and adaptive immune cells, which serve as critical regulatory factors for maintaining adipose tissue homeostasis. The substantial enrichment of anti-inflammatory immune cells, including M2-like macrophages, eosinophils, innate lymphoid cells (ILCs), regulatory T cells (Treg cells), invariant natural killer T cells (iNKT cells), and $\gamma\delta$ T cells in mouse WAT, establishes an anti-inflammatory microenvironment favorable for adipose tissue homeostasis [84, 85].

The role of adipose tissue macrophages (ATMs) as representative immune cells in adipose tissue deserves further investigation. As fundamental components of the innate immune system, ATMs are traditionally categorized into classically activated (M1-like) and alternatively activated (M2-like) macrophages [86, 87]. The M2-like phenotype of ATMs has been reported to control inflammation in the adipose tissue and regulate adaptive thermogenesis and adipogenesis. In response to cold exposure, macrophages are recruited to iWAT and alternatively activated by interleukin (IL)-4 (IL-4)/IL-13 produced by local eosinophils [88]. These alternatively activated macrophages induce the expression of tyrosine hydroxylase and promote catecholamine production, thereby promoting lipolysis and thermogenesis in adipose tissue [88]. In addition, macrophages regulate adipocyte biogenesis and ECM remodeling during adipogenesis. In $\beta_3$-adrenergic receptor ($\beta_3$-AR)-induced de novo adipogenesis, dying adipocytes would recruit M2-polarized macrophages to facilitate adipocyte clearance and express high levels of osteopontin, attracting a subpopulation of $CD4^+$/platelet-derived growth factor receptor alpha $(PDGFR\alpha)^+$ progenitors [89]. In mice, M2-polarized macrophages exhibit physiological endocytosis of collagen, which reduces proinflammatory collagen fragments and is also associated with the physiological expansion of sWAT [90]. Notably, ATMs undergo massive changes in composition, distribution, and function during diet-induced obesity and aging, and novel ATM subpopulations have recently been identified [91–99]. Obesity and aging induce ATMs toward a proinflammatory phenotype. In both aged mice and older humans [100–102], M1-like macrophages were found to secrete more proinflammatory cytokines such as IL-6, tumor necrosis factor $\alpha$ (TNF$\alpha$), and IL-1$\beta$ in adipose tissue, engaging in localized and systemic inflammation to influence insulin sensitivity and metabolic homeostasis [87].

## Innervation in Adipose Tissue

The presence of sympathetic innervation in adipose tissue, which locally releases noradrenaline and drives lipolysis and thermogenesis in brown and beige adipocytes, is well-established [103–105]. Sympathetic nerves express tyrosine hydroxylase, a marker of noradrenergic neurons, and catalyze the synthesis of norepinephrine (NE) under conditions such as cold exposure and fasting [28, 104]. NE induces adipocyte lipolysis through the activation of the $\beta$3-ARs-cyclic AMP (cAMP)-protein kinase A (PKA) signaling pathway, which in turn phosphorylates hormone-sensitive lipase (HSL), NE induce adipocyte lipolysis [106]. By subsequently coupling both Gαs and Gαi, subunits of the G protein-coupled receptors for transmembrane signaling, NE interacts with β-ARs to activate the MAPK cascade to stimulate the gene expression of UCP1, the primary driver of uncoupled mitochondrial respiration and adaptive nonshivering thermogenesis in brown and beige adipocytes [107–109].

Abundant sympathetic innervation is observed in the iBAT of mice and supraclavicular fat pad of humans. It is now well-known that BAT is much more densely innervated by sympathetic nerves than WAT [110–112]. Sympathetic innervation also differs significantly between the subcutaneous and visceral depots. Mice iWAT exhibits a more abundant parenchymal sympathetic innervation than eWAT [113]. In addition, after 48 h of cold exposure, there is a more marked increase in the proportion of $UCP1^+$ beige cells and higher sympathetic innervation density in the inguinal region than dorsolumbar regions [113].

The sympathetic innervation of thermogenic adipose tissue is typically regulated by a multitude of factors derived from adipocytes, including the transcription regulator PRDM16 and the neurotrophic factor S100b. The expression level of PRDM16 is associated with the density of sympathetic projections within the adipose tissue. Researchers found that there were no significant cold-induced regional differences in sympathetic innervation density in adipocyte-specific knockdown of PRDM16 mice [113], whereas overexpression of PRDM16 led to increased sympathetic innervation of beige adipocytes in iWAT in mice [114]. This is further supported by the abundant sympathetic innervation observed in regions with high levels of PRDM16 in iBAT in mice and the supraclavicular fat pad in humans [110, 111]. As a neurotrophic factor, S100b can stimulate sympathetic axon growth [115]. Recently, Zeng et al. showed that a previously unknown protein, calsyntenin 3β (CLSTN3β), can bind to and enhance the protein expression of S100b, thereby promoting the sympathetic innervation of both brown and beige adipocytes [115].

## Extracellular Matrix (ECM) of Adipose Tissue

Constituted by a complex amalgam of proteins, polysaccharides, and proteoglycans, ECM provides a dynamic mechanical and structural scaffold that simultaneously serves as a biochemical framework to facilitate intercellular communication [116]. The interactions between the peri-adipocyte ECM and its cognate cell surface receptors have been reviewed systematically [117].

The ECM is remodeled during adipogenesis. During the differentiation of human mesenchymal stem cells (hMSC) to adipocytes, significant elevation in laminin-α4 and laminin-ß4 expression levels was observed, and specific laminin was integrated into the basement membrane [118]. In mice, laminin-α4 was found to negatively regulate the thermogenic phenotype and pathways involved in mitochondrial biogenesis in adipocytes through the suppression of the AMP-activated protein kinase alpha subunit (AMPKα) [119].

High levels of collagen (Col) VI (Col VI) expression have been observed in obese adipocytes, and pericellular fibrosis is present in omental adipose tissue and subcutaneous adipose tissue of obese individuals [120–123]. Fibrosis restricts adipocyte hypertrophy but is associated with a reciprocal increase in adipocyte hyperplasia and may be linked to decreased insulin sensitivity, adverse outcomes of bariatric surgery, and difficulty in weight loss [124].

The ECM may determine the fate of progenitor cells in the adipose tissue. In human WAT, the ECM can be subdivided into septa and stroma, with the former exhibiting a higher concentration of fiber components, such as Col I, Col III, and elastin than the latter [125]. These two compartments constitute different niches for the $CD45^-/CD34^+/CD31^-$ progenitors within the adipose lobules. In the stroma, there is enrichment of the mesenchymal stem cell antigen-1 $(MSCA1)^+$ adipogenic subset, whereas in the septa, the $CD271^{high}$ subsets with a high tendency for myofibroblast conversion are more abundant [125].

Adipose tissue is widely distributed throughout the body and exhibits a rich heterogeneity. Previously, adipose tissue was considered uniform, but current research has revealed differences across various fat depots and even within individual depots, including cellular origin, metabolic functions, and interactions with other organs. Next, we will elaborate on the anatomy, functions, and physiological roles of adipose tissue based on the location of the fat deposit.

## Visceral White Adipose Tissue (vWAT)

The definition of vWAT remains unclear. Some researchers define vWAT as the adipose tissue surrounding intra-abdominal organs, whereas others consider retroperitoneal fat, including perirenal fat, as part of vWAT [126, 127]. One classification method defines genuine vWAT based on whether the fat depot drains into the portal vein, represented by fat depots around the omentum and mesentery [3]. The portal vein theory posits that the delivery of FFAs and proinflammatory cytokines from vWAT to the liver via the portal vein contributes to insulin resistance and metabolic deterioration observed in obese individuals [128]. However, in the broad sense, "visceral" should include the gonads and kidneys, and this section introduces the broader concept of vWAT.

### *Omental Adipose Tissue*

The omentum is rich in adipose tissue and can be subdivided into greater and lesser omenta. The greater omentum has received more attention and been studied extensively. Although considered as WAT [129, 130], omental adipose tissue can undergo beige adipogenesis in vitro through a gut-derived peptide called glucose-dependent insulinotropic polypeptide (GIP) [131], and by a thiazolidinedione (TZD) named rosiglitazone [132].

The omentum is microscopically composed of translucent membranous areas and highly vascularized adipose-rich regions, with different proportions and anatomical locations in humans and rodents with greater omenta [133]. The human greater omentum predominantly consists of fat cells [133, 134], while murine omental adipose tissue accounts for only 40% of the omentum [133]. The omental adipose tissue is enclosed by two layers of mesothelial cells to represent a substantial, flattened layer in human omentum [134–136] but is restricted as an adipose-rich rim of murine omentum [133]. Macroscopically, human omental fat resembles an apron hanging from the greater gastric curvature down to the level of the pubis symphysis with a surface area of 300 $cm^2$ to 1500 $cm^2$ [134, 137]. However, murine omental adipose tissue presents as a much smaller fat strip that extends along the anterior greater curvature of the stomach [133]. Nonetheless, human and mouse omenta share similar surface structures via electron microscopy [133], indicating that mice may be justified as experimental models to study omental adipose tissue to some extent.

Current studies on omental adipocytes have often been conducted in pathological contexts, and the physiological roles of omental fat depots remain to be elucidated. Clinical evidence has demonstrated that obesity-induced omental adipocyte hypertrophy is strongly associated with impaired systemic insulin sensitivity and hepatic steatosis in obese adults [138, 139], indicating that omental adipose tissue may contribute to obesity-related metabolic disorders. In addition, omental adipose tissue is found to promote the development of malignant tumors owing to its highly vascularized microenvironment and the biochemical signals it releases [140, 141].

## *Mesenteric Adipose Tissue*

Mesenteric adipose tissue (MAT) extends proximally from the mesenteric attachment toward the mesentery base [142], referring specifically to the adipose tissue attached to the intestinal walls [143]. MAT can be regionally differentiated into small intestinal MAT and large intestinal MAT. Small intestinal MAT is not only characterized by smaller adipocytes but also contains a higher prevalence of fat-associated lymphoid clusters compared to large intestinal MAT [144]. Evidence from inflammatory bowel studies suggests that MAT may play a role in bowel immune regulation. In patients with Crohn's disease, bowel inflammation penetrates the adjacent adipose tissue along the mesentery [145]. Liu et al. [146] suggested that MAT is an independent determinant of metabolic syndrome and is associated with the thickness of the intima and media in human carotid arteries [146]. Furthermore, MAT also exhibits browning potential, as demonstrated by the ability of both human MAT and primary mesenteric adipocytes from Crohn's disease patients and healthy controls to induce browning [147]. However, it should be noted that research on MAT is currently limited. The close anatomical proximity of

MAT to surrounding tissues complicates sample collection, often resulting in vascular contamination, which has impeded comprehensive investigations into the role of the mesenteric fat pad in metabolic health [148, 149].

## *Perirenal Adipose Tissue*

Perirenal adipose tissue (PRAT) surrounds the kidneys on both sides and is enclosed by a thick multilayered fibrous tissue known as the renal fascia [150]. The medial part of PRAT, referred to as medial PRAT (mPRAT), is found around the helium of the kidney, where studies have discovered UCP1-expressing brown adipocytes [150]. In contrast, the lateral portion of the PRAT, termed lateral PRAT (lPRAT), mainly consists of white adipocytes [21]. Observations of age-related changes in human PRAT suggest gradual substitution of brown adipocytes with white adipocytes over time [151]. However, analysis of PRAT obtained from individuals in Siberia at the time of autopsy showed up to a 40% composition of brown adipocytes, demonstrating the capacity of PRAT to adapt to colder environments [152]. Additionally, dormant brown adipocytes, characterized by unilocular lipid droplets and a gene expression pattern that shares features with white adipocytes, yet with positive expression of UCP1, have been identified in the PRAT of adult kidney donors [153]. Therefore, PRAT is a unique visceral fat depot containing a mixture of brown and white adipocytes. Further studies are needed to investigate the origin and plasticity of cellular heterogeneity in PRAT.

## *Gonadal White Adipose Tissue*

Gonadal white adipose tissue (gWAT) is sometimes termed periovarian WAT in females and epididymal WAT (eWAT) in males. gWAT, a specialized form of adipose tissue in mammals, envelops reproductive organs, specifically encompassing the uterus and ovaries in females and the epididymis and testes in males [149]. Periovarian WAT is situated in proximity to the ovaries, uterus, and bladder and plays a crucial role in the local endocrine milieu. In instances of obesity, observations indicate that gonadal and perirenal fat depots exhibit continuity, with fat accumulation surrounding the uterine horns and ovarian coalescing [149]. eWAT, on the other hand, is intimately associated with the epididymis, testicular tissues, and prominent vasculature supplying the epididymis, highlighting its integral role in structural support and potential metabolic influence on these reproductive structures. However, in humans, no corresponding gonadal fat region has been identified as that in mice. Therefore, whether gWAT in mice accurately represents vWAT in humans for research purposes remains a subject of ongoing debate.

## Subcutaneous Adipose Tissue

Located beneath the skin, encapsulated by connective tissue and supported by blood vessels and nerves, subcutaneous adipose tissue (sAT) is widely distributed. Given the significant differences in the distribution patterns of sATs between mice and humans, this section will discuss sATs in mice and humans separately.

### *Mice Subcutaneous Adipose Tissue*

In mice, sAT can be divided into anterior and posterior sATs [3]. The anterior sAT is primarily located along the midline and on both sides of the mouse, including fat depots located at the deep cervical, superficial cervical, interscapular, subscapular, and axillo-thoracic areas [154]. Posterior sAT is primarily found in the inguinal, dorsolumbar, and gluteal regions [154]. Typically, the term sWAT is used to refer to sAT; however, this oversimplifies the functional diversity of sAT. Indeed, because inguinal adipose tissue is the largest and most easily isolated fat depot among mouse sATs, it is considered the archetype of sATs, exhibiting typical WAT characteristics at room temperature and is commonly referred to as iWAT [155]. However, iWAT does not represent the entirety of sAT, as it exhibits much greater heterogeneity than previously thought. Upon cold exposure and exercise, iWAT activates thermogenesis, which is attributed to the reversible browning process of beige adipocytes [156–158]. Wu et al. isolated clonal beige adipocyte cell lines from mouse iWAT, which could be induced to activate thermogenesis [24]. Unlike posterior sAT, mouse anterior sAT appears brown, and interscapular fat depot displays visible brown and white components even at room temperature [154, 159]. The brown component of interscapular adipose tissue, commonly referred to as interscapular BAT (iBAT), is the largest brown adipose depot in mice and is the primary focus of BAT research [155].

Despite primarily exhibiting the characteristics of WAT, mouse iWAT demonstrates features distinct from those of vWAT. With the exception of inducible thermogenic activation under physiological stimuli such as cold exposure and exercise, iWAT appears less geared toward proliferation. In an HFD regimen, vWAT demonstrates an earlier onset of lipogenesis, observable as early as 4 weeks, whereas iWAT exhibits hypertrophy over an extended duration, typically not until 12 weeks [160]. Additionally, iWAT is distinguished by a reduced rate of macrophage infiltration, contributing to its diminished release of proinflammatory cytokines compared to vWAT [161]. Further studies have confirmed the critical anti-inflammatory role of CXCL14-expressing adipose stem cells (ASCs) from iWAT in obesity. This effect was achieved by suppressing the inflammatory CXCL12-CXCR4 pathway in monocytes, consistent with previous findings [162, 163]. This aligns with emerging evidence that iWAT ASCs inhibit monocyte infiltration in obesity [164], accentuating the specialized role that ASCs play in diverse adipose tissues under metabolic stress.

### *Human Subcutaneous Adipose Tissue*

sAT is widely distributed throughout the human body. In infancy, interscapular adipose tissue comprises adipocytes that express genes characteristic of classical brown adipocytes [165]. However, with growth, BAT in the interscapular region eventually regresses [165]. In adults, the main sATs are primarily located in the abdominal region, buttocks, and thighs [166]. Specifically, adipose tissue found in the buttocks and thighs is collectively referred to as the gluteofemoral depot [167, 168]. Research has demonstrated that gluteofemoral sAT exhibits a reduced level of lipolytic activity compared to sAT in the abdominal region [169]. In humans, sAT may possess inducible thermogenic activation capabilities. Histological and molecular evidence suggests that human thigh adipose tissue upregulates thermogenesis-related genes such as UCP1 in response to cold stimulation, indicating the presence of browning [170, 171]. However, PET/CT imaging did not reveal an increase in glucose uptake in the sAT in response to cold stimulation [172, 173]. In one study, acute cold exposure decreased the local blood flow in human sAT, resulting in reduced glucose uptake [174]. Apart from the lower sensitivity of PET/CT, potentially leading to false-negative results, one possible hypothesis is that, under cold stimulation, sAT primarily serves as a thermal insulator through vasoconstriction, with the activation of thermogenesis requiring longer exposure time to manifest [16].

## Perivascular Adipose Tissue

### *Anatomy of the Perivascular Adipose Tissue*

Perivascular adipose tissue (PVAT) is defined as a unique fat depot surrounding blood vessels throughout the body (except for the cerebral vasculature), without any anatomical structural barrier [175–177]. According to this definition, PVAT can be subdivided into periaortic adipose tissue and organ-specific fat, encircling distinct organ vasculature. Therefore, broadly speaking, epicardial adipose tissue (EAT) and adipose tissue of the mesenteric bed and renal sinus should fall under the category of PVAT [178, 179]. However, EAT will be discussed separately, while the latter two are currently considered parts of mesenteric and perirenal adipose tissue and classified as vWAT. Moreover, in the existing literature, the term PVAT primarily refers to periaortic adipose tissue, with special emphasis on thoracic periaortic PVAT (tPVAT) and abdominal periaortic PVAT (aPVAT). Therefore, this section predominantly focuses on periaortic adipose tissue.

### Thoracic Periaortic PVAT

The tPVAT extends from the aortic arch at the T4 vertebra to the T10–T11 vertebrae, namely, to the aortic hiatus of the diaphragm [180]. Microscopically, the majority of rodent tPVAT exhibit morphological characteristics of BAT, with adipocytes having round nuclei, abundant mitochondria, and small multilocular lipid droplets, whereas the remaining small areas structurally resemble WAT [181, 182]. A series of human autopsies validated that multilocular BAT-like cells in tPVAT immunochemically express high levels of UCP1, a unique marker of brown adipocytes [183]. Using microarray analysis of mice, tPVAT shares approximately identical global gene expression patterns with classical iBAT, with disparity lying in only 228 significantly different genes (0.79%) [181]. PVAT is reported to exhibit active thermogenesis similar to BAT in response to cold exposure in mice [35]. This is further supported by $^{18}$F-FDG PET/CT studies in adult humans showing that cold-induced $^{18}$F-FDG intake is markedly upregulated in the periaortic area [184].

Although sharing some similarities with BAT, it exhibits notable differences from classical BAT when exposed to external stimuli. In mice fed a HFD for 13 or more weeks, tPVAT and iBAT both showed resistance to the extent of inflammation. However, the appearance of BAT appeared "whiter," and an unexpected distortion in the architecture of BAT lipid droplets was observed, a phenomenon not seen in tPVAT. Nevertheless, tPVAT developed a dissimilar manifestation, with swollen mitochondria and unfolding of cristae. These findings are supported by alterations in gene expression of HFD [181].

Intriguingly, further efforts were made in an experimental investigation to subdivide tPVAT into lateral tPVAT (L-tPVAT) and anterior tPVAT (A-tPVAT), presenting a quite discrepant picture [185]. More UCP1$^{+}$ adipocytes are situated in A-tPVAT than in L-tPVAT. In addition, disparity lies in the biological activity. L-tPVAT had a significantly improved vasorelaxation effect on the aorta compared to A-tPVAT in the vascular ring experiment, whose biological basis and physiological regulatory role remain to be further explored.

### Abdominal Periaortic PVAT

The aPVAT encompasses the abdominal aorta, from the aortic hiatus of the diaphragm to the femoral bifurcation [180]. Morphologically, adipocytes within the aPVAT, as observed via light and electron microscopy in both rodents and humans, are primarily unilocular, resembling those in WAT [182, 186, 187]. Human aPVAT has chemoattractant properties and secretes comparable amounts of chemokines as sWAT [187]. Given the growing evidence for the physiological and pathological roles of WAT, clarifying the similarities and differences between aPVAT and WAT may reveal the bioactivity and metabolic mechanisms of aPVAT, which warrant further study.

### Difference Between tPVAT and aPVAT

These findings have encouraged researchers to hypothesize that tPVAT and aPVAT represent BAT and WAT, respectively, which are thought to act differently in their bioactive patterns. In response to stimuli, WAT tends to initiate the inflammatory pathway more than BAT. Corroborating this, an experimental study in an HFD-induced obese mouse model revealed that aPVAT explants were capable of promoting more macrophage migration than tPVAT explants, accompanied by a higher elevation in the mRNA abundance of macrophage markers and proinflammatory chemokines in aPVAT than in tPVAT [182]. In addition, aPVAT in rats was more inflamed than tPVAT, with a proinflammatory profile of gene expression and immune cell infiltration [186]. Further insights are awaited from clinical and experimental data to investigate the biological basis and functional links between the PVAT, WAT, and BAT.

In general, periaortic PVAT has attracted considerable attention owing to its unique anatomical location and relative convenience. PVAT is analogous to the comparatively well-studied WAT and BAT. tPVAT exhibits a BAT-like phenotype, capable of thermogenesis, whereas aPVAT resembles WAT, presenting a more pro-inflammatory pattern. Although questions remain to be answered, exploration of PVAT opens a window into the external environment of the vasculature and provides novel opportunities to identify the cross talk between PVAT and homeostasis of the cardiovascular system.

## *Origin of Perivascular Adipose Tissue*

### The Debate of Perivascular Adipose Tissue Origin

Evidence regarding the developmental origin of PVAT has yielded controversial results. Some studies have shown that progenitor cells of PVAT share a common developmental origin with vascular smooth muscle cells (VSMCs) [35, 185]. Smooth muscle protein 22-alpha (SM22α), a putative calcium-binding protein, is expressed in mesenchymal populations around the dorsal aorta during mouse embryogenesis at embryonic day (E) 9.0 and is considered one of the earliest markers of VSMC [188]. A lineage-tracing study elucidated that adipocytes within the tPVAT primarily originate from $SM22\alpha^+$ progenitors, providing powerful support for this hypothesis [185]. Research has discovered that PPARγ is an essential factor for adipogenesis in vitro and in vivo [189, 190]. Progenitor cells prenatally expressing PPARγ are capable of proliferating and are the main source of the rapid growth of adipocyte development from birth to 1 month old in mice [191]. In an experimental study using a smooth muscle cell (SMC)-specific *Pparg* knockout mouse model with a genotype of $\textit{Pparg}^{flox/flox}$*;SM22α-Cre;*$\textit{ApoE}^{-/-}$, there was no PVAT supposedly surrounding the aortic arch, thoracic and abdominal aorta, along with a significant reduction in mesenteric adipose tissue [35], further corroborating this hypothesis.

However, the dissenting voice is that PVAT develops from a fibroblastic lineage [192]. Single-cell RNA sequencing (scRNA-seq) analysis for stromal vascular fraction (SVF) separated from thoracic aortas at postnatal day (P) 3 (with nearly identical integration with E18) revealed several cell populations including SMCs, adipocytes, and a connective array of fibroblast cells. Fibroblast cell populations can be further classified into progenitor cells, intermediate cells, and preadipocytes. To verify whether SMCs serve as the origin of periaortic fat cells, Angueira et al. [192] used Myh11-Cre;tdTomato reporter mice as genetic lineage-tracing models. TdTomato$^+$ adipocytes were not detected around the aorta at P3 in reporter mice, challenging the notion that PVAT shares a common origin with SMCs, as previously discussed. Furthermore, ex vivo experiments to estimate adipogenic potential showed that SMCs present nearly no ability to undergo adipocyte differentiation. In contrast, progenitor cells and preadipocytes of the fibroblast cell populations are capable of developing adipocytes, and preadipocytes display the strongest adipogenic potential. The supposition of periaortic PVAT arising from fibroblastic lineage was further demonstrated using genetic lineage-tracing models.

As Chang et al. [35] utilized the SM22α-Cre expressed as early as E9.0 in mice in their studies, Angueira et al. [192] investigated SVF of periaortic PVAT at E18 and mainly at P3 in mice. The disparity in progenitor cells of PVAT may lie in the stages chosen differently to explore development during the embryonic period and after birth. This possibility cannot be ruled out because, in adult mice of 13 weeks old, the SVF of tPVAT encompasses a new cluster of SMCs contributing to PVAT formation [192]. More comprehensive research is necessary to define these conditions more clearly, emphasizing the need for more precise descriptions of the developmental trajectory.

### Regional Differences in Perivascular Adipose Tissue Origin

The phenotypic and bioactive disparity along the aorta discussed above suggests that PVAT may be a heterogeneous tissue derived from different progenitor cells with local specificity. Administration of thiazolidinediones (TZDs), known as PPARγ agonists, markedly augments the differentiation of preadipocytes with a high affinity to PPARγ [193]. Precursor cells stemming from aortic explant from aortic rings in C57Bl/6 J mice were used ex vivo for adipogenesis using rosiglitazone as one of the TZDs, depicting different gene expression profiles of tPVAT and aPVAT [194]. Moreover, the results of transcriptome analysis and aortic ring array suggested that progenitor cells from tPVAT and aPVAT maintained their inherent identity to develop into transcriptionally more thermogenic and WAT-like adipocytes, respectively, with cell-autonomous traits, even ex vivo [194].

Except for what was mentioned above, developmental settings may be different in PVAT, even at different anatomical locations of the same arterial segment. Transcriptome analysis of L-tPVAT and A-tPVAT showed a massive discrepancy in gene expression for regulating development [185]. At the same time, a lineage-tracing study further validated that adipocytes derived from Myf5$^+$ are located in

L-tPVAT rather than in A-tPVAT, reflecting that L-tPVAT is present at an earlier development stage [185].

Taken together, PVAT is a highly heterogeneous tissue derived from multiple progenitor and precursor cells, which remains to be precisely determined. It is crucial to acknowledge that current studies on the origins of PVAT are mainly conducted in experimental animal models, underlining that these findings have yet to be determined in humans.

## *Endocrine and Physical Function of Perivascular Adipose Tissue*

The long-standing notion that PVAT provides structural support for the vascular system has led to its importance being overlooked. It was not until 1991 that PVAT was first demonstrated to attenuate the contractile effects of NE in the rat aorta ex vivo [195]. Given its close anatomical proximity to the vascular wall and the fact that human aortic arch PVAT was even observed with adipocytes infiltrating the adventitia under light microscopy [196], the physiological characteristics of PVAT have attracted enormous interest. Recently, PVAT has garnered considerable attention because of its role in vascular tone by mediating both vasodilation and vasoconstriction. It has been reported that PVAT secretes PVAT-derived relaxing factors (PVRFs) to blunt the agonist-induced vasocontractile impact in vascular protection, such as NO, palmitic acid methyl ester (PAME), and prostacyclin [35, 197, 198]. Vasocontractile effects may be modulated by PVAT-derived contractile factors (PVCFs), including catecholamines, angiotensin II, and chemerin [199–201]. Notably, the pinpoint mechanisms of PVAT in performing and balancing both converse functions remain unclear. Apart from the effects on vascular tone, studies have provided new insights into the role of PVAT in VSMC growth and migration by releasing growth factors and influencing the formation of neointima [202, 203].

As aging is an inevitable and natural biological process, its effect on PVAT remains unclear. Bioinformatics analysis revealed a decrease in the adipogenic differentiation competence of resident stromal cells within the PVAT into endothelial and brown fat cells, promoting vascular remodeling [204]. In line with this finding, the number of brown-like adipocytes within tPVAT was observed to decline in aged mice [185], which can be seen in the periaortic PVAT of aged rats with a decrement in anti-constrictive function [205]. In addition, aging results in older mice being more vulnerable to obesity by enhancing macrophage infiltration in the PVAT and exacerbating the alteration in the vascular wall to a more proinflammatory pattern than in younger mice [206]. In terms of inflammation, which plays a core role in atherosclerosis, PVAT has the potential to be a new predictor and therapeutic target for cardiovascular aging-related diseases, which requires further investigation.

Apart from the heterogeneous morphological, original, and functional disparities in PVAT discussed above, emerging evidence reveals that a significant gender gap lies in PVAT as well. The Framingham Heart Study with 3246 participants concluded that the volume of tPVAT in women was higher than that in men, in line with a higher prevalence of excess tPVAT in women as aging [207]. Experimental data from stroke-prone spontaneously hypertensive rats (SHRSP) suggest that sex differences can be attributed to the decreasing vasodilation function of male PVAT [208]. Because the vasoprotective factor estrogen works as a vasodilator, an anti-inflammatory and antioxidant factor, a regulator of the RAAS system, and an inhibitor of the sympathetic nervous system [209–211], further research is required to identify the clinical significance and mechanisms of metabolic sexual disparity in PVAT.

## Epicardial Adipose Tissue (EAT)

### *Anatomy of Epicardial Adipose Tissue*

Epicardial adipose tissue (EAT) is characterized by unique properties that differentiate it from classical WAT and BAT. It is found between the myocardium (heart muscle) and epicardium (outer layer of the heart) and is supplied by the coronary arteries [212]. Approximately 80% of the heart surface is covered by epicardial fat, which accounts for approximately 20% of the total weight of the heart [213]. EAT is localized around the major coronary arteries, ventricular surface, and apex in humans, but it is primarily concentrated in the atrioventricular groove in mice and rats [214]. EAT can be categorized into two types: coronary EAT, which is directly adjacent to the coronary artery adventitia, and myocardial EAT, which is located over the myocardium [215]. The EAT surrounding the left atrium shows high expression of genes encoding arrhythmogenic factors [216], whereas the coronary artery EAT is characterized by high expression of genes that promote inflammatory adipokines and regulate glucose and lipid metabolism [216].

EAT possesses unique functionalities owing to its distinct anatomical location and direct contact with the heart but without fascial separation from the myocardium [212]. EAT safeguards the heart and coronary arteries by preventing mechanical deformation and shielding the heart from the impacts of low body temperature [217]. In neonates, the EAT has thermogenic characteristics and functions similar to BAT. However, as age increases, the function of EAT changes from thermogenesis to energy storage, and the BAT-like activity of EAT significantly decreases with age [218]. Additionally, the release of FFAs from adipocytes in the EAT may serve as a local and rapidly mobilizable energy supply for cardiomyocytes [219, 220]. Moreover, EAT protects myocardial cells from lipotoxicity by converting excess FFAs into fatty acyl-coenzyme A (acetyl-CoA), which is then stored as triacylglycerol in lipid droplets of adipocytes [221, 222].

## *Adipogenesis of Epicardial Adipose Tissue*

Research on the origin of EAT is limited. A study involving embryos from miscarriages and ectopic pregnancies found that the precursor tissue of EAT appears as early as the beginning of the second gestational month (days 33–35 of gestation), and unilocular adipocytes are present by the eighth gestational month (week 32 of gestation) [223]. EAT formation begins in the atrioventricular grooves and progresses to the interatrial groove and right ventricle, eventually reaching the interventricular groove [223]. In mice, EAT typically originates in the atrioventricular groove, which is derived from the epicardial component in embryos [224]. The EAT in the atrioventricular groove of mice appears between the first and second weeks after birth, and its formation depends on the expression of PPARγ [224]. In addition, studies have shown the involvement of Wt1 (Wilms tumor 1) and Msln (mesothelin) cells in the development of EAT adipocytes [129, 225]. In a previous study using *Wt1*$^{\mathrm{CreERT2}}$;*Rosa26-mTmG* reporter mice, it was found that approximately 66% of epicardial adipocytes were originated from Wt1-expressing mesothelial cells during E14.5 to E16.5 [129]. By injection of the *Msln-Cre* adenovirus into *Rosa26-mTmG* embryonic hearts, a study revealed that the embryonic epicardium contributes to the formation of epicardial fat during development [225].

## *Endocrine Function of Epicardial Adipose Tissue*

Similar to other adipose tissues, EAT possesses endocrine functions and secretes adiponectin, leptin, adrenomedullin, omentin, visfatin, resistin, and other adipokines [226]. For example, adiponectin secreted by EAT can alleviate cardiac dysfunction by regulating PPARγ [227]. Furthermore, adiponectin directly reduces myocardial NADPH oxidase activity through endocrine or paracrine mechanisms, and its expression in the EAT is regulated by the paracrine effects of heart-released oxidation products [228]. Adiponectin can also protect cardiomyocytes from enlargement and reduce inflammation and fibrosis in the heart, thereby lowering the risk of adverse clinical events [229].

However, in terms of endocrine function, EAT behaves more like the so-called bad adipose tissue. Numerous studies have suggested that EAT promotes disease progression by secreting proinflammatory and pro-fibrotic mediators [216]. EAT-derived leptin induces metabolic syndrome-related myocardial injury through two pathways: mitochondrial oxidative stress, which leads to apoptosis, and inflammation via AP1 translocation [230]. High levels of IL-1β, IL-6, monocyte chemoattractant protein 1 (MCP-1), and TNFα mRNA and protein have been observed in this tissue [231]. These inflammatory factors can contribute to the formation of atrial fibrillation [232]. Patients with coronary artery disease demonstrate mild inflammation of the EAT due to macrophage infiltration [231]. The proinflammatory, profibrotic, and proarrhythmic effects of EAT-derived EVs may play a role in the development of cardiovascular diseases associated with obesity and abnormal adipose tissue accumulation [233].

## Dermal Adipose Tissue

### *Anatomy of Dermal Adipose Tissue*

Dermal adipose tissue (dAT) is a loosely structured dermal component composed of adipocytes that are functionally integrated into the skin [234, 235]. In rodents, it typically manifests as a substantial layer of adipose cells ranging from 2 to 15 cells in thickness, situated immediately within the reticular dermis, demarcated by a discrete muscle layer known as the panniculus carnosus [236, 237]. The human skin is distinguished by a prominent dAT layer that lacks a well-defined boundary with sAT, a feature attributed to an underdeveloped skeletal muscle infrastructure [234, 238]. During the growth phase, hair follicles interact closely with dermal adipocytes [237]. Additionally, sweat glands, ubiquitously present across the human skin, are embedded in the dermis and occasionally come into contact with the interface of dAT [239].

Over the past few decades, increasing attention has been paid to the importance of dAT in various biological processes. Presently, dermal adipocytes are acknowledged as a discrete, exceptional population distinguished by their unique biological characteristics that extend beyond mere metabolic functions [235, 240]. Many contemporary studies have elucidated the multifaceted involvement of dAT in an extensive array of physiological processes, including the regulation of hair follicle cycling [241], modulation of local immune responses during infections [242], and facilitation of wound-healing mechanisms [243, 244]. Segalla et al. [240] categorized dAT into at least three or four functional compartments within the dermis: near pilosebaceous units, sweat glands, fibroblasts/immune cells/connective tissue, and a diffuse, unclustered form with undefined roles. Recent studies have shown that dAT in mice undergoes reversible dedifferentiation and displays high plasticity depending on environmental changes [245]. In short, dAT exhibits fascinating, non-metabolic attributes, underscoring its specialized and intriguing nature [235, 240].

### *Origin of Dermal Adipose Tissue*

During mouse skin development, adipose precursors emerge on E16.5 [246]. Multilocular adipocytes first emerge between E17.5 and E18.5 and rapidly multiply in number and size [247–249], peaking around P12. At P17, the dAT around the hair follicles began to collapse with a decrease in cell size rather than substantial cell apoptosis [249]. By P19–P23, dermal adipocytes shrink to less than half their size

[247, 249]. Mature dermal adipocytes dedifferentiate into smaller fibroblast-like adipocyte-derived precursor cells during the degenerative phase of hair follicles, which can re-differentiate back into mature adipocytes during the active growth phase of hair follicle [250]. The cyclical growth of dAT is a consequence of hypertrophy, involving an increase in the size of preexisting adipocytes, and hyperplasia, a process driven by the proliferation and differentiation of adipocyte precursors into mature, lipid-filled, unilocular adipocytes [241, 251, 252].

During the embryonic period, early mouse adipose precursors can be labelled with *Dlk1*-CreER [241]. Following E16.5, there is a dermal lineage specification in which cells expressing Dlk1 are restricted to give rise to lower dermal derivatives including adipocytes [246]. In the adult mouse skin, a large portion of adipose progenitors can be marked using *Pdgfrα*-CreER [241]. PDGFRα$^+$ progenitor cells also have the ability to differentiate into fibroblasts, which can participate in wound-healing [246]. The proliferation of both adipocyte stem cells and preadipocytes critically relies on PDGFα/PDGFRα signaling, and mesenchymal cell-specific deletion of PDGFα in mice impedes dAT expansion during this period [252]. This highlights the importance of PDGFα/PDGFRα signaling in regulating the hyperplastic growth of dAT.

The origin and development of dAT are distinct from those of the traditional WAT and BAT. However, multiple adipocyte precursor populations have been identified in murine dAT, all sharing common cellular surface markers (CD34, CD29, and Sca1) with adipocyte precursors from other WAT depots [241, 253]. Lin$^-$, CD24$^+$, CD29$^+$, CD34$^+$, and Atxn1$^+$, a marker set originally developed for visceral and subcutaneous adipose stem cells [253, 254], can also be used to sort dermal adipose progenitors for quantification and gene expression profiling [252]. Similar to vWAT, CD24$^+$ skin progenitors represent the early adipose stem cells necessary for the long-term maintenance of dAT [252].Within this framework, CD24$^+$ cells denote an adipocyte stem cell pool, whereas CD24$^-$ cells represent committed pre-adipocytes [252, 255]. Both CD24$^+$ and CD24$^-$ adipocyte precursor subsets can proliferate; however, the CD24$^+$ adipocyte stem cell population experiences preferential expansion, giving rise to approximately 20% of new adipocytes [241, 252].

## Bone Marrow Adipose Tissue (BMAT)

### *Anatomy of Bone Marrow Adipose Tissue*

Bone marrow adipose tissue (BMAT) contributes to over 10% of total body fat and represents a distinctive category of adipose tissue with unique functional characteristics [256, 257]. BMAT functions as a modulator of skeletal homeostasis and energy metabolism, which fills 50–70% of the bone marrow cavity in the human body [258]. Both murine and human bone marrow adipocytes exhibit a large lipid droplet, enveloped by a very thin layer of cytoplasm, with the cell nucleus situated peripherally, and they express comparable levels of adiponectin and leptin as white adipocytes [259]. The average diameter of these marrow adipocytes is generally less than that of subcutaneous and visceral white adipocytes, usually measuring between 40 and 65 micrometers [260].

There are two types of BMAT: constitutive BMAT (cBMAT), which is located within the yellow marrow of distal bones and has little to no hematopoiesis, and regulated BMAT (rBMAT), which is found in proximal bones with active hematopoiesis and bone remodeling [261–263]. cBMAT exists in the medullary canal at the junction of the sacrum and ankle, whereas rBMAT develops within the bone marrow of long bones after birth, extending from beneath the growth plate to the metaphysis and entering the diaphysis [264]. cBMAT plays a suppressive role in hematopoiesis, potentially by preserving hematopoietic stem cells (HSCs) in a quiescent state, as elegantly demonstrated by research from the Daley laboratory [265]. In relation to the skeletal system, the volume of cBMAT, as assessed by techniques such as osmium staining and micro-CT in rodents or *magnetic resonance imaging* (MRI) in humans, is inversely proportional to bone mass [266, 267]. Conversely, rBMAT exhibits a more adaptable nature, functioning as an adipose tissue with endocrine and paracrine effects on hematopoietic and skeletal remodeling processes [256]. A high rBMAT volume is frequently, albeit not universally, correlated with reduced bone mass [268, 269]. However, these findings regarding BMAT have only been confirmed in mice, and whether they hold true for human BMAT remains to be determined.

## *Adipogenesis of Bone Marrow Adipose Tissue*

Previous studies have shown that the source of adipocytes in the bone marrow differs from that in WAT [270]. The bone marrow exists in two forms: red and yellow. The red marrow is active in blood cell formation, and its red color is derived from abundant erythrocytes [271]. The yellow marrow is mostly filled with fat cells, defining its color, as it also contains some primitive blood cells [271]. During growth, bone marrow transitions from red to yellow as fat accumulates, particularly in long bone shafts and trabecular regions by early adulthood, with fat comprising 20–25% of the adult skeletal mass [271]. BMAT typically expands first at the ends of long bones and then progresses toward the shaft, moving toward the ends [272, 273]. As this occurs, the red marrow reduces and eventually occupies only specific sites such as the ribs, sternum, certain vertebrae, sacrum, and coccyx by approximately 40 years of age [272, 273].

Bone marrow adipocytes originates from mesenchymal stem cells (MSCs), which have the potential to differentiate into both adipocytes and osteoblasts [274]. *PPARG* is recognized as a key regulator that governs the process of adipocyte differentiation during the development of bone marrow adipocytes. Deletion of *Pparg* results in a decrease in the number of bone marrow adipocytes [275]. Additionally, haploinsufficiency of *Pparg* has been associated with a notable increase in bone mass [276]. Bone marrow adipocytes originate from heterogeneous lineages. Lineage-tracing studies have revealed the origins of MSCs, showing that they derive from Osterix1-positive cells found in neonatal bone marrow [277]. However, in adult but not young mice, MSCs have been associated with cells that express LepR [274]. Notably, the leptin signaling pathway is instrumental in managing both the development of bone marrow adipose tissue and bone formation in adult mice [278].

Bone marrow adipocytes are undeniably derived from the mesenchymal progenitor cells that inhabit the bone marrow. Nonetheless, the exact lineage hierarchy of these adipocytes within BMAT, as well as the molecular pathways governing their differentiation, is yet to be fully elucidated.

## *Endocrine Function and Physiology of Bone Marrow Adipose Tissue*

BMAT plays a pivotal role in the maintenance of bone marrow homeostasis. It is also involved in regulating hematopoiesis and osteoblastogenesis. In contrast, BMAT significantly contributes to energy metabolism, highlighting its multifaceted influence on both the hematological and skeletal systems, as well as overall metabolic health.

BMAT is a complex and dynamically active endocrine adipose depot located in non-hematopoietic bone marrow niches that interacts with hematopoietic cells and bone tissue [279]. Nguyen et al. proposed that adipocytes might facilitate hematopoiesis by secreting leptin [280]. Notably, in the early stages of postnatal development, endothelial and stromal cells expressing LepR are demonstrated to exhibit the highest levels of stem cell factor [281], which is a crucial component for the development and maintenance of hematopoietic stem cells. Furthermore, crucial molecules, such as CXCL12, IL-3, IL-8, CSF3, and LIF, are expressed in bone marrow adipocytes, directly promoting the survival of HSCs [282]. According to a report by Corre et al. [283], bone marrow adipocytes can produce cytokines such as IL-7, IL-15, EPO, and TPO, which directly promote hematopoiesis through the induction of lymphoid and myeloid differentiation of HSC clonogenic progenitors. Bone marrow adipocytes express or produce factors that participate in regulating the differentiation of HSCs into lymphocytes, granulocytes, megakaryocytes, and erythrocytes, but their specific role is not yet clear [284]. For instance, the reduction in B lymphocyte production is associated with an increase in adipose tissue in the bone marrow, a part of which promotes the generation of myeloid-derived suppressor cells [285]. These cells inhibit the production of B lymphocytes via the secretion of IL-1. During in vitro co-culture with HSCs in vitro, bone marrow adipocytes can support megakaryocyte maturation and promote platelet production [286].

To some extent, marrow adipocytes have been shown to exert a suppressive effect on bone formation. Indeed, there is evidence suggesting that the deletion or reduction of marrow adipocytes leads to enhanced bone formation [287]. In vitro studies have shown that adipocytes derived from MSCs can induce differentiation of MSC-derived osteoblasts into adipocyte phenotypes [288]. The lipotoxicity of FFAs produced by bone marrow adipocytes is mediated by their action as natural ligands for PPARγ, inducing activation of PPARγ in MSCs, pre-osteoblasts, and osteoblasts, promoting adipogenesis and transforming osteoblasts into adipocytes by inhibiting *Runx2* activity [279, 289]. Furthermore, within the FFAs produced by bone marrow adipocytes, palmitic acid impairs osteoblast formation and mineralization by inhibiting the Wnt/β-catenin and BMP2/Runx2/Smads pathways and also induces apoptosis and autophagy dysfunction in bone cells [279, 290, 291].

Notably, BMAT possesses endocrine functions and is also subject to regulation by endocrine factors. It exhibits endocrine/paracrine roles by secreting peptides, such as leptin and adiponectin, as well as cytokines, including IL-6, IL-1β, and TNFα [292]. In addition, compared to epididymal adipocytes, there is an upregulation of IL-6, IL-1β, and TNFα gene expression in mouse bone marrow adipocytes, indicating that BMAT has inflammatory potential [293]. In 2014, Cawthorn et al. [256] demonstrated that BMAT acts as an endocrine organ and contributes significantly to circulating adiponectin levels under various conditions in control mice, genetically modified mice (Ocn-Wnt10 mice), rabbits, and humans. Research conducted on human cell cultures has demonstrated that marrow adipocytes exhibit high levels of leptin expression and secretion [294]. Further research is needed to explore the effects of adipokines such as leptin and adiponectin on bone. Expansion of BMAT is regulated by growth hormone, with a marked increase in BMAT occurring in pituitary-ectomized rats, an effect that is mitigated by the introduction of external growth hormone [295].

Furthermore, the BMAT is innervated by sympathetic and sensory axons, which can potentially influence its adaptation to different skeletal sites [296]. It is directly responsive to adrenergic signals and can be modified by the neural regulation of surrounding cells, indicating multiple potential mechanisms for the neural regulation of BMAT metabolism [296].

## Mammary Adipose Tissue

Primarily composed of epithelial cells and adipocytes, the mammary gland functions as an exocrine organ responsible for milk production and secretion [297, 298]. Enveloping the ducts and alveoli, the mammary fat pad is a crucial component of the mammary gland, occupying most of the breast [299]. Because the mammary gland is an appendage of the skin, mammary adipose tissue is considered by some researcher to be a type of sAT [298, 299]. An early ultrastructural study suggested that adipocytes within the mammary gland are derived directly from perivascular mesenchymal cells. Throughout the differentiation process, these cells gradually accumulate lipids, ultimately acquiring the morphological characteristics of typical white adipocytes [300]. However, the remarkable functional and morphological changes in the mammary gland during the cyclical lactation process indicate that mammary adipose tissue exhibits characteristics distinct from those of typical sAT, such as iWAT. Although mammary adipose tissue constitutes the primary component of the breast stroma, it has received much less attention from researchers than the mammary parenchyma in studies targeting the mammary gland. In adipose tissue-focused research, mammary adipose tissue is often excluded because of concerns regarding the potential confounding effects of mammary epithelial cells. Nevertheless, there is growing recognition of the role of mammary adipose tissue in mammary development and lactation.

Mammary adipose tissue plays a critical role in postnatal development of the mammary gland. In lipodystrophic mice, mammary gland development remains impaired, even with exogenous estradiol supplementation, as evidenced by the absence of duct terminal end buds [301]. Loss of adipose tissues at 2 weeks and 7 weeks after birth results in decreased duct branching points and terminal end buds, as well as excessive lobulation in mice, respectively [302]. These findings suggest that adipose tissues are essential for the formation of the ductal network during puberty and for maintaining proper lobular architecture. Mammary adipose tissue, a crucial component of the breast stromal microenvironment, has been primed to contribute significantly. Signaling molecules secreted by mammary adipose tissue are crucial for regulating mammary epithelial cells. Conditioned media derived from fully differentiated adipocytes inhibits apoptosis of mammary epithelial cells [303]. The genetic deletion of functional leptin or lack of leptin receptors results in the complete loss of ductal epithelial development, and this defect can be reversed by restoring leptin signaling [304]. Mice overexpressing adiponectin have exhibited abnormal mammary growth and incomplete ductal development [305]. Interestingly, the "white" phenotype of adipocytes may be essential for mammary gland development. As mentioned above, mammary adipose tissue is primarily composed of white adipocytes, while the presence of typical brown adipocytes remains uncertain. An in vivo experiment showed that browning of mammary adipose tissue by a β3-AR agonist is associated with inhibited mammary development [306]. Additionally, proliferation of mammary epithelial cells is suppressed when co-cultured with brown adipocytes in vitro [306].

During lactation, most of the mammary fat pad is replaced by the expanded ductal system and developed alveoli [307]. A plausible hypothesis for the disappearance of mammary adipose tissue during lactation is that adipocytes undergo delipidation to provide the necessary substrates and energy for milk production and secretion. The anabolic processes of mammary adipose tissue in lactating mice were inhibited [308]. The number of lactating offspring and the duration of lactation are determining factors for the extent of mammary adipose tissue depletion, with adipose tissue proximal to the mammary epithelial tissue being consumed more extensively [309]. Adipocytes may be an alternative source of secretory alveoli in addition to ductal stem cells. In late-pregnancy mice, a single large lipid droplet appears in the cytoplasm of alveolar cells, which displays characteristics of both epithelial and adipocyte cells [310]. Immunohistochemical analysis using perilipins 1 and 2 revealed a population of cells with an intermediate adipoepithelial phenotype in the mammary fat pad of mice during pregnancy and lactation [311]. These cells highly express E74-like factor 5 (Elf5), a key regulator of alveologenesis, and both transient and stable ectopic expressions of Elf5 in cultured adipocytes induce the expression of whey acidic protein [311]. Furthermore, using the Cre-loxP recombination system, it has been shown that mouse mammary adipocytes may directly transdifferentiate into secretory epithelial cells during pregnancy [312]. These temporary lipid-containing secretory alveolar cells that are derived from adipocytes during lactation are defined as "pink adipocytes" by Cinti et al. [310]. Mammary adipose tissue also promotes gland involution. Lipid tracking and lipidomics results indicate that adipocytes within mammary adipose tissue receive lipids from alveoli, and removal of mammary adipose tissue impedes mammary gland involution [313].

## Summary

This chapter briefly introduces the anatomy and physiology of the adipose tissue. The classification of adipose tissue by "color" has evolved from simply reflecting its appearance to now representing the metabolic and thermogenic states of adipocytes, establishing a paradigm in adipose tissue research. Unilocular white adipocytes primarily store energy, whereas multilocular, mitochondria-rich brown adipocytes are mainly involved in thermogenesis. Accumulating evidence indicates the existence of beige adipocytes, which lie between white and brown adipocytes and can be inducibly activated for thermogenesis. Adipose tissue also functions as an endocrine organ, playing essential roles in metabolic regulation and other physiological functions by secreting polypeptides, lipokines, EVs, and exosomal miRNAs. Immune cells, neural innervation, and ECM collectively influence the adipocyte phenotype and local microenvironment in adipose tissue.

Adipose tissue, distributed widely throughout the body, demonstrates substantial heterogeneity, with adipocytes in various fat depots and even within the same depot, showing notable disparities in origin, phenotype, and physiological functions. Furthermore, adipose tissue is dynamic, manifesting structural and functional alterations in response to physiological stimuli or at specific stages of the life cycle. A comprehensive understanding of adipose tissue biology is important for the development of future therapies.

**Sources of Funding** This work was sponsored by the National Natural Science Foundation of China (Grant Numbers: 82300489, 82130012, and 81830010), the Shanghai Pujiang Program (No. 23PJD084), and the Nurture projects for basic research of Shanghai Chest Hospital (Grant Number: 2022YNJCQ03).

**Conflict of Interest** The authors have no relevant financial or nonfinancial interests to disclose.

## References

1. Hassan M, Latif N, Yacoub M. Adipose tissue: friend or foe? Nat Rev Cardiol. 2012;9:689–702.
2. Cannon B, Nedergaard J. Brown adipose tissue: function and physiological significance. Physiol Rev. 2004;84:277–359.
3. Chun KH. Mouse model of the adipose organ: the heterogeneous anatomical characteristics. Arch Pharm Res. 2021;44:857–75.
4. Zwick RK, Guerrero-Juarez CF, Horsley V, Plikus MV. Anatomical, physiological, and functional diversity of adipose tissue. Cell Metab. 2018;27:68–83.
5. Knittle JL, Timmers K, Ginsberg-Fellner F, Brown RE, Katz DP. The growth of adipose tissue in children and adolescents. Cross-sectional and longitudinal studies of adipose cell number and size. J Clin Invest. 1979;63:239–46.
6. Spalding KL, Arner E, Westermark PO, Bernard S, Buchholz BA, Bergmann O, Blomqvist L, Hoffstedt J, Näslund E, Britton T, Concha H, Hassan M, Rydén M, Frisén J, Arner P. Dynamics of fat cell turnover in humans. Nature. 2008;453:783–7.
7. Shamsi F, Wang CH, Tseng YH. The evolving view of thermogenic adipocytes—ontogeny, niche and function. Nat Rev Endocrinol. 2021;17:726–44.
8. Virtanen KA, Lidell ME, Orava J, Heglind M, Westergren R, Niemi T, Taittonen M, Laine J, Savisto NJ, Enerbäck S, Nuutila P. Functional brown adipose tissue in healthy adults. N Engl J Med. 2009;360:1518–25.
9. van Marken Lichtenbelt WD, Vanhommerig JW, Smulders NM, Drossaerts JM, Kemerink GJ, Bouvy ND, Schrauwen P, Teule GJ. Cold-activated brown adipose tissue in healthy men. N Engl J Med. 2009;360:1500–8.
10. Heaton JM. The distribution of brown adipose tissue in the human. J Anat. 1972;112:35–9.
11. Rogers NH. Brown adipose tissue during puberty and with aging. Ann Med. 2015;47:142–9.
12. Cohade C, Osman M, Pannu HK, Wahl RL. Uptake in supraclavicular area fat ("USA-fat"): description on 18F-FDG PET/CT. J Nucl Med. 2003;44:170–6.
13. Yeung HW, Grewal RK, Gonen M, Schöder H, Larson SM. Patterns of (18)F-FDG uptake in adipose tissue and muscle: a potential source of false-positives for PET. J Nucl Med. 2003;44:1789–96.
14. Levy SB, Leonard WR. The evolutionary significance of human brown adipose tissue: integrating the timescales of adaptation. Evol Anthropol. 2021;31:75–91.
15. Scheja L, Heeren J. Metabolic interplay between white, beige, brown adipocytes and the liver. J Hepatol. 2016;64:1176–86.
16. Hu Y, Huang Y, Jiang Y, Weng L, Cai Z, He B. The different shades of thermogenic adipose tissue. Curr Obes Rep. 2024;13:440.
17. Lin CS, Klingenberg M. Isolation of the uncoupling protein from brown adipose tissue mitochondria. FEBS Lett. 1980;113:299–303.
18. Seale P, Kajimura S, Yang W, Chin S, Rohas LM, Uldry M, Tavernier G, Langin D, Spiegelman BM. Transcriptional control of brown fat determination by PRDM16. Cell Metab. 2007;6:38–54.
19. Puigserver P, Wu Z, Park CW, Graves R, Wright M, Spiegelman BM. A cold-inducible coactivator of nuclear receptors linked to adaptive thermogenesis. Cell. 1998;92:829–39.
20. Gesta S, Blüher M, Yamamoto Y, Norris AW, Berndt J, Kralisch S, Boucher J, Lewis C, Kahn CR. Evidence for a role of developmental genes in the origin of obesity and body fat distribution. Proc Natl Acad Sci U S A. 2006;103:6676–81.
21. de Jong JM, Larsson O, Cannon B, Nedergaard J. A stringent validation of mouse adipose tissue identity markers. Am J Physiol Endocrinol Metab. 2015;308:E1085–105.
22. Vitali A, Murano I, Zingaretti MC, Frontini A, Ricquier D, Cinti S. The adipose organ of obesity-prone C57BL/6J mice is composed of mixed white and brown adipocytes. J Lipid Res. 2012;53:619–29.

23. Cao L, Choi EY, Liu X, Martin A, Wang C, Xu X, During MJ. White to brown fat phenotypic switch induced by genetic and environmental activation of a hypothalamic-adipocyte axis. Cell Metab. 2011;14:324–38.
24. Wu J, Boström P, Sparks LM, Ye L, Choi JH, Giang AH, Khandekar M, Virtanen KA, Nuutila P, Schaart G, Huang K, Tu H, van Marken Lichtenbelt WD, Hoeks J, Enerbäck S, Schrauwen P, Spiegelman BM. Beige adipocytes are a distinct type of thermogenic fat cell in mouse and human. Cell. 2012;150:366–76.
25. Lehnig AC, Stanford KI. Exercise-induced adaptations to white and brown adipose tissue. J Exp Biol. 2018;221:jeb161570.
26. Rosenwald M, Perdikari A, Rülicke T, Wolfrum C. Bi-directional interconversion of brite and white adipocytes. Nat Cell Biol. 2013;15:659–67.
27. Pilkington AC, Paz HA, Wankhade UD. Beige adipose tissue identification and marker specificity—overview. Front Endocrinol (Lausanne). 2021;12:599134.
28. Wang W, Seale P. Control of brown and beige fat development. Nat Rev Mol Cell Biol. 2016;17:691–702.
29. Harms M, Seale P. Brown and beige fat: development, function and therapeutic potential. Nat Med. 2013;19:1252–63.
30. Sharp LZ, Shinoda K, Ohno H, Scheel DW, Tomoda E, Ruiz L, Hu H, Wang L, Pavlova Z, Gilsanz V, Kajimura S. Human BAT possesses molecular signatures that resemble beige/brite cells. PLoS One. 2012;7:e49452.
31. Trayhurn P, Beattie JH. Physiological role of adipose tissue: white adipose tissue as an endocrine and secretory organ. Proc Nutr Soc. 2001;60:329–39.
32. Kershaw EE, Flier JS. Adipose tissue as an endocrine organ. J Clin Endocrinol Metab. 2004;89:2548–56.
33. Mechanick JI, Zhao S, Garvey WT. The Adipokine-cardiovascular-lifestyle network: translation to clinical practice. J Am Coll Cardiol. 2016;68:1785–803.
34. Pestel J, Blangero F, Watson J, Pirola L, Eljaafari A. Adipokines in obesity and metabolic-related-diseases. Biochimie. 2023;212:48–59.
35. Chang L, Villacorta L, Li R, Hamblin M, Xu W, Dou C, Zhang J, Wu J, Zeng R, Chen YE. Loss of perivascular adipose tissue on peroxisome proliferator-activated receptor-γ deletion in smooth muscle cells impairs intravascular thermoregulation and enhances atherosclerosis. Circulation. 2012;126:1067–78.
36. Zhang Y, Proenca R, Maffei M, Barone M, Leopold L, Friedman JM. Positional cloning of the mouse obese gene and its human homologue. Nature. 1994;372:425–32.
37. Hube F, Lietz U, Igel M, Jensen PB, Tornqvist H, Joost HG, Hauner H. Difference in leptin mRNA levels between omental and subcutaneous abdominal adipose tissue from obese humans. Horm Metab Res. 1996;28:690–3.
38. Pérez-Pérez A, Toro A, Vilariño-García T, Maymó J, Guadix P, Dueñas JL, Fernández-Sánchez M, Varone C, Sánchez-Margalet V. Leptin action in normal and pathological pregnancies. J Cell Mol Med. 2018;22:716–27.
39. Vilariño-García T, Polonio-González ML, Pérez-Pérez A, Ribalta J, Arrieta F, Aguilar M, Obaya JC, Gimeno-Orna JA, Iglesias P, Navarro J, Durán S, Pedro-Botet J, Sánchez-Margalet V. Role of leptin in obesity, cardiovascular disease, and type 2 diabetes. Int J Mol Sci. 2024;25:2338.
40. Margetic S, Gazzola C, Pegg GG, Hill RA. Leptin: a review of its peripheral actions and interactions. Int J Obes Relat Metab Disord. 2002;26:1407–33.
41. Friedman J. The long road to leptin. J Clin Invest. 2016;126:4727–34.
42. Scherer PE, Williams S, Fogliano M, Baldini G, Lodish HF. A novel serum protein similar to C1q, produced exclusively in adipocytes. J Biol Chem. 1995;270:26746–9.
43. Berner HS, Lyngstadaas SP, Spahr A, Monjo M, Thommesen L, Drevon CA, Syversen U, Reseland JE. Adiponectin and its receptors are expressed in bone-forming cells. Bone. 2004;35:842–9.

44. AlM D, Jonas J-C, Bauche IB, Cornu O, Brichard SM. Induction of adiponectin in skeletal muscle by inflammatory cytokines: in vivo and in vitro studies. Endocrinology. 2004;145:5589–97.
45. Wang Y, Xu A, Knight C, Xu LY, Cooper GJS. Hydroxylation and glycosylation of the four conserved lysine residues in the collagenous domain of adiponectin. Potential role in the modulation of its insulin-sensitizing activity. J Biol Chem. 2002;277:19521–9.
46. Yamauchi T, Kamon J, Ito Y, Tsuchida A, Yokomizo T, Kita S, Sugiyama T, Miyagishi M, Hara K, Tsunoda M, Murakami K, Ohteki T, Uchida S, Takekawa S, Waki H, Tsuno NH, Shibata Y, Terauchi Y, Froguel P, Tobe K, Koyasu S, Taira K, Kitamura T, Shimizu T, Nagai R, Kadowaki T. Cloning of adiponectin receptors that mediate antidiabetic metabolic effects. Nature. 2003;423:762–9.
47. Straub LG, Scherer PE. Metabolic messengers: adiponectin. Nat Metab. 2019;1:334–9.
48. Maeda N, Shimomura I, Kishida K, Nishizawa H, Matsuda M, Nagaretani H, Furuyama N, Kondo H, Takahashi M, Arita Y, Komuro R, Ouchi N, Kihara S, Tochino Y, Okutomi K, Horie M, Takeda S, Aoyama T, Funahashi T, Matsuzawa Y. Diet-induced insulin resistance in mice lacking adiponectin/ACRP30. Nat Med. 2002;8:731–7.
49. Kim J-Y, van de Wall E, Laplante M, Azzara A, Trujillo ME, Hofmann SM, Schraw T, Durand JL, Li H, Li G, Jelicks LA, Mehler MF, Hui DY, Deshaies Y, Shulman GI, Schwartz GJ, Scherer PE. Obesity-associated improvements in metabolic profile through expansion of adipose tissue. J Clin Invest. 2007;117:2621–37.
50. Fang H, Judd RL. Adiponectin regulation and function. Compr Physiol. 2018;8:1031–63.
51. Mandal P, Pratt BT, Barnes M, McMullen MR, Nagy LE. Molecular mechanism for adiponectin-dependent M2 macrophage polarization: link between the metabolic and innate immune activity of full-length adiponectin. J Biol Chem. 2011;286:13460–9.
52. Caligiuri A, Bertolani C, Guerra CT, Aleffi S, Galastri S, Trappoliere M, Vizzutti F, Gelmini S, Laffi G, Pinzani M, Marra F. Adenosine monophosphate-activated protein kinase modulates the activated phenotype of hepatic stellate cells. Hepatology. 2008;47:668–76.
53. Okamoto Y, Kihara S, Ouchi N, Nishida M, Arita Y, Kumada M, Ohashi K, Sakai N, Shimomura I, Kobayashi H, Terasaka N, Inaba T, Funahashi T, Matsuzawa Y. Adiponectin reduces atherosclerosis in apolipoprotein E-deficient mice. Circulation. 2002;106:2767–70.
54. Qiao L, Wattez J-S, Lee S, Nguyen A, Schaack J, Hay WW, Shao J. Adiponectin deficiency impairs maternal metabolic adaptation to pregnancy in mice. Diabetes. 2017;66:1126–35.
55. Hu X, Jogasuria A, Wang J, Kim C, Han Y, Shen H, Wu J, You M. MitoNEET deficiency alleviates experimental alcoholic steatohepatitis in mice by stimulating endocrine adiponectin-Fgf15 axis. J Biol Chem. 2016;291:22482–95.
56. Scheja L, Heeren J. The endocrine function of adipose tissues in health and cardiometabolic disease. Nat Rev Endocrinol. 2019;15:507–24.
57. Cao H, Gerhold K, Mayers JR, Wiest MM, Watkins SM, Hotamisligil GS. Identification of a lipokine, a lipid hormone linking adipose tissue to systemic metabolism. Cell. 2008;134:933–44.
58. Yore MM, Syed I, Moraes-Vieira PM, Zhang T, Herman MA, Homan EA, Patel RT, Lee J, Chen S, Peroni OD, Dhaneshwar AS, Hammarstedt A, Smith U, McGraw TE, Saghatelian A, Kahn BB. Discovery of a class of endogenous mammalian lipids with anti-diabetic and anti-inflammatory effects. Cell. 2014;159:318–32.
59. Dimopoulos N, Watson M, Sakamoto K, Hundal HS. Differential effects of palmitate and palmitoleate on insulin action and glucose utilization in rat L6 skeletal muscle cells. Biochem J. 2006;399:473–81.
60. Diakogiannaki E, Dhayal S, Childs CE, Calder PC, Welters HJ, Morgan NG. Mechanisms involved in the cytotoxic and cytoprotective actions of saturated versus monounsaturated long-chain fatty acids in pancreatic beta-cells. J Endocrinol. 2007;194:283–91.
61. Kuda O, Brezinova M, Rombaldova M, Slavikova B, Posta M, Beier P, Janovska P, Veleba J, Kopecky J Jr, Kudova E, Pelikanova T, Kopecky J. Docosahexaenoic acid-derived fatty

acid esters of Hydroxy fatty acids (FAHFAs) with anti-inflammatory properties. Diabetes. 2016;65:2580–90.
62. Ma Y, Kind T, Vaniya A, Gennity I, Fahrmann JF, Fiehn O. An in silico MS/MS library for automatic annotation of novel FAHFA lipids. J Cheminform. 2015;7:53.
63. Syed I, Rubin de Celis MF, Mohan JF, Moraes-Vieira PM, Vijayakumar A, Nelson AT, Siegel D, Saghatelian A, Mathis D, Kahn BB. PAHSAs attenuate immune responses and promote β cell survival in autoimmune diabetic mice. J Clin Invest. 2019;129:3717–31.
64. Colombo M, Raposo G, Théry C. Biogenesis, secretion, and intercellular interactions of exosomes and other extracellular vesicles. Annu Rev Cell Dev Biol. 2014;30:255–89.
65. Harding C, Heuser J, Stahl P. Receptor-mediated endocytosis of transferrin and recycling of the transferrin receptor in rat reticulocytes. J Cell Biol. 1983;97:329–39.
66. Thomou T, Mori MA, Dreyfuss JM, Konishi M, Sakaguchi M, Wolfrum C, Rao TN, Winnay JN, Garcia-Martin R, Grinspoon SK, Gorden P, Kahn CR. Adipose-derived circulating miRNAs regulate gene expression in other tissues. Nature. 2017;542:450–5.
67. Yu Y, Du H, Wei S, Feng L, Li J, Yao F, Zhang M, Hatch GM, Chen L. Adipocyte-derived exosomal MiR-27a induces insulin resistance in skeletal muscle through repression of PPARγ. Theranostics. 2018;8:2171–88.
68. Müller G, Schneider M, Biemer-Daub G, Wied S. Microvesicles released from rat adipocytes and harboring glycosylphosphatidylinositol-anchored proteins transfer RNA stimulating lipid synthesis. Cell Signal. 2011;23:1207–23.
69. Mori MA, Thomou T, Boucher J, Lee KY, Lallukka S, Kim JK, Torriani M, Yki-Järvinen H, Grinspoon SK, Cypess AM, Kahn CR. Altered miRNA processing disrupts brown/white adipocyte determination and associates with lipodystrophy. J Clin Invest. 2014;124:3339–51.
70. Xing X, Li Z, Yang X, Li M, Liu C, Pang Y, Zhang L, Li X, Liu G, Xiao Y. Adipose-derived mesenchymal stem cells-derived exosome-mediated microRNA-342-5p protects endothelial cells against atherosclerosis. Aging. 2020;12:3880–98.
71. Scheele C, Wolfrum C. Brown adipose crosstalk in tissue plasticity and human metabolism. Endocr Rev. 2020;41:53–65.
72. Villarroya F, Cereijo R, Villarroya J, Giralt M. Brown adipose tissue as a secretory organ. Nat Rev Endocrinol. 2017;13:26–35.
73. BonDurant LD, Ameka M, Naber MC, Markan KR, Idiga SO, Acevedo MR, Walsh SA, Ornitz DM, Potthoff MJ. FGF21 regulates metabolism through adipose-dependent and -independent mechanisms. Cell Metab. 2017;25:935.
74. Schlein C, Talukdar S, Heine M, Fischer AW, Krott LM, Nilsson SK, Brenner MB, Heeren J, Scheja L. FGF21 lowers plasma triglycerides by accelerating lipoprotein catabolism in White and Brown adipose tissues. Cell Metab. 2016;23:441–53.
75. Lin Z, Tian H, Lam KSL, Lin S, Hoo RCL, Konishi M, Itoh N, Wang Y, Bornstein SR, Xu A, Li X. Adiponectin mediates the metabolic effects of FGF21 on glucose homeostasis and insulin sensitivity in mice. Cell Metab. 2013;17:779–89.
76. Geng L, Lam KSL, Xu A. The therapeutic potential of FGF21 in metabolic diseases: from bench to clinic. Nat Rev Endocrinol. 2020;16:654–67.
77. Deshmukh AS, Peijs L, Beaudry JL, Jespersen NZ, Nielsen CH, Ma T, Brunner AD, Larsen TJ, Bayarri-Olmos R, Prabhakar BS, Helgstrand C, Severinsen MCK, Holst B, Kjaer A, Tang-Christensen M, Sanfridson A, Garred P, Privé GG, Pedersen BK, Gerhart-Hines Z, Nielsen S, Drucker DJ, Mann M, Scheele C. Proteomics-based comparative mapping of the secretomes of human Brown and White adipocytes reveals EPDR1 as a novel Batokine. Cell Metab. 2019;30:963.
78. Cataldo LR, Gao Q, Argemi-Muntadas L, Hodek O, Cowan E, Hladkou S, Gheibi S, Spégel P, Prasad RB, Eliasson L, Scheele C, Fex M, Mulder H, Moritz T. The human batokine EPDR1 regulates β-cell metabolism and function. Mol Metab. 2022;66:101629.
79. Stanford KI, Lynes MD, Takahashi H, Baer LA, Arts PJ, May FJ, Lehnig AC, Middelbeek RJW, Richard JJ, So K, Chen EY, Gao F, Narain NR, Distefano G, Shettigar VK, Hirshman MF, Ziolo MT, Kiebish MA, Tseng YH, Coen PM, Goodyear LJ. 12,13-diHOME: an

exercise-induced Lipokine that increases skeletal muscle fatty acid uptake. Cell Metab. 2018;27:1111–1120.e3.
80. Lynes MD, Leiria LO, Lundh M, Bartelt A, Shamsi F, Huang TL, Takahashi H, Hirshman MF, Schlein C, Lee A, Baer LA, May FJ, Gao F, Narain NR, Chen EY, Kiebish MA, Cypess AM, Blüher M, Goodyear LJ, Hotamisligil GS, Stanford KI, Tseng YH. The cold-induced lipokine 12,13-diHOME promotes fatty acid transport into brown adipose tissue. Nat Med. 2017;23:631–7.
81. Pinckard KM, Shettigar VK, Wright KR, Abay E, Baer LA, Vidal P, Dewal RS, Das D, Duarte-Sanmiguel S, Hernández-Saavedra D, Arts PJ, Lehnig AC, Bussberg V, Narain NR, Kiebish MA, Yi F, Sparks LM, Goodpaster BH, Smith SR, Pratley RE, Lewandowski ED, Raman SV, Wold LE, Gallego-Perez D, Coen PM, Ziolo MT, Stanford KI. A novel endocrine role for the BAT-released Lipokine 12,13-diHOME to mediate cardiac function. Circulation. 2021;143:145–59.
82. Zhao H, Chen X, Hu G, Li C, Guo L, Zhang L, Sun F, Xia Y, Yan W, Cui Z, Guo Y, Guo X, Huang C, Fan M, Wang S, Zhang F, Tao L. Small extracellular vesicles from Brown adipose tissue mediate exercise cardioprotection. Circ Res. 2022;130:1490–506.
83. Rossi AB, Vergnanini AL. Cellulite: a review. J Eur Acad Dermatol Venereol. 2000;14:251–62.
84. Kane H, Lynch L. Innate immune control of adipose tissue homeostasis. Trends Immunol. 2019;40:857–72.
85. Vohralik EJ, Psaila AM, Knights AJ, Quinlan KGR. EoTHINophils: eosinophils as key players in adipose tissue homeostasis. Clin Exp Pharmacol Physiol. 2020;47:1495–505.
86. Hill AA, Reid Bolus W, Hasty AH. A decade of progress in adipose tissue macrophage biology. Immunol Rev. 2014;262:134–52.
87. Cai Z, Huang Y, He B. New insights into adipose tissue macrophages in obesity and insulin resistance. Cells. 2022;11:1424.
88. Qiu Y, Nguyen Khoa D, Odegaard Justin I, Cui X, Tian X, Locksley Richard M, Palmiter Richard D, Chawla A. Eosinophils and type 2 cytokine signaling in macrophages orchestrate development of functional beige fat. Cell. 2014;157:1292–308.
89. Lee Y-H, Petkova AP, Granneman JG. Identification of an adipogenic niche for adipose tissue remodeling and restoration. Cell Metab. 2013;18:355–67.
90. Vujičić M, Broderick I, Salmantabar P, Perian C, Nilsson J, Sihlbom Wallem C, Wernstedt AI. A macrophage-collagen fragment axis mediates subcutaneous adipose tissue remodeling in mice. Proc Natl Acad Sci U S A. 2024;121:e2313185121.
91. Hill DA, Lim HW, Kim YH, Ho WY, Foong YH, Nelson VL, Nguyen HCB, Chegireddy K, Kim J, Habertheuer A, Vallabhajosyula P, Kambayashi T, Won KJ, Lazar MA. Distinct macrophage populations direct inflammatory versus physiological changes in adipose tissue. Proc Natl Acad Sci U S A. 2018;115:E5096–105.
92. Burl RB, Ramseyer VD, Rondini EA, Pique-Regi R, Lee YH, Granneman JG. Deconstructing adipogenesis induced by beta3-adrenergic receptor activation with single-cell expression profiling. Cell Metab. 2018;28(2):300–309.e4.
93. Sarvari AK, Van Hauwaert EL, Markussen LK, Gammelmark E, Marcher AB, Ebbesen MF, Nielsen R, Brewer JR, Madsen JGS, Mandrup S. Plasticity of Epididymal adipose tissue in response to diet-induced obesity at single-nucleus resolution. Cell Metab. 2021;33:437–453.e5.
94. Harasymowicz NS, Rashidi N, Savadipour A, Wu CL, Tang R, Bramley J, Buchser W, Guilak F. Single-cell RNA sequencing reveals the induction of novel myeloid and myeloid-associated cell populations in visceral fat with long-term obesity. FASEB J. 2021;35:e21417.
95. Emont MP, Jacobs C, Essene AL, Pant D, Tenen D, Colleluori G, Di Vincenzo A, Jorgensen AM, Dashti H, Stefek A, McGonagle E, Strobel S, Laber S, Agrawal S, Westcott GP, Kar A, Veregge ML, Gulko A, Srinivasan H, Kramer Z, De Filippis E, Merkel E, Ducie J, Boyd CG, Gourash W, Courcoulas A, Lin SJ, Lee BT, Morris D, Tobias A, Khera AV, Claussnitzer

M, Pers TH, Giordano A, Ashenberg O, Regev A, Tsai LT, Rosen ED. A single-cell atlas of human and mouse white adipose tissue. Nature. 2022;603:926–33.

96. Jaitin DA, Adlung L, Thaiss CA, Weiner A, Li B, Descamps H, Lundgren P, Bleriot C, Liu Z, Deczkowska A, Keren-Shaul H, David E, Zmora N, Eldar SM, Lubezky N, Shibolet O, Hill DA, Lazar MA, Colonna M, Ginhoux F, Shapiro H, Elinav E, Amit I. Lipid-associated macrophages control metabolic homeostasis in a Trem2-dependent manner. Cell. 2019;178:686–698.e14.
97. Cox N, Crozet L, Holtman IR, Loyher PL, Lazarov T, White JB, Mass E, Stanley ER, Elemento O, Glass CK, Geissmann F. Diet-regulated production of PDGFcc by macrophages controls energy storage. Science. 2021;373:eabe9383.
98. Pirzgalska RM, Seixas E, Seidman JS, Link VM, Sanchez NM, Mahu I, Mendes R, Gres V, Kubasova N, Morris I, Arus BA, Larabee CM, Vasques M, Tortosa F, Sousa AL, Anandan S, Tranfield E, Hahn MK, Iannacone M, Spann NJ, Glass CK, Domingos AI. Sympathetic neuron-associated macrophages contribute to obesity by importing and metabolizing norepinephrine. Nat Med. 2017;23:1309–18.
99. Andersson O, Korach-Andre M, Reissmann E, Ibanez CF, Bertolino P. Growth/differentiation factor 3 signals through ALK7 and regulates accumulation of adipose tissue and diet-induced obesity. Proc Natl Acad Sci U S A. 2008;105:7252–6.
100. Lumeng CN, Liu J, Geletka L, Delaney C, Delproposto J, Desai A, Oatmen K, Martinez-Santibanez G, Julius A, Garg S, Yung RL. Aging is associated with an increase in T cells and inflammatory macrophages in visceral adipose tissue. J Immunol. 2011;187:6208–16.
101. Wu D, Ren Z, Pae M, Guo W, Cui X, Merrill AH, Meydani SN. Aging up-regulates expression of inflammatory mediators in mouse adipose tissue. J Immunol. 2007;179:4829–39.
102. Morrisette-Thomas V, Cohen AA, Fulop T, Riesco E, Legault V, Li Q, Milot E, Dusseault-Belanger F, Ferrucci L. Inflamm-aging does not simply reflect increases in pro-inflammatory markers. Mech Ageing Dev. 2014;139:49–57.
103. Bachman ES, Dhillon H, Zhang CY, Cinti S, Bianco AC, Kobilka BK, Lowell BB. betaAR signaling required for diet-induced thermogenesis and obesity resistance. Science. 2002;297:843–5.
104. Zeng W, Pirzgalska RM, Pereira MM, Kubasova N, Barateiro A, Seixas E, Lu YH, Kozlova A, Voss H, Martins GG, Friedman JM, Domingos AI. Sympathetic neuro-adipose connections mediate leptin-driven lipolysis. Cell. 2015;163:84–94.
105. Madden KS. Sympathetic neural-immune interactions regulate hematopoiesis, thermoregulation and inflammation in mammals. Dev Comp Immunol. 2017;66:92–7.
106. Collins S, Cao W, Robidoux J. Learning new tricks from old dogs: beta-adrenergic receptors teach new lessons on firing up adipose tissue metabolism. Mol Endocrinol. 2004;18:2123–31.
107. Soeder KJ, Snedden SK, Cao W, Della Rocca GJ, Daniel KW, Luttrell LM, Collins S. The beta3-adrenergic receptor activates mitogen-activated protein kinase in adipocytes through a Gi-dependent mechanism. J Biol Chem. 1999;274:12017–22.
108. Cao W, Medvedev AV, Daniel KW, Collins S. Beta-adrenergic activation of p38 MAP kinase in adipocytes: cAMP induction of the uncoupling protein 1 (UCP1) gene requires p38 MAP kinase. J Biol Chem. 2001;276:27077–82.
109. Collins S. β-Adrenergic receptors and adipose tissue metabolism: evolution of an old story. Annu Rev Physiol. 2022;84:1–16.
110. Mori S, Beyer RS, Bernardes de Souza B, Sorg JM, Hoover DB, Sacks HS, Fishbein MC, Chang G, Peacock WJ, St John MA, Law J, Symonds ME, Ajijola OA, Shivkumar K, Srikanthan P. Sympathetic innervation of the supraclavicular brown adipose tissue: a detailed anatomical study. PLoS One. 2023;18:e0290455.
111. François M, Torres H, Huesing C, Zhang R, Saurage C, Lee N, Qualls-Creekmore E, Yu S, Morrison CD, Burk D, Berthoud HR, Münzberg H. Sympathetic innervation of the interscapular brown adipose tissue in mouse. Ann N Y Acad Sci. 2019;1454:3–13.

112. Wang P, Loh KH, Wu M, Morgan DA, Schneeberger M, Yu X, Chi J, Kosse C, Kim D, Rahmouni K, Cohen P, Friedman J. A leptin-BDNF pathway regulating sympathetic innervation of adipose tissue. Nature. 2020;583:839–44.
113. Chi J, Wu Z, Choi CHJ, Nguyen L, Tegegne S, Ackerman SE, Crane A, Marchildon F, Tessier-Lavigne M, Cohen P. Three-dimensional adipose tissue imaging reveals regional variation in beige fat biogenesis and PRDM16-dependent sympathetic neurite density. Cell Metab. 2018;27:226–236.e3.
114. Seale P, Conroe HM, Estall J, Kajimura S, Frontini A, Ishibashi J, Cohen P, Cinti S, Spiegelman BM. Prdm16 determines the thermogenic program of subcutaneous white adipose tissue in mice. J Clin Invest. 2011;121:96–105.
115. Zeng X, Ye M, Resch JM, Jedrychowski MP, Hu B, Lowell BB, Ginty DD, Spiegelman BM. Innervation of thermogenic adipose tissue via a calsyntenin 3beta-S100b axis. Nature. 2019;569:229–35.
116. Divoux A, Clément K. Architecture and the extracellular matrix: the still unappreciated components of the adipose tissue. Obes Rev. 2011;12:e494–503.
117. Lin D, Chun TH, Kang L. Adipose extracellular matrix remodelling in obesity and insulin resistance. Biochem Pharmacol. 2016;119:8–16.
118. Noro A, Sillat T, Virtanen I, Ingerpuu S, Bäck N, Konttinen YT, Korhonen M. Laminin production and basement membrane deposition by mesenchymal stem cells upon adipogenic differentiation. J Histochem Cytochem. 2013;61:719–30.
119. Goddi A, Carmona A, Park SY, Dalgin G, Gonzalez Porras MA, Brey EM, Cohen RN. Laminin-α4 negatively regulates adipocyte Beiging through the suppression of AMPKα in male mice. Endocrinology. 2022;163:bqac154.
120. Sun K, Li X, Scherer PE. Extracellular matrix (ECM) and fibrosis in adipose tissue: overview and perspectives. Compr Physiol. 2023;13:4387–407.
121. Khan T, Muise ES, Iyengar P, Wang ZV, Chandalia M, Abate N, Zhang BB, Bonaldo P, Chua S, Scherer PE. Metabolic dysregulation and adipose tissue fibrosis: role of collagen VI. Mol Cell Biol. 2009;29:1575–91.
122. Jones JEC, Rabhi N, Orofino J, Gamini R, Perissi V, Vernochet C, Farmer SR. The adipocyte acquires a fibroblast-like transcriptional signature in response to a high fat diet. Sci Rep. 2020;10:2380.
123. Divoux A, Tordjman J, Lacasa D, Veyrie N, Hugol D, Aissat A, Basdevant A, Guerre-Millo M, Poitou C, Zucker JD, Bedossa P, Clément K. Fibrosis in human adipose tissue: composition, distribution, and link with lipid metabolism and fat mass loss. Diabetes. 2010;59:2817–25.
124. DeBari MK, Abbott RD. Adipose tissue fibrosis: mechanisms, models, and importance. Int J Mol Sci. 2020;21:6030.
125. Estève D, Boulet N, Belles C, Zakaroff-Girard A, Decaunes P, Briot A, Veeranagouda Y, Didier M, Remaury A, Guillemot JC, Ledoux S, Dani C, Bouloumié A, Galitzky J. Lobular architecture of human adipose tissue defines the niche and fate of progenitor cells. Nat Commun. 2019;10:2549.
126. Nauli AM, Matin S. Why do men accumulate abdominal visceral fat? Front Physiol. 2019;10:1486.
127. Cohen CA, Shea AA, Heffron CL, Schmelz EM, Roberts PC. Intra-abdominal fat depots represent distinct immunomodulatory microenvironments: a murine model. PLoS One. 2013;8:e66477.
128. Björntorp P. "Portal" adipose tissue as a generator of risk factors for cardiovascular disease and diabetes. Arteriosclerosis. 1990;10:493–6.
129. Chau YY, Bandiera R, Serrels A, Martínez-Estrada OM, Qing W, Lee M, Slight J, Thornburn A, Berry R, McHaffie S, Stimson RH, Walker BR, Chapuli RM, Schedl A, Hastie N. Visceral and subcutaneous fat have different origins and evidence supports a mesothelial source. Nat Cell Biol. 2014;16:367–75.
130. Massier L, Jalkanen J, Elmastas M, Zhong J, Wang T, Nono Nankam PA, Frendo-Cumbo S, Bäckdahl J, Subramanian N, Sekine T, Kerr AG, Tseng BTP, Laurencikiene J, Buggert M,

Lourda M, Kublickiene K, Bhalla N, Andersson A, Valsesia A, Astrup A, Blaak EE, Ståhl PL, Viguerie N, Langin D, Wolfrum C, Blüher M, Rydén M, Mejhert N. An integrated single cell and spatial transcriptomic map of human white adipose tissue. Nat Commun. 2023;14:1438.
131. Chen X, He X, Guo Y, Liu L, Li H, Tan J, Feng W, Guan H, Cao X, Xiao H, Li Y. Glucose-dependent insulinotropic polypeptide modifies adipose plasticity and promotes beige adipogenesis of human omental adipose-derived stem cells. FASEB J. 2021;35:e21534.
132. Lee MJ, Jash S, Jones JEC, Puri V, Fried SK. Rosiglitazone remodels the lipid droplet and britens human visceral and subcutaneous adipocytes ex vivo. J Lipid Res. 2019;60:856–68.
133. Wilkosz S, Ireland G, Khwaja N, Walker M, Butt R, de Giorgio-Miller A, Herrick SE. A comparative study of the structure of human and murine greater omentum. Anat Embryol. 2005;209:251–61.
134. Platell C, Cooper D, Papadimitriou JM, Hall JC. The omentum. World J Gastroenterol. 2000;6:169–76.
135. Collins D, Hogan AM, O'Shea D, Winter DC. The omentum: anatomical, metabolic, and surgical aspects. J Gastrointest Surg. 2009;13:1138–46.
136. Gama JFG, Pinheiro DF, Da Silva RF, Quirico-Santos T, Lagrota-Candido J. The omentum harbors unique conditions in the peritoneal cavity to promote healing and regeneration for diaphragm muscle repair in mdx mice. Cell Tissue Res. 2020;382:447–55.
137. Liebermann-Meffert D. The greater omentum. Anatomy, embryology, and surgical applications. Surg Clin North Am. 2000;80(1):275–93.
138. Liu F, He J, Liu B, Zhang P, Wang H, Sun X, Chu X, Guan W, Feng W, Bi Y, Zhu D. Association of Omental Adipocyte Hypertrophy and Fibrosis with human obesity and type 2 diabetes. Obesity (Silver Spring). 2021;29:976–84.
139. O'Connell J, Lynch L, Cawood TJ, Kwasnik A, Nolan N, Geoghegan J, McCormick A, O'Farrelly C, O'Shea D. The relationship of omental and subcutaneous adipocyte size to metabolic disease in severe obesity. PLoS One. 2010;5:e9997.
140. Chkourko Gusky H, Diedrich J, MacDougald OA, Podgorski I. Omentum and bone marrow: how adipocyte-rich organs create tumour microenvironments conducive for metastatic progression. Obes Rev. 2016;17:1015–29.
141. Gerber SA, Rybalko VY, Bigelow CE, Lugade AA, Foster TH, Frelinger JG, Lord EM. Preferential attachment of peritoneal tumor metastases to omental immune aggregates and possible role of a unique vascular microenvironment in metastatic survival and growth. Am J Pathol. 2006;169:1739–52.
142. Morshedzadeh N, Rahimlou M, Asadzadeh Aghdaei H, Shahrokh S, Reza Zali M, Mirmiran P. Association between adipokines levels with inflammatory bowel disease (IBD): systematic reviews. Dig Dis Sci. 2017;62:3280–6.
143. Gu P, Dube S, McGovern DPB. Medical and surgical implications of mesenteric adipose tissue in Crohn's disease: a review of the literature. Inflamm Bowel Dis. 2023;29:458–69.
144. Zhang H, Ding Y, Zeng Q, Wang D, Liu G, Hussain Z, Xiao B, Liu W, Deng T. Characteristics of mesenteric adipose tissue attached to different intestinal segments and their roles in immune regulation. Am J Physiol Gastrointest Liver Physiol. 2022;322:G310–26.
145. Karaskova E, Velganova-Veghova M, Geryk M, Foltenova H, Kucerova V, Karasek D. Role of adipose tissue in inflammatory bowel disease. Int J Mol Sci. 2021;22:4226.
146. Liu KH, Chan YL, Chan WB, Chan JCN, Chu CWW. Mesenteric fat thickness is an independent determinant of metabolic syndrome and identifies subjects with increased carotid intima-media thickness. Diabetes Care. 2006;29:379–84.
147. Zuo L, Geng Z, Song X, Li J, Wang Y, Zhang X, Ge S, Wang L, Liu G, Zhang L, Xu M, Zhao Y, Shen L, Ge Y, Wu R, Hu J. Browning of mesenteric White adipose tissue in Crohn's disease: a new pathological change and therapeutic target. J Crohns Colitis. 2023;17:1179–92.
148. Chusyd DE, Wang D, Huffman DM, Nagy TR. Relationships between rodent White adipose fat pads and human White adipose fat depots. Front Nutr. 2016;3:10.
149. Bagchi DP, MacDougald OA. Identification and dissection of diverse mouse adipose depots. J Vis Exp. 2019;149:10.3791/59499.

150. Liu B-X, Sun W, Kong X-Q. Perirenal fat: a unique fat pad and potential target for cardiovascular disease. Angiology. 2019;70:584–93.
151. Tanuma Y, Ohata M, Ito T, Yokochi C. Possible function of human brown adipose tissue as suggested by observation on perirenal brown fats from necropsy cases of variable age groups. Arch Histol Jpn. 1976;39:117–45.
152. Efremova A, Senzacqua M, Venema W, Isakov E, Di Vincenzo A, Zingaretti MC, Protasoni M, Thomski M, Giordano A, Cinti S. A large proportion of mediastinal and perirenal visceral fat of Siberian adult people is formed by UCP1 immunoreactive multilocular and paucilocular adipocytes. J Physiol Biochem. 2020;76:185–92.
153. Jespersen NZ, Feizi A, Andersen ES, Heywood S, Hattel HB, Daugaard S, Peijs L, Bagi P, Feldt-Rasmussen B, Schultz HS, Hansen NS, Krogh-Madsen R, Pedersen BK, Petrovic N, Nielsen S, Scheele C. Heterogeneity in the perirenal region of humans suggests presence of dormant brown adipose tissue that contains brown fat precursor cells. Mol Metab. 2019;24:30–43.
154. Frontini A, Cinti S. Distribution and development of brown adipocytes in the murine and human adipose organ. Cell Metab. 2010;11:253–6.
155. Waldén TB, Hansen IR, Timmons JA, Cannon B, Nedergaard J. Recruited vs. nonrecruited molecular signatures of brown, "brite," and white adipose tissues. Am J Physiol Endocrinol Metab. 2012;302:E19–31.
156. Wu MV, Bikopoulos G, Hung S, Ceddia RB. Thermogenic capacity is antagonistically regulated in classical brown and white subcutaneous fat depots by high fat diet and endurance training in rats: impact on whole-body energy expenditure. J Biol Chem. 2014;289:34129–40.
157. Guo Y, Zhang Q, Zheng L, Shou J, Zhuang S, Xiao W, Chen P. Depot-specific adaption of adipose tissue for different exercise approaches in high-fat diet/streptozocin-induced diabetic mice. Front Physiol. 2023;14:1189528.
158. Ye L, Wu J, Cohen P, Kazak L, Khandekar MJ, Jedrychowski MP, Zeng X, Gygi SP, Spiegelman BM. Fat cells directly sense temperature to activate thermogenesis. Proc Natl Acad Sci U S A. 2013;110:12480–5.
159. Negron SG, Xu B, Lin Z. Isolating Brown adipocytes from murine interscapular Brown adipose tissue for gene and protein expression analysis. J Vis Exp. 2021;169. https://doi.org/10.3791/62332.
160. Wang QA, Tao C, Gupta RK, Scherer PE. Tracking adipogenesis during white adipose tissue development, expansion and regeneration. Nat Med. 2013;19:1338–44.
161. Cancello R, Tordjman J, Poitou C, Guilhem G, Bouillot JL, Hugol D, Coussieu C, Basdevant A, Bar Hen A, Bedossa P, Guerre-Millo M, Clément K. Increased infiltration of macrophages in omental adipose tissue is associated with marked hepatic lesions in morbid human obesity. Diabetes. 2006;55:1554–61.
162. Kim D, Kim J, Yoon JH, Ghim J, Yea K, Song P, Park S, Lee A, Hong C-P, Jang MS, Kwon Y, Park S, Jang MH, Berggren P-O, Suh P-G, Ryu SH. CXCL12 secreted from adipose tissue recruits macrophages and induces insulin resistance in mice. Diabetologia. 2014;57:1456–65.
163. Tanegashima K, Suzuki K, Nakayama Y, Tsuji K, Shigenaga A, Otaka A, Hara T. CXCL14 is a natural inhibitor of the CXCL12–CXCR4 signaling axis. FEBS Lett. 2013;587:1731–5.
164. Hwang I, Jo K, Shin KC, Kim JI, Ji Y, Park YJ, Park J, Jeon YG, Ka S, Suk S, Noh HL, Choe SS, Alfadda AA, Kim JK, Kim S, Kim JB. GABA-stimulated adipose-derived stem cells suppress subcutaneous adipose inflammation in obesity. Proc Natl Acad Sci. 2019;116:11936–45.
165. Lidell ME, Betz MJ, Dahlqvist Leinhard O, Heglind M, Elander L, Slawik M, Mussack T, Nilsson D, Romu T, Nuutila P, Virtanen KA, Beuschlein F, Persson A, Borga M, Enerbäck S. Evidence for two types of brown adipose tissue in humans. Nat Med. 2013;19:631–4.
166. Gesta S, Tseng Y-H, Kahn CR. Developmental origin of fat: tracking obesity to its source. Cell. 2007;131:242–56.
167. Bjørndal B, Burri L, Staalesen V, Skorve J, Berge RK. Different adipose depots: their role in the development of metabolic syndrome and mitochondrial response to hypolipidemic agents. J Obes. 2011;2011:490650.

168. Tchkonia T, Thomou T, Zhu Y, Karagiannides I, Pothoulakis C, Jensen Michael D, Kirkland JL. Mechanisms and metabolic implications of regional differences among fat depots. Cell Metab. 2013;17:644–56.
169. Lafontan M, Langin D. Lipolysis and lipid mobilization in human adipose tissue. Prog Lipid Res. 2009;48:275–97.
170. Kern PA, Finlin BS, Zhu B, Rasouli N, McGehee RE Jr, Westgate PM, Dupont-Versteegden EE. The effects of temperature and seasons on subcutaneous white adipose tissue in humans: evidence for thermogenic gene induction. J Clin Endocrinol Metab. 2014;99:E2772–9.
171. Finlin BS, Memetimin H, Confides AL, Kasza I, Zhu B, Vekaria HJ, Harfmann B, Jones KA, Johnson ZR, Westgate PM, Alexander CM, Sullivan PG, Dupont-Versteegden EE, Kern PA. Human adipose beiging in response to cold and mirabegron. JCI Insight. 2018;3:e121510.
172. Ouellet V, Labbé SM, Blondin DP, Phoenix S, Guérin B, Haman F, Turcotte EE, Richard D, Carpentier AC. Brown adipose tissue oxidative metabolism contributes to energy expenditure during acute cold exposure in humans. J Clin Invest. 2012;122:545–52.
173. Muzik O, Mangner TJ, Leonard WR, Kumar A, Janisse J, Granneman JG. 15O PET measurement of blood flow and oxygen consumption in cold-activated human brown fat. J Nucl Med. 2013;54:523–31.
174. Muzik O, Mangner TJ, Leonard WR, Kumar A, Granneman JG. Sympathetic innervation of cold-activated Brown and White fat in lean young adults. J Nucl Med. 2017;58:799–806.
175. Horimatsu T, Kim HW, Weintraub NL. The role of perivascular adipose tissue in non-atherosclerotic vascular disease. Front Physiol. 2017;8:969.
176. Hildebrand S, Stümer J, Pfeifer A. PVAT and its relation to Brown, beige, and White adipose tissue in development and function. Front Physiol. 2018;9:70.
177. Mancio J, Oikonomou EK, Antoniades C. Perivascular adipose tissue and coronary atherosclerosis. Heart. 2018;104:1654–62.
178. Mendizábal Y, Llorens S, Nava E. Vasoactive effects of prostaglandins from the perivascular fat of mesenteric resistance arteries in WKY and SHROB rats. Life Sci. 2013;93:1023–32.
179. Chughtai HL, Morgan TM, Rocco M, Stacey B, Brinkley TE, Ding J, Nicklas B, Hamilton C, Hundley WG. Renal sinus fat and poor blood pressure control in middle-aged and elderly individuals at risk for cardiovascular events. Hypertension. 2010;56:901–6.
180. Li X, Ma Z, Zhu YZ. Regional heterogeneity of perivascular adipose tissue: morphology, origin, and secretome. Front Pharmacol. 2021;12:697720.
181. Fitzgibbons TP, Kogan S, Aouadi M, Hendricks GM, Straubhaar J, Czech MP. Similarity of mouse perivascular and brown adipose tissues and their resistance to diet-induced inflammation. Am J Physiol Heart Circ Physiol. 2011;301:H1425–37.
182. Police SB, Thatcher SE, Charnigo R, Daugherty A, Cassis LA. Obesity promotes inflammation in periaortic adipose tissue and angiotensin II-induced abdominal aortic aneurysm formation. Arterioscler Thromb Vasc Biol. 2009;29:1458–64.
183. Kortelainen ML, Pelletier G, Ricquier D, Bukowiecki LJ. Immunohistochemical detection of human brown adipose tissue uncoupling protein in an autopsy series. J Histochem Cytochem. 1993;41:759–64.
184. Nedergaard J, Bengtsson T, Cannon B. Unexpected evidence for active brown adipose tissue in adult humans. Am J Physiol Endocrinol Metab. 2007;293:E444–52.
185. Ye M, Ruan CC, Fu M, Xu L, Chen D, Zhu M, Zhu D, Gao P. Developmental and functional characteristics of the thoracic aorta perivascular adipocyte. Cell Mol Life Sci. 2019;76:777–89.
186. Padilla J, Jenkins NT, Vieira-Potter VJ, Laughlin MH. Divergent phenotype of rat thoracic and abdominal perivascular adipose tissues. Am J Physiol Regul Integr Comp Physiol. 2013;304:R543–52.
187. Henrichot E, Juge-Aubry CE, Pernin A, Pache JC, Velebit V, Dayer JM, Meda P, Chizzolini C, Meier CA. Production of chemokines by perivascular adipose tissue: a role in the pathogenesis of atherosclerosis? Arterioscler Thromb Vasc Biol. 2005;25:2594–9.

188. Li L, Miano JM, Mercer B, Olson EN. Expression of the SM22alpha promoter in transgenic mice provides evidence for distinct transcriptional regulatory programs in vascular and visceral smooth muscle cells. J Cell Biol. 1996;132:849–59.
189. Chawla A, Schwarz EJ, Dimaculangan DD, Lazar MA. Peroxisome proliferator-activated receptor (PPAR) gamma: adipose-predominant expression and induction early in adipocyte differentiation. Endocrinology. 1994;135:798–800.
190. Barak Y, Nelson MC, Ong ES, Jones YZ, Ruiz-Lozano P, Chien KR, Koder A, Evans RM. PPAR gamma is required for placental, cardiac, and adipose tissue development. Mol Cell. 1999;4:585–95.
191. Tang W, Zeve D, Suh JM, Bosnakovski D, Kyba M, Hammer RE, Tallquist MD, Graff JM. White fat progenitor cells reside in the adipose vasculature. Science. 2008;322:583–6.
192. Angueira AR, Sakers AP, Holman CD, Cheng L, Arbocco MN, Shamsi F, Lynes MD, Shrestha R, Okada C, Batmanov K, Susztak K, Tseng YH, Liaw L, Seale P. Defining the lineage of thermogenic perivascular adipose tissue. Nat Metab. 2021;3:469–84.
193. Adams M, Montague CT, Prins JB, Holder JC, Smith SA, Sanders L, Digby JE, Sewter CP, Lazar MA, Chatterjee VK, O'Rahilly S. Activators of peroxisome proliferator-activated receptor gamma have depot-specific effects on human preadipocyte differentiation. J Clin Invest. 1997;100:3149–53.
194. Tran KV, Fitzgibbons T, Min SY, DeSouza T, Corvera S. Distinct adipocyte progenitor cells are associated with regional phenotypes of perivascular aortic fat in mice. Mol Metab. 2018;9:199–206.
195. Soltis EE, Cassis LA. Influence of perivascular adipose tissue on rat aortic smooth muscle responsiveness. Clin Exp Hypertens A. 1991;13:277–96.
196. Chatterjee TK, Stoll LL, Denning GM, Harrelson A, Blomkalns AL, Idelman G, Rothenberg FG, Neltner B, Romig-Martin SA, Dickson EW, Rudich S, Weintraub NL. Proinflammatory phenotype of perivascular adipocytes: influence of high-fat feeding. Circ Res. 2009;104:541–9.
197. Gao YJ, Lu C, Su LY, Sharma AM, Lee RM. Modulation of vascular function by perivascular adipose tissue: the role of endothelium and hydrogen peroxide. Br J Pharmacol. 2007;151:323–31.
198. Lee YC, Chang HH, Chiang CL, Liu CH, Yeh JI, Chen MF, Chen PY, Kuo JS, Lee TJ. Role of perivascular adipose tissue-derived methyl palmitate in vascular tone regulation and pathogenesis of hypertension. Circulation. 2011;124:1160–71.
199. Ayala-Lopez N, Martini M, Jackson WF, Darios E, Burnett R, Seitz B, Fink GD, Watts SW. Perivascular adipose tissue contains functional catecholamines. Pharmacol Res Perspect. 2014;2:e00041.
200. Lu C, Su LY, Lee RM, Gao YJ. Mechanisms for perivascular adipose tissue-mediated potentiation of vascular contraction to perivascular neuronal stimulation: the role of adipocyte-derived angiotensin II. Eur J Pharmacol. 2010;634:107–12.
201. Watts SW, Dorrance AM, Penfold ME, Rourke JL, Sinal CJ, Seitz B, Sullivan TJ, Charvat TT, Thompson JM, Burnett R, Fink GD. Chemerin connects fat to arterial contraction. Arterioscler Thromb Vasc Biol. 2013;33:1320–8.
202. Wang P, Xu TY, Guan YF, Su DF, Fan GR, Miao CY. Perivascular adipose tissue-derived visfatin is a vascular smooth muscle cell growth factor: role of nicotinamide mononucleotide. Cardiovasc Res. 2009;81:370–80.
203. Takaoka M, Nagata D, Kihara S, Shimomura I, Kimura Y, Tabata Y, Saito Y, Nagai R, Sata M. Periadventitial adipose tissue plays a critical role in vascular remodeling. Circ Res. 2009;105:906–11.
204. Pan XX, Ruan CC, Liu XY, Kong LR, Ma Y, Wu QH, Li HQ, Sun YJ, Chen AQ, Zhao Q, Wu F, Wang XJ, Wang JG, Zhu DL, Gao PJ. Perivascular adipose tissue-derived stromal cells contribute to vascular remodeling during aging. Aging Cell. 2019;18:e12969.
205. Gómez-Serrano M, Camafeita E, López JA, Rubio MA, Bretón I, García-Consuegra I, García-Santos E, Lago J, Sánchez-Pernaute A, Torres A, Vázquez J, Peral B. Differential pro-

teomic and oxidative profiles unveil dysfunctional protein import to adipocyte mitochondria in obesity-associated aging and diabetes. Redox Biol. 2017;11:415–28.
206. Bailey-Downs LC, Tucsek Z, Toth P, Sosnowska D, Gautam T, Sonntag WE, Csiszar A, Ungvari Z. Aging exacerbates obesity-induced oxidative stress and inflammation in perivascular adipose tissue in mice: a paracrine mechanism contributing to vascular redox dysregulation and inflammation. J Gerontol A Biol Sci Med Sci. 2013;68:780–92.
207. Britton KA, Pedley A, Massaro JM, Corsini EM, Murabito JM, Hoffmann U, Fox CS. Prevalence, distribution, and risk factor correlates of high thoracic periaortic fat in the Framingham Heart Study. J Am Heart Assoc. 2012;1:e004200.
208. Small HY, McNeilly S, Mary S, Sheikh AM, Delles C. Resistin mediates sex-dependent effects of perivascular adipose tissue on vascular function in the Shrsp. Sci Rep. 2019;9:6897.
209. Chambliss KL, Shaul PW. Estrogen modulation of endothelial nitric oxide synthase. Endocr Rev. 2002;23:665–86.
210. Ramesh SS, Christopher R, Indira Devi B, Bhat DI. The vascular protective role of oestradiol: a focus on postmenopausal oestradiol deficiency and aneurysmal subarachnoid haemorrhage. Biol Rev Camb Philos Soc. 2019;94:1897–917.
211. Maric-Bilkan C, Manigrasso MB. Sex differences in hypertension: contribution of the renin-angiotensin system. Gend Med. 2012;9:287–91.
212. Iacobellis G, Corradi D, Sharma AM. Epicardial adipose tissue: anatomic, biomolecular and clinical relationships with the heart. Nat Clin Pract Cardiovasc Med. 2005;2:536–43.
213. Iacobellis G, Bianco AC. Epicardial adipose tissue: emerging physiological, pathophysiological and clinical features. Trends Endocrinol Metab. 2011;22:450–7.
214. Lu Z, Jiang Z, Tang J, Lin C-P, Zhang H. Functions and origins of cardiac fat. FEBS J. 2023;290:1705–18.
215. Company JM, Booth FW, Laughlin MH, Arce-Esquivel AA, Sacks HS, Bahouth SW, Fain JN. Epicardial fat gene expression after aerobic exercise training in pigs with coronary atherosclerosis: relationship to visceral and subcutaneous fat. J Appl Physiol (1985). 2010;109:1904–12.
216. Iacobellis G. Epicardial adipose tissue in contemporary cardiology. Nat Rev Cardiol. 2022;19:593–606.
217. Li C, Liu X, Adhikari BK, Chen L, Liu W, Wang Y, Zhang H. The role of epicardial adipose tissue dysfunction in cardiovascular diseases: an overview of pathophysiology, evaluation, and management. Front Endocrinol. 2023;14:1167952.
218. Fainberg HP, Birtwistle M, Alagal R, Alhaddad A, Pope M, Davies G, Woods R, Castellanos M, May ST, Ortori CA, Barrett DA, Perry V, Wiens F, Stahl B, van der Beek E, Sacks H, Budge H, Symonds ME. Transcriptional analysis of adipose tissue during development reveals depot-specific responsiveness to maternal dietary supplementation. Sci Rep. 2018;8:9628.
219. Marchington JM, Mattacks CA, Pond CM. Adipose tissue in the mammalian heart and pericardium: structure, foetal development and biochemical properties. Comp Biochem Physiol B. 1989;94:225–32.
220. Lopaschuk GD, Ussher JR, Folmes CDL, Jaswal JS, Stanley WC. Myocardial fatty acid metabolism in health and disease. Physiol Rev. 2010;90:207–58.
221. Iacobellis G. Local and systemic effects of the multifaceted epicardial adipose tissue depot. Nat Rev Endocrinol. 2015;11:363–71.
222. Ernault AC, Meijborg VMF, Coronel R. Modulation of cardiac arrhythmogenesis by epicardial adipose tissue: JACC state-of-the-art review. J Am Coll Cardiol. 2021;78:1730–45.
223. Perez-Miguelsanz J, Jiménez-Ortega V, Cano-Barquilla P, Garaulet M, Esquifino AI, Varela-Moreiras G, Fernández-Mateos P. Early appearance of epicardial adipose tissue through human development. Nutrients. 2021;13:2906.
224. Yamaguchi Y, Cavallero S, Patterson M, Shen H, Xu J, Kumar SR, Sucov HM. Adipogenesis and epicardial adipose tissue: a novel fate of the epicardium induced by mesenchymal transformation and PPARγ activation. Proc Natl Acad Sci. 2015;112:2070–5.

225. Liu Q, Huang X, Oh J-H, Lin R-Z, Duan S, Yu Y, Yang R, Qiu J, Melero-Martin JM, Pu WT, Zhou B. Epicardium-to-fat transition in injured heart. Cell Res. 2014;24:1367–9.
226. Doukbi E, Soghomonian A, Sengenès C, Ahmed S, Ancel P, Dutour A, Gaborit B. Browning epicardial adipose tissue: friend or foe? Cells. 2022;11:991.
227. Mukherjee AG, Renu K, Gopalakrishnan AV, Jayaraj R, Dey A, Vellingiri B, Ganesan R. Epicardial adipose tissue and cardiac lipotoxicity: a review. Life Sci. 2023;328:121913.
228. Antonopoulos AS, Margaritis M, Verheule S, Recalde A, Sanna F, Herdman L, Psarros C, Nasrallah H, Coutinho P, Akoumianakis I, Brewer AC, Sayeed R, Krasopoulos G, Petrou M, Tarun A, Tousoulis D, Shah AM, Casadei B, Channon KM, Antoniades C. Mutual regulation of epicardial adipose tissue and myocardial redox state by PPAR-γ/adiponectin signalling. Circ Res. 2016;118:842–55.
229. Packer M. Epicardial adipose tissue May mediate deleterious effects of obesity and inflammation on the myocardium. J Am Coll Cardiol. 2018;71:2360–72.
230. Chen H, Liu L, Li M, Zhu D, Tian G. Epicardial adipose tissue-derived leptin promotes myocardial injury in metabolic syndrome rats through PKC/NADPH oxidase/ROS pathway. J Am Heart Assoc. 2023;12:e029415.
231. Mazurek T, Zhang L, Zalewski A, Mannion JD, Diehl JT, Arafat H, Sarov-Blat L, O'Brien S, Keiper EA, Johnson AG, Martin J, Goldstein BJ, Shi Y. Human epicardial adipose tissue is a source of inflammatory mediators. Circulation. 2003;108:2460–6.
232. Frederiksen TC, Dahm CC, Preis SR, Lin H, Trinquart L, Benjamin EJ, Kornej J. The bidirectional association between atrial fibrillation and myocardial infarction. Nat Rev Cardiol. 2023;20:631–44.
233. Shaihov-Teper O, Ram E, Ballan N, Brzezinski RY, Naftali-Shani N, Masoud R, Ziv T, Lewis N, Schary Y, Levin-Kotler LP, Volvovitch D, Zuroff EM, Amunts S, Regev-Rudzki N, Sternik L, Raanani E, Gepstein L, Leor J. Extracellular vesicles from epicardial fat facilitate atrial fibrillation. Circulation. 2021;143:2475–93.
234. Driskell RR, Jahoda CAB, Chuong C-M, Watt FM, Horsley V. Defining dermal adipose tissue. Exp Dermatol. 2014;23:629–31.
235. Chen SX, Zhang LJ, Gallo RL. Dermal White adipose tissue: a newly recognized layer of skin innate defense. J Invest Dermatol. 2019;139:1002–9.
236. Kasza I, Suh Y, Wollny D, Clark RJ, Roopra A, Colman RJ, MacDougald OA, Shedd TA, Nelson DW, Yen MI, Yen CL, Alexander CM. Syndecan-1 is required to maintain intradermal fat and prevent cold stress. PLoS Genet. 2014;10:e1004514.
237. Guerrero-Juarez CF, Plikus MV. Emerging nonmetabolic functions of skin fat. Nat Rev Endocrinol. 2018;14:163–73.
238. Walker GE, Marzullo P, Verti B, Guzzaloni G, Maestrini S, Zurleni F, Liuzzi A, Di Blasio AM. Subcutaneous abdominal adipose tissue subcompartments: potential role in rosiglitazone effects. Obesity. 2008;16:1983–91.
239. J K K. Some histological aspects of the palmar digital pads in the vervet monkey. Folia Primatol (Basel). 1983;41:41.
240. Segalla L, Chirumbolo S, Sbarbati A. Dermal white adipose tissue: much more than a metabolic, lipid-storage organ? Tissue Cell. 2021;71:101583.
241. Festa E, Fretz J, Berry R, Schmidt B, Rodeheffer M, Horowitz M, Horsley V. Adipocyte lineage cells contribute to the skin stem cell niche to drive hair cycling. Cell. 2011;146:761–71.
242. Zhang L-j, Guerrero-Juarez CF, Hata T, Bapat SP, Ramos R, Plikus MV, Gallo RL. Innate immunity. Dermal adipocytes protect against invasive Staphylococcus aureus skin infection. Science. 2015;347:67–71.
243. Plikus MV, Guerrero-Juarez CF, Ito M, Li YR, Dedhia PH, Zheng Y, Shao M, Gay DL, Ramos R, Hsi T-C, Oh JW, Wang X, Ramirez A, Konopelski SE, Elzein A, Wang A, Supapannachart RJ, Lee H-L, Lim CH, Nace A, Guo A, Treffeisen E, Andl T, Ramirez RN, Murad R, Offermanns S, Metzger D, Chambon P, Widgerow AD, Tuan T-L, Mortazavi A, Gupta RK, Hamilton BA, Millar SE, Seale P, Pear WS, Lazar MA, Cotsarelis G. Regeneration of fat cells from myofibroblasts during wound healing. Science (New York, NY). 2017;355:748–52.

244. Rivera-Gonzalez G, Shook B, Horsley V. Adipocytes in skin health and disease. Cold Spring Harb Perspect Med. 2014;4:a015271.
245. Zhang Z, Shao M, Hepler C, Zi Z, Zhao S, An YA, Zhu Y, Ghaben AL, Wang M-Y, Li N, Onodera T, Joffin N, Crewe C, Zhu Q, Vishvanath L, Kumar A, Xing C, Wang QA, Gautron L, Deng Y, Gordillo R, Kruglikov I, Kusminski CM, Gupta RK, Scherer PE. Dermal adipose tissue has high plasticity and undergoes reversible dedifferentiation in mice. J Clin Invest. 2019;129:5327–42.
246. Driskell RR, Lichtenberger BM, Hoste E, Kretzschmar K, Simons BD, Charalambous M, Ferron SR, Herault Y, Pavlovic G, Ferguson-Smith AC, Watt FM. Distinct fibroblast lineages determine dermal architecture in skin development and repair. Nature. 2013;504:277–81.
247. Donati G, Proserpio V, Lichtenberger BM, Natsuga K, Sinclair R, Fujiwara H, Watt FM. Epidermal Wnt/β-catenin signaling regulates adipocyte differentiation via secretion of adipogenic factors. Proc Natl Acad Sci U S A. 2014;111:E1501–9.
248. Wojciechowicz K, Gledhill K, Ambler CA, Manning CB, Jahoda CA. Development of the mouse dermal adipose layer occurs independently of subcutaneous adipose tissue and is marked by restricted early expression of FABP4. PLoS One. 2013;8:e59811.
249. Wojciechowicz K, Markiewicz E, Jahoda CAB. C/EBPalpha identifies differentiating preadipocytes around hair follicles in foetal and neonatal rat and mouse skin. Exp Dermatol. 2008;17:675–80.
250. Kruglikov IL, Zhang Z, Scherer PE. The role of immature and mature adipocytes in hair cycling. Trends Endocrinol Metab. 2019;30:93–105.
251. Zhang B, Tsai P-C, Gonzalez-Celeiro M, Chung O, Boumard B, Perdigoto CN, Ezhkova E, Hsu Y-C. Hair follicles' transit-amplifying cells govern concurrent dermal adipocyte production through sonic hedgehog. Genes Dev. 2016;30:2325–38.
252. Rivera-Gonzalez GC, Shook BA, Andrae J, Holtrup B, Bollag K, Betsholtz C, Rodeheffer MS, Horsley V. Skin adipocyte stem cell self-renewal is regulated by a PDGFA/AKT-signaling Axis. Cell Stem Cell. 2016;19:738–51.
253. Rodeheffer MS, Birsoy K, Friedman JM. Identification of White adipocyte progenitor cells in vivo. Cell. 2008;135:240–9.
254. Berry R, Rodeheffer MS. Characterization of the adipocyte cellular lineage in vivo. Nat Cell Biol. 2013;15:302–8.
255. Berry DC, Stenesen D, Zeve D, Graff JM. The developmental origins of adipose tissue. Development (Cambridge). 2013;140:3939–49.
256. Cawthorn William P, Scheller Erica L, Learman Brian S, Parlee Sebastian D, Simon Becky R, Mori H, Ning X, Bree Adam J, Schell B, Broome David T, Soliman Sandra S, DelProposto JL, Lumeng Carey N, Mitra A, Pandit Sandeep V, Gallagher Katherine A, Miller Joshua D, Krishnan V, Hui Susanta K, Bredella Miriam A, Fazeli Pouneh K, Klibanski A, Horowitz Mark C, Rosen Clifford J, MacDougald OA. Bone marrow adipose tissue is an endocrine organ that contributes to increased circulating adiponectin during caloric restriction. Cell Metab. 2014;20:368–75.
257. Suchacki KJ, Tavares AAS, Mattiucci D, Scheller EL, Papanastasiou G, Gray C, Sinton MC, Ramage LE, McDougald WA, Lovdel A, Sulston RJ, Thomas BJ, Nicholson BM, Drake AJ, Alcaide-Corral CJ, Said D, Poloni A, Cinti S, Macpherson GJ, Dweck MR, Andrews JPM, Williams MC, Wallace RJ, van Beek EJR, MacDougald OA, Morton NM, Stimson RH, Cawthorn WP. Bone marrow adipose tissue is a unique adipose subtype with distinct roles in glucose homeostasis. Nat Commun. 2020;11:3097.
258. Custer RP, Ahlfeldt FE. Studies on the structure and function of bone marrow. II. Variations in cellularity in various bones with advancing years of life and their relative response to stimuli. J Lab Clin Med. 1932;17:960–2.
259. Attané C, Estève D, Chaoui K, Iacovoni JS, Corre J, Moutahir M, Valet P, Schiltz O, Reina N, Muller C. Human bone marrow is comprised of adipocytes with specific lipid metabolism. Cell Rep. 2020;30:949–958.e6.
260. Marinelli Busilacchi E, Morsia E, Poloni A. Bone marrow adipose tissue. Cells. 2024;13:724.

261. Pachón-Peña G, Bredella MA. Bone marrow adipose tissue in metabolic health. Trends Endocrinol Metab. 2022;33:401–8.
262. Fazeli PK, Horowitz MC, MacDougald OA, Scheller EL, Rodeheffer MS, Rosen CJ, Klibanski A. Marrow fat and bone-new perspectives. J Clin Endocrinol Metab. 2013;98:935–45.
263. Scheller EL, Doucette CR, Learman BS, Cawthorn WP, Khandaker S, Schell B, Wu B, Ding SY, Bredella MA, Fazeli PK, Khoury B, Jepsen KJ, Pilch PF, Klibanski A, Rosen CJ, MacDougald OA. Region-specific variation in the properties of skeletal adipocytes reveals regulated and constitutive marrow adipose tissues. Nat Commun. 2015;6:7808.
264. Scheller EL, Rosen CJ. What's the matter with MAT? Marrow adipose tissue, metabolism, and skeletal health. Ann N Y Acad Sci. 2014;1311:14–30.
265. Naveiras O, Nardi V, Wenzel PL, Hauschka PV, Fahey F, Daley GQ. Bone-marrow adipocytes as negative regulators of the haematopoietic microenvironment. Nature. 2009;460:259–63.
266. Scheller EL, Troiano N, Vanhoutan JN, Bouxsein MA, Fretz JA, Xi Y, Nelson T, Katz G, Berry R, Church CD, Doucette CR, Rodeheffer MS, Macdougald OA, Rosen CJ, Horowitz MC. Use of osmium tetroxide staining with microcomputerized tomography to visualize and quantify bone marrow adipose tissue in vivo. Methods Enzymol. 2014;537:123–39.
267. Bredella MA, Lin E, Gerweck AV, Landa MG, Thomas BJ, Torriani M, Bouxsein ML, Miller KK. Determinants of bone microarchitecture and mechanical properties in obese men. J Clin Endocrinol Metab. 2012;97:4115–22.
268. Bornstein S, Brown SA, Le PT, Wang X, DeMambro V, Horowitz MC, MacDougald O, Baron R, Lotinun S, Karsenty G, Wei W, Ferron M, Kovacs CS, Clemmons D, Wan Y, Rosen CJ. FGF-21 and skeletal remodeling during and after lactation in C57BL/6J mice. Endocrinology. 2014;155:3516–26.
269. Ackert-Bicknell CL, Shockley KR, Horton LG, Lecka-Czernik B, Churchill GA, Rosen CJ. Strain-specific effects of rosiglitazone on bone Mass, body composition, and serum insulin-like growth factor-I. Endocrinology. 2009;150:1330–40.
270. Tavassoli M. Ultrastructural development of bone marrow adipose cell. Acta Anat (Basel). 1976;94:65–77.
271. Basic anatomical and physiological data for use in radiological protection: reference values. A report of age- and gender-related differences in the anatomical and physiological characteristics of reference individuals. ICRP Publication 89. Ann ICRP; 2002. p. 32.
272. Justesen J, Stenderup K, Ebbesen EN, Mosekilde L, Steiniche T, Kassem M. Adipocyte tissue volume in bone marrow is increased with aging and in patients with osteoporosis. Biogerontology. 2001;2:165–71.
273. Hindorf C, Glatting G, Chiesa C, Lindén O, Flux G. EANM dosimetry committee guidelines for bone marrow and whole-body dosimetry. Eur J Nucl Med Mol Imaging. 2010;37:1238–50.
274. Zhou BO, Yue R, Murphy MM, Peyer JG, Morrison SJ. Leptin-receptor-expressing mesenchymal stromal cells represent the main source of bone formed by adult bone marrow. Cell Stem Cell. 2014;15:154–68.
275. Li Z, Bagchi DP, Zhu J, Bowers E, Yu H, Hardij J, Mori H, Granger K, Skjaerlund J, Mandair G, Abrishami S, Singer K, Hankenson KD, Rosen CJ, MacDougald OA. Constitutive bone marrow adipocytes suppress local bone formation. JCI Insight. 2022;7:7.
276. Akune T, Ohba S, Kamekura S, Yamaguchi M, Chung U-I, Kubota N, Terauchi Y, Harada Y, Azuma Y, Nakamura K, Kadowaki T, Kawaguchi H. PPARgamma insufficiency enhances osteogenesis through osteoblast formation from bone marrow progenitors. J Clin Invest. 2004;113:846–55.
277. Mizoguchi T, Pinho S, Ahmed J, Kunisaki Y, Hanoun M, Mendelson A, Ono N, Kronenberg HM, Frenette PS. Osterix marks distinct waves of primitive and definitive stromal progenitors during bone marrow development. Dev Cell. 2014;29:340–9.
278. Yue R, Zhou BO, Shimada IS, Zhao Z, Morrison SJ. Leptin receptor promotes adipogenesis and reduces osteogenesis by regulating mesenchymal stromal cells in adult bone marrow. Cell Stem Cell. 2016;18:782–96.

279. Li J, Chen X, Lu L, Yu X. The relationship between bone marrow adipose tissue and bone metabolism in postmenopausal osteoporosis. Cytokine Growth Factor Rev. 2020;52:88–98.
280. Nguyen TT, Corvera S. Adipose tissue as a linchpin of organismal ageing. Nat Metab. 2024;6:793–807.
281. Kara N, Xue Y, Zhao Z, Murphy MM, Comazzetto S, Lesser A, Du L, Morrison SJ. Endothelial and leptin receptor+ cells promote the maintenance of stem cells and hematopoiesis in early postnatal murine bone marrow. Dev Cell. 2023;58:348.
282. Mattiucci D, Maurizi G, Izzi V, Cenci L, Ciarlantini M, Mancini S, Mensà E, Pascarella R, Vivarelli M, Olivieri A, Leoni P, Poloni A. Bone marrow adipocytes support hematopoietic stem cell survival. J Cell Physiol. 2018;233:1500–11.
283. Corre J, Planat-Benard V, Corberand JX, Pénicaud L, Casteilla L, Laharrague P. Human bone marrow adipocytes support complete myeloid and lymphoid differentiation from human CD34+ cells. Br J Haematol. 2004;127:344–7.
284. Labella R, Vujačić M, Trivanović D. Bone marrow adipose tissue: regulation of osteoblastic niche, hematopoiesis and hematological malignancies. Stem Cell Rev Rep. 2023;19:1135–51.
285. Kennedy DE, Knight KL. Inhibition of B Lymphopoiesis by adipocytes and IL-1-producing myeloid-derived suppressor cells. J Immunol. 2015;195:2666–74.
286. Valet C, Batut A, Vauclard A, Dortignac A, Bellio M, Payrastre B, Valet P, Severin S. Adipocyte fatty acid transfer supports megakaryocyte maturation. Cell Rep. 2020;32:107875.
287. Zou W, Rohatgi N, Brestoff JR, Li Y, Barve RA, Tycksen E, Kim Y, Silva MJ, Teitelbaum SL. Ablation of fat cells in adult mice induces massive bone gain. Cell Metab. 2020;32:801–813.e6.
288. Clabaut A, Delplace S, Chauveau C, Hardouin P, Broux O. Human osteoblasts derived from mesenchymal stem cells express adipogenic markers upon coculture with bone marrow adipocytes. Differentiation. 2010;80:40–5.
289. Li J, Liu X, Zuo B, Zhang L. The role of bone marrow microenvironment in governing the balance between osteoblastogenesis and adipogenesis. Aging Dis. 2016;7:514–25.
290. Gunaratnam K, Vidal C, Gimble JM, Duque G. Mechanisms of palmitate-induced lipotoxicity in human osteoblasts. Endocrinology. 2014;155:108–16.
291. Al Saedi A, Bermeo S, Plotkin L, Myers DE, Duque G. Mechanisms of palmitate-induced lipotoxicity in osteocytes. Bone. 2019;127:353–9.
292. Caers J, Deleu S, Belaid Z, De Raeve H, Van Valckenborgh E, De Bruyne E, DeFresne MP, Van Riet I, Van Camp B, Vanderkerken K. Neighboring adipocytes participate in the bone marrow microenvironment of multiple myeloma cells. Leukemia. 2007;21:1580–4.
293. Liu LF, Shen WJ, Ueno M, Patel S, Kraemer FB. Characterization of age-related gene expression profiling in bone marrow and epididymal adipocytes. BMC Genomics. 2011;12:212.
294. Laharrague P, Larrouy D, Fontanilles AM, Truel N, Campfield A, Tenenbaum R, Galitzky J, Corberand JX, Pénicaud L, Casteilla L. High expression of leptin by human bone marrow adipocytes in primary culture. FASEB J. 1998;12:747–52.
295. Menagh PJ, Turner RT, Jump DB, Wong CP, Lowry MB, Yakar S, Rosen CJ, Iwaniec UT. Growth hormone regulates the balance between bone formation and bone marrow adiposity. J Bone Miner Res. 2010;25:757–68.
296. Zhang X, Hassan MG, Scheller EL. Neural regulation of bone marrow adipose tissue. Best Pract Res Clin Endocrinol Metab. 2021;35:101522.
297. Hennighausen L, Robinson GW. Information networks in the mammary gland. Nat Rev Mol Cell Biol. 2005;6:715–25.
298. Colleluori G, Perugini J, Barbatelli G, Cinti S. Mammary gland adipocytes in lactation cycle, obesity and breast cancer. Rev Endocr Metab Disord. 2021;22:241–55.
299. Kothari C, Diorio C, Durocher F. The importance of breast adipose tissue in breast cancer. Int J Mol Sci. 2020;21:5760.
300. Bani-Sacchi T, Bianchi S, Bani G, Bigazzi M. Ultrastructural studies on white adipocyte differentiation in the mouse mammary gland following estrogen and relaxin. Acta Anat (Basel). 1987;129:1–9.

301. Brenot A, Hutson I, Harris C. Epithelial-adipocyte interactions are required for mammary gland development, but not for milk production or fertility. Dev Biol. 2020;458:153–63.
302. Landskroner-Eiger S, Park J, Israel D, Pollard JW, Scherer PE. Morphogenesis of the developing mammary gland: stage-dependent impact of adipocytes. Dev Biol. 2010;344:968–78.
303. Dzięgelewska-Sokołowska Ż, Majewska A, Prostek A, Gajewska M. Adipocyte-derived paracrine factors regulate the in vitro development of bovine mammary epithelial cells. Int J Mol Sci. 2023;24:13348.
304. Hu X, Juneja SC, Maihle NJ, Cleary MP. Leptin—a growth factor in normal and malignant breast cells and for normal mammary gland development. J Natl Cancer Inst. 2002;94:1704–11.
305. Landskroner-Eiger S, Qian B, Muise ES, Nawrocki AR, Berger JP, Fine EJ, Koba W, Deng Y, Pollard JW, Scherer PE. Proangiogenic contribution of adiponectin toward mammary tumor growth in vivo. Clin Cancer Res. 2009;15:3265–76.
306. Lang L, Zheng J, Liang S, Zhang F, Fu Y, Deng K, Li F, Yang X, Wang J, Luo Y, Zhang S, Zhu X, Wang L, Gao P, Zhu C, Shu G, Xi Q, Zhang Y, Jiang Q, Wang S. Browning of mammary fat suppresses pubertal mammary gland development of mice via elevation of serum phosphatidylcholine and inhibition of PI3K/Akt pathway. Int J Mol Sci. 2023;24:16171.
307. Fata JE, Werb Z, Bissell MJ. Regulation of mammary gland branching morphogenesis by the extracellular matrix and its remodeling enzymes. Breast Cancer Res. 2004;6:1–11.
308. Bartley JC, Emerman JT, Bissell MJ. Metabolic cooperativity between epithelial cells and adipocytes of mice. Am J Physiol. 1981;241:C204–8.
309. Elias JJ, Pitelka DR, Armstrong RC. Changes in fat cell morphology during lactation in the mouse. Anat Rec. 1973;177:533–47.
310. Cinti S. Pink adipocytes. Trends Endocrinol Metab. 2018;29:651–66.
311. Prokesch A, Smorlesi A, Perugini J, Manieri M, Ciarmela P, Mondini E, Trajanoski Z, Kristiansen K, Giordano A, Bogner-Strauss JG, Cinti S. Molecular aspects of adipoepithelial transdifferentiation in mouse mammary gland. Stem Cells. 2014;32:2756–66.
312. Morroni M, Giordano A, Zingaretti MC, Boiani R, De Matteis R, Kahn BB, Nisoli E, Tonello C, Pisoschi C, Luchetti MM, Marelli M, Cinti S. Reversible transdifferentiation of secretory epithelial cells into adipocytes in the mammary gland. Proc Natl Acad Sci U S A. 2004;101:16801–6.
313. Zwick RK, Rudolph MC, Shook BA, Holtrup B, Roth E, Lei V, Van Keymeulen A, Seewaldt V, Kwei S, Wysolmerski J, Rodeheffer MS, Horsley V. Adipocyte hypertrophy and lipid dynamics underlie mammary gland remodeling after lactation. Nat Commun. 2018;9:3592.

# Chapter 3
# Metabolic Function of the Adipose Tissue

**Gergana Deevska**

## Abbreviations

| | |
|---|---|
| β-AR | β-Adrenergic receptor |
| AA | Amino acid |
| ABCA1 | ATP-binding cassette transporter protein A1 |
| ABCG1 | ATP-binding cassette transporter protein G1 |
| ABHD5 | Alpha/beta hydrolase domain-containing protein 5 |
| ACAT | Acyl-CoA:cholesterol acyltransferase |
| ACC | Acetyl-CoA carboxylase |
| acyl-CoA | Acyl coenzyme A |
| AdipoR | Adiponectin receptor |
| AGPAT (*s.* LPAAT) | 1-Acylglycerol-3-phosphate O-acyltransferase |
| AMP | Adenosine monophosphate |
| AMPK | Adenosine monophosphate-activated protein kinase |
| apoA-I | Adipolipoprotein A-I |
| aSMase | Acid sphingomyelinase |
| ATGL | Adipose triacylglycerol lipase |
| ATRA | All-trans retinoic acid |
| BAT | Brown adipose tissue |
| BCAA | Branch chain amino acids |
| CCT | CTP:phosphocholine cytidylyltransferase |
| CDP | Cytidine diphosphate |
| CDS | Chanarin-Dorfman syndrome |
| CE | Cholesterol ester |

G. Deevska (✉)
Sam Houston State University College of Osteopathic Medicine, Conroe, TX, USA
e-mail: g.deevska@shsu.edu



D. Avtanski (ed.), *Adipose Tissue*, Contemporary Endocrinology,
https://doi.org/10.1007/978-3-031-72570-8_3

| | |
|---|---|
| C/EBP | CCAAT/enhancer-binding proteins |
| CERS | Ceramide synthase |
| ChREBP | Carbohydrate-response element-binding protein |
| CPT | CDP-choline:1,2-diacylglycerol choline phosphotransferase |
| DAG | Diacylglycerol |
| DES1 | Dihydroceramide desaturase 1 |
| DGAT | Diacylglycerol acyltransferase |
| DHAP | Dihydroxyacetone phosphate |
| DNL | De novo lipogenesis |
| eNOS | Endothelial nitric oxide synthase |
| ER | Endoplasmic reticulum |
| ETC | Electron transport chain |
| FAS | Fatty acid synthase |
| FA | Fatty acid |
| FC | Free cholesterol |
| FFA | Free fatty acid |
| G3P | Glycerol-3-phosphate |
| G6P | Glucose-6-phosphate |
| GAPDH | Glyceraldehyde-3-phosphate dehydrogenase |
| GCS (*s.* UGCG) | Glucosylceramide synthase |
| GLP | Glucagon-like peptide |
| GLUT | Glucose transporter |
| GPAT | Glycerol-3-phosphate acyltransferase |
| GPCR | G protein-coupled receptor |
| GPD1 | Glycerol 3-phosphate dehydrogenase |
| GPDH | 3-Glycerol phosphate dehydrogenase |
| HDL | High-density lipoprotein |
| HK | Hexokinase |
| HMP | Hexose-monophosphate |
| HSL | Hormone-sensitive lipase |
| iNOS | Inducible nitric oxide synthase |
| iPFK2 | Inducible 6-phosphofructo-2-kinase |
| IR | Insulin receptor |
| IR-A | Insulin receptor isoform A |
| IR-B | Insulin receptor isoform B |
| KSR | 3-Ketosphinganine reductase |
| LCFA | Long-chain fatty acid |
| LD | Lipid droplet |
| LDH | Lactate dehydrogenase |
| LDLR | Low-density lipoprotein receptor |
| LPA | Lysophosphatidic acid |
| LPAAT (*s.* AGPAT) | Lysophosphatidic acid acyltransferase |
| LRP1 | Low-density lipoprotein receptor-related protein 1 |
| MDH | NAD-dependent malate dehydrogenase |
| MGL | Monoacylglycerol lipase |

| | |
|---|---|
| NO | Nitric oxide |
| PA | Phosphatidic acid |
| PAP (*s.* Lipin) | Phosphatidic acid phosphatase |
| PC | Phosphatidylcholine |
| PDC | Pyruvate dehydrogenase complex |
| PDK | Pyruvate dehydrogenase kinase |
| PDP | Pyruvate dehydrogenase phosphatase |
| PE | Phosphatidylethanolamine |
| PEPCK-C | Phosphoenolpyruvate carboxykinase |
| PFK | Phosphofructokinase 1 |
| PI | Phosphatidylinositol |
| PK | Pyruvate kinase |
| PKA | cAMP-dependent protein kinase |
| PKL | Liver-specific pyruvate kinase |
| PKM | Pyruvate kinase muscle isoenzyme |
| PKR | Red blood cells-specific pyruvate kinase |
| PL | Phospholipid |
| PLTP | Phospholipid transfer protein |
| PPAR | Peroxisome proliferator-activated receptor |
| RA | Retinoic acid |
| RAR | Retinoic acid receptor |
| ROS | Reactive oxygen species |
| S1P | Sphingosine-1-phosphate |
| SM | Sphingomyelin |
| SMS1 | Sphingomyelin synthase |
| SPHK | Sphingosine kinase |
| SPT | Serine palmitoyltransferase |
| SR-BI | Scavenger receptor class B type 1 |
| TAG | Triacylglycerol |
| TCA | Tricarboxylic acid |
| TG | Triglyceride |
| TR | Thyroid receptor |
| UCP1 (*s.* thermogenin) | Uncoupling protein-1 |
| UGCG (*s.* GCS) | UDP-glucose:ceramide glucosyltransferase |
| VLDLR | Very low-density lipoprotein receptor |
| WAT | White adipose tissue |

## Introduction

Adipose tissue is a complex metabolically active organ consisting of two functionally distinct tissue types, white adipose tissue (WAT) and brown adipose tissue (BAT), which differ in size and body distribution. WAT is the largest organ distributed throughout the body and is designed to store considerable amounts of energy in the form of triacylglycerols (TAGs). Nearly all TAGs in the body are found in adipocytes, the primary cell type of adipose tissue, with less than 5% located in other tissues such as skeletal muscle cells, liver, and pancreas. WAT adipocytes are unique cells in terms of size and expansion capacity. They contain a single centrally located LD) consisting of TAGs and cholesteryl esters (CEs) surrounded by a phospholipid (PL) monolayer and LD-specific proteins. These unilocular cells are dynamically altered in response to the nutrient condition. During periods of nutrient excess, extra nutrients are converted into TAGs through lipogenesis. TAGs accumulate within LD, causing adipocyte to expand by several folds in size. During nutrient deprivation, stored TAGs are broken down to release free fatty acids (FFAs) via lipolysis, and different body tissues are used to fulfill their energy needs. This is referred to as TAG turnover, which is a slow process central to maintaining whole body energy homeostasis in a healthy adult individual [1].

Conversely, BAT comprises only a small percentage of adult human adipose tissue, localized primarily to the supraclavicular and paravertebral spaces [2, 3]. However, it does not play a role in energy storage. Instead, it dissipates stored energy to produce heat and regulate body temperature, thereby playing a role in thermogenesis [4]. Brown adipocytes are morphologically different from white adipocytes and contain numerous tiny LDs and well-developed mitochondria that express high levels of uncoupling protein-1 (UCP1 or thermogenin). A third type of adipocytes, namely "beige" or "brite," has been described in recent years [5]. Although located within WAT, they are morphologically and metabolically similar to brown adipocytes.

Moreover, white adipocytes can be converted into beige adipocytes in response to specific environmental, neuronal, and hormonal stimuli [6]. This unique morphological and metabolic plasticity of adipose tissue underlies its crucial role in maintaining energy homeostasis and ultimately in regulating metabolism and body weight. In addition, the function of adipose tissue as an endocrine organ requires a highly active protein synthesis machinery to produce and secrete hundreds of small peptide hormones and other metabolites (collectively known as "adipokines" from WAT and "batokines" from BAT). They further contribute to the crosstalk between cells and tissue types and ensure adequate whole-body homeostatic control. This chapter focuses on the critical role of adipose tissue in whole-body energy homeostasis by exploring the major metabolic pathways in different adipocyte types, summarizing their regulatory mechanisms and potential impact on human health and disease.

## Role of Adipose Tissue in Energy Homeostasis

The primary role of the adipose organ in regulating whole-body energy homeostasis is to (1) act as a dynamic energy reservoir, ensuring that an adequate supply of energy substrates for other tissues is maintained within healthy ranges under different physiological states; (2) convert chemical energy into heat via non-shivering thermogenesis under certain stress conditions; and (3) communicate with the rest of the body via endocrine signals to achieve a higher level of regulatory control of energy homeostasis.

The role of adipose tissue in dynamic energy storage and release within the human body is attained by WAT. This tissue type plays an integral role in maintaining energy homeostasis, ensuring that energy is stored during periods of excess caloric intake and released when the body needs it, such as during fasting or times of increased energy demand. The consumption of more calories than required for immediate energy expenditure leads to the conversion of extra energy into TAG molecules via lipogenesis and their storage in adipocytes as LDs. These LDs can expand in size and number as more energy is stored, allowing the WAT to adapt to the changing energy needs of the body. During fasting or with increased energy demand, the body relies on the energy stored in white adipocytes to meet its energy needs. The release of stored energy from adipose tissue occurs through lipolysis. TAG molecules are hydrolyzed to FFAs and glycerol, which are released into the bloodstream for use as energy substrates by other tissues, such as the muscle and liver. Lipogenesis and lipolysis metabolic pathways are described in greater detail in section "Central Metabolic Pathways in Adipocytes" of this chapter. These processes are regulated by a complex signaling network that involves various hormones and enzymes. They ensure metabolic flexibility and a consistent energy supply, which are essential for survival.

Non-shivering thermogenesis, a heat-generating process that prevents hypothermia without muscle shivering, is a unique feature of brown and beige adipocytes [7]. Thermogenesis is regulated by environmental factors such as cold exposure as and the sympathetic nervous system. In addition to its central role in regulating body temperature, it supports the maintenance of energy balance within the body by dissipating excess energy as heat, thus enhancing energy expenditure. Notably, recent studies have provided compelling evidence that the thermogenic-mediated functions of brown/beige fat extend beyond heat generation and play an active role in regulating glucose and lipid metabolism, insulin sensitivity, and overall adipose tissue homeostasis [8]. Hence, regulation of the number and function of brown/beige adipocytes is emerging as a promising strategy for developing novel approaches that support the management of excess body adiposity and diabetes [9]. An in-depth overview of the process of non-shivering thermogenesis and its underlying regulatory mechanisms is provided in section "Regulation of Adipose Tissue Metabolism" of this chapter.

Efficient crosstalk between adipose tissues and other tissues and organs in the human body is essential for maintaining a continuous, well-balanced homeostatic state. This is achieved by the secretion of a large number of factors with hormonal, autocrine, and paracrine properties. These factors include peptides, hormones, growth factors, binding proteins and precursors, a variety of novel signaling lipids and lipid mediators, certain inflammatory mediators, and exosomal miRNAs [10]. They are named adipokines and batokines and represent a complex and poorly understood network of biochemical mediators produced in WAT and BAT, respectively, in response to metabolic, environmental, and genetic triggers aimed at communicating with other tissues to maintain homeostasis and adapt to changes [11]. The endocrine roles of adipose tissue are discussed more comprehensively in other chapters of this book.

## Central Metabolic Pathways in Adipocytes

### *Lipid Metabolism in Adipose Tissue*

Lipids comprise a diverse group of molecules with unique structures, functions, and metabolic pathways that govern their synthesis and turnover. The primary energy storage lipid molecule in adipose tissue is the glycerolipid TAG. It is the primary neutral lipid that accumulates inside LD and comprises the hydrophobic core of dynamic cellular organelles. The LD surface is composed of polar, amphipathic lipids, which are relatively minor, but important lipid intermediates with specific functions. These include cholesterol, several different phospholipids, and sphingolipids.

#### TAG Synthesis and Turnover

TAG synthesis occurs via two distinct pathways: the glycerol-3-phosphate (G3P) pathway and the acyl coenzyme A (acyl-CoA): monoacylglycerol (MAG) pathway. Both pathways involve multiple steps taking place at the endoplasmic reticulum (ER) membranes. The activity of the MGAT pathway in humans is limited to small intestinal enterocytes during the process of active absorption of dietary fat due to the abundance of the digestion product sn-2 MAG. In adipocytes, the major pathway that takes place is the G3P pathway. It comprises four sequential steps, in which three FFAs are added to a glycerol backbone to form TAG. This pathway was originally named the "Kennedy pathway" in honor of Kennedy and his colleagues, who identified the enzymes of the TAG pathway in the 1950s and 1960s [12, 13]. It begins with acylation of G3P at the sn-1 position by a long-chain acyl-CoA thioester, a step catalyzed by the enzyme glycerol-3-phosphate acyltransferase (GPAT). The product of this first step is lysophosphatidic acid (LPA). The GPAT-catalyzed reaction is the rate-limiting step of the de novo TAG synthesis pathway. In mammals, four isoforms of GPATs (GPAT1–4) have been identified and characterized as having different subcellular localizations, substrate preferences, and tissue-specific expression profiles [14]. GPAT1 and GPAT2 are localized in the mitochondrial outer membrane, whereas GPAT3 and GPAT4 are localized in the ER membrane.

Moreover, GPTA1 and GPTA2, although expressed to some degree in WAT, appear to have limited contribution to the total activity of G3P conversion to LPA [15–17]. GPAT3 and GPAT4, on the other hand, are both expressed in WAT. GPAT3 is the predominant isoform expressed in WAT, accounting for as much as 80% of total GPAT activity [14, 18]. GPAT4, in addition to its expression in WAT, has also been found in BAT. Although the mRNA abundance of this isoform is high in adipose depots, the data suggest that most of the GPAT activity in adipocytes is provided by GPAT3 [19]. Its expression and activity are further significantly upregulated during adipocyte maturation [18].

A second acylation at the sn-2 position of LPA by acyl-CoA:1-acylglycerol-3-phosphate O-acyltransferase (AGPAT or LPA acyltransferase (LPAAT)) produces a second intermediate of this pathway, phosphatidic acid (PA). As many as 13 genes encoding proteins with similar structures and sequence similarities to previously characterized glycerolipid acyltransferases have been identified and named *Agpat 1–13*; however, only AGPAT1, 2, and 3 have been shown to catalyze the esterification of LPA to form PA [13, 20]. WAT expresses all three AGPAT isoforms; however, AGPAT-2 and -3 are barely detectable. AGPAT2 is critical for TAG synthesis in this tissue type and its expression is twofold greater than AGPAT1. Deficiency of AGPAT2 in humans results in the Berardinelli-Siep form of congenital lipodystrophy, indicating that the different isoforms cannot compensate for its loss [21].

In step 3 of this pathway, PA is hydrolyzed by the enzyme phosphatidic acid phosphatase (PAP, also called Lipin) to form diacylglycerol (DAG). PAP catalytic activity is $Mg^{2+}$-dependent, and in mammalian cells, there are three isoforms (PAP1–3) with distinct structural similarities and tissue-specific expression patterns. All three PAP isoforms are expressed in human fat, with PAP1 being the primary lipin enzyme in both WAT and BAT [22]. Mutations in this isoform lead to a marked lipodystrophy phenotype in mice, whereas humans with mutations in the gene encoding PAP1 exhibit no defects in adipogenesis or reduction in adiposity, suggesting the existence of a yet-to-be-identified compensatory mechanism for DAG synthesis in adipocytes [23]. DAG is not only the immediate precursor for the synthesis of TAG but also a vital branch-point intermediate that intersects the synthesis of TAG and the major phospholipids that compose the phospholipid monolayer of LD, phosphatidylcholine (PC), and phosphatidylethanolamine (PE).

The fourth and final esterification step in the TAG synthesis pathway is catalyzed by diacylglycerol acyltransferase (DGAT). Two genes have been identified in mammalian cells that encode structurally distinct enzymes that catalyze the same biochemical reaction: *DGAT1* and *DGAT2* [24, 25]. Both enzymes are highly expressed in cells and tissue types that actively produce TAG, including white and brown adipose tissues [26]. It appears that the two DGAT enzymes can compensate for each other in adipose tissue, suggesting that they do not possess completely distinct functions and are most likely not essential for TAG storage in adipocytes [27].

Lipolysis is the metabolic pathway through which TAG is completely catabolized to FFAs and glycerol. It is the primary turnover pathway for TAG molecules that occurs in the cytosol of adipocytes. In white adipose tissue, the primary function of lipolysis is to generate FFAs that are excreted into the circulation and used for energy production in other tissues, therefore participating in the maintenance of whole-body energy homeostasis. Conversely, fatty acids (FAs) generated via lipolysis are intracellularly used as substrates for β-oxidation and energy production in brown adipocytes. Lipolyzed FAs may also be re-esterified back to TAG or activated for use in the synthesis of other FA-containing intracellular lipids, such as phospholipids, cholesteryl esters, and sphingolipids.

The central pathway of adipocyte lipolysis is catalyzed by three main lipases, namely adipose triacylglycerol lipase, hormone-sensitive lipase, and monoacylglycerol lipase, which work coordinatively to remove the FA moieties of TAGs, sequentially releasing the glycerol backbone.

The removal of the first FA from the TAG molecule producing DAG is catalyzed by the enzyme adipose triacylglycerol lipase (ATGL, also known as desnutrin, PNPLA2, and calcium-independent phospholipase A2ζ), which is the major TAG-specific hydrolase highly expressed in adipose tissue [28–30]. This is the rate-limiting step in TAG degradation in vivo [31]. ATGL is a LD-associated enzyme that possesses the highest substrate specificity towards TAG, but can also hydrolyze DAG, although at a much lower level. In addition, it is not currently known whether ATGL may have specificity towards the sn position of the FAs in TAG, their chain length, or saturation level. Consistent with the key role of this enzyme in TAG lipolysis, ATGL-null mice show increased white adiposity, and silencing ATGL in human adipocytes significantly reduces lipolysis, whereas overexpression has the opposite effect [31, 32]. Humans lacking functional ATGL, however, do not have increased adiposity despite the presence of multisystem TAG storage underlying severe pathologies [33].

For optimal activity, ATGL requires the presence of a specific activator protein known as alpha/beta hydrolase domain-containing protein 5 (ABHD5, also known as CGI-58). CGI-58 is an essential component of the lipolytic system capable of potently inducing the activity of ATGL in tissue extracts from adipose tissue [34]. Genetic defects in this activator protein result in the neutral lipid storage disease Chanarin-Dorfman syndrome (CDS), characterized by intracellular accumulation of TAG in LDs of most tissues and ichthyosis [35].

The second step in TAG turnover involves the release of another FA from DAG to produce MAG under the catalytic activity of the enzyme hormone-sensitive lipase (HSL). This name originates from the fact that lipase is under hormonal regulation [36]. HSL was historically the first identified neutral lipase in adipose tissue and, for many years, was considered the sole key player in TAG hydrolysis, catalyzing the rate-limiting reaction until the discovery and characterization of ATGL [37]. Recent advances in our understanding of the role of HSL in TAG hydrolysis indicate that although it has broad substrate specificity towards tri-, di-, and monoacylglycerols, as well as cholesterol and retinyl esters, the rates of hydrolysis of DAG are much higher than those of other substrates and at least tenfold higher than for TAG [38]. The structural and functional properties of HSLs and their different modes of regulation have been extensively studied over the past five decades and are well characterized (reviewed in [38]). Briefly, during lipolysis, HSL translocates to the surface of the LD in response to hormonal stimuli that reversibly activate phosphorylation sites located within the regulatory domain of the enzyme via cAMP-dependent protein kinase (PKA) and intracellular levels of cAMP. In adipose tissue, in addition to post-translational modifications, protein-protein interactions with specific LD-associated proteins, such as perilipins, vimentin, and FABP, have also been to play an important role in regulating HSL activity [39–41]. In humans, mutations in the gene encoding the HSL protein result in more than 50% decreased expression and lead to adipose tissue dysfunction presenting with features of lipodystrophic phenotype [42, 43].

Lipase catalyzes the final step in TAG hydrolysis, in which the third FA is released from MAG to yield free glycerol monoacylglycerol lipase (MGL). This enzyme is ubiquitously expressed in various tissues, with high levels in adipose tissue. In contrast to the other two lipases, MGL does not hydrolyze other glycerolipids and possesses specificity towards the hydrolysis of monoacylglycerols at both the sn-1- and sn-2-position of the FA residue [44, 45]. Although this enzyme has been extensively studied for its critical role in the brain endocannabinoid signaling system [46], its physiological role in adipose tissue lipolysis remains to be elucidated.

A shared noteworthy characteristic of ATGL, HSL, and MGL is the existence of regulatory mechanisms at the level of protein interaction partners that modulate enzyme activity, localization, stability, or affinity to other proteins. Deficiency of many of these proteins with regulatory and/or supporting functions leads to pronounced alterations in the metabolism of TAG and other neutral lipids, often resulting in adipose tissue dysfunction and consequent pathologies [47].

### Cholesterol Storage and Efflux

Cholesterol is the second most abundant neutral lipid accumulating in adipocytes LDs. As much as one-fourth of total body cholesterol is stored in adipose tissue, which could increase to one-half of the adipose tissue hypertrophy and hyperplasia associated with obesity [48, 49]. Interestingly, nearly all LD-associated cholesterol is free cholesterol (FC), with less than 6% cholesterol esters (CE) [50]. This is consistent with the finding that acyl-CoA:cholesterol acyltransferase (ACAT), an enzyme that catalyzes the esterification of cholesterol with long-chain FAs (LCFAs), has relatively low expression levels of in mouse adipose tissue compared to other tissues [51]. Furthermore, the rate of de novo cholesterol synthesis in adipocytes has been reported to be negligent. Instead, these cells acquire cholesterol via lipoprotein uptake via several different receptor-mediated mechanisms, including the low-density lipoprotein receptor (LDLR), scavenger receptor class B type 1 (SR-BI), low-density lipoprotein receptor-related protein 1 (LRP1), and very low-density lipoprotein receptor (VLDLR) [52–55]. Diet is the primary source of adipocyte cholesterol, as evidenced by earlier studies demonstrating that dietary cholesterol levels are more strongly correlated with adipocyte cholesterol content than with plasma cholesterol concentrations [48, 56]. Circulating triglyceride (TG)-rich chylomicrons deliver dietary cholesterol to the adipose tissue, where uptake occurs in association with TG accrual through the above-mentioned membrane-anchored lipoprotein receptors.

The removal of cholesterol from adipocytes is achieved through efflux via two ATP-binding cassette transporter proteins, ABCA1 and ABCG1, and SR-BI [57].

ABCA1 acts to efflux excess cellular FC to adipolipoprotein A-I (apoA-I), forming nascent high-density lipoprotein (HDL) particles in circulation, contributing to systemic HDL biogenesis [58]. SR-BI, on the other hand, transfers cholesterol to the mature circulating HDL. It was initially shown that ABCA1 and SR-BI are the primary adipocyte-expressed transport systems for cholesterol efflux in adipose tissue [59]. However, more recent studies, implicated ABCG1 as a significant player in adipose tissue cholesterol homeostasis and overall lipid storage capacity [60, 61]. Both ABCA1 and ABCG1 expression levels in adipose tissue and FC efflux appeared to be proportional to adipocyte size and cellular TAG content. However, the mechanism underlying this association remains to be elucidated.

Adipocyte phospholipid transfer protein (PLTP) has recently been identified as an additional contributor to cholesterol efflux in adipose tissue. It stimulates adipocyte FC efflux and contributes a small but significant pool of plasma HDL cholesterol, which illustrates a necessary but not yet well-understood link between phospholipids and cholesterol efflux in this cell type [62].

## Synthesis and Remodeling of Major Phospholipids in Adipose Tissue

The major LD-associated surface phospholipids in mammalian cells are PC, PE, and phosphatidylinositol (PI). Compared with other membranes, LDs are relatively deficient in phosphatidylserine and phosphatidic acid but are enriched in some lysophospholipids [63]. The metabolic steps for synthesizing the three major phospholipids found in adipocyte LDs—PC, PE, and PI –are summarized below. Notably, these phospholipids share a common intermediate with the TAG biosynthetic pathway. DAG is a precursor for the synthesis of PC and PE, whereas phosphatidic acid is used to synthesize PI.

The initial step in PC synthesis begins with the import of extracellular choline inside adipocytes and its rapid phosphorylation to phosphocholine by the cytosolic enzyme, choline kinase. In the subsequent reaction, phosphocholine is converted to cytidine diphosphate (CDP)-choline by CTP:phosphocholine cytidylyltransferase (CCT), the rate-limiting reaction for PC biosynthesis. The final reaction in this pathway is the transfer of choline from CDP-choline to DAG in a reaction catalyzed by CDP-choline:1,2-diacylglycerol choline phosphotransferase (CPT), an integral membrane protein localized primarily to the ER [64, 65].

PE is synthesized via the CDP-ethanolamine pathway, which is composed of several reactions that closely parallel those of the CDP-choline pathway for PC synthesis. The cytosolic enzyme ethanolamine kinase first activates the amino alcohol ethanolamine to phospho-ethanolamine. Subsequently, CTP reacts with the phosphorylated amine to produce CDP-ethanolamine in a reaction catalyzed by the cytosolic enzyme CTP: phosphoethanolamine cytidylyltransferase. In the final step of PE synthesis via this pathway, the integral ER membrane protein CDP-ethanolamine:1,2-diacylglycerol ethanolaminephosphotransferase converts CDP-ethanolamine and DAG to PE [65, 66].

PI is produced from phosphatidic acid (PA) via conversion to CDP-diacylglycerol, which then serves as a precursor for the biosynthesis of several acidic phospholipids, including phosphatidylinositol (PI). In the first step, CDP-diacylglycerol synthase catalyzes the reaction between CTP and PA, resulting in the formation of CDP-diacylglycerol, which is then used in a subsequent reaction to provide the DAG skeleton for the synthesis of PI. This step is catalyzed by the enzyme PI synthase [65].

The degradation and remodeling of phospholipids are achieved by different enzymatic activities of a large group of intracellular phospholipases that vary considerably in structure, function, and tissue expression. They hydrolyze phospholipid substrates at specific ester bonds and play a vital role in the remodeling of existing PL, as well as in the generation of a wide array of bioactive second messenger molecules that play a role in cell signaling [67].

## Sphingolipid Metabolism in Adipose Tissue

Sphingolipids (SLs) are a diverse class of lipid molecules that contain the long-chain amino alcohol sphingosine as the backbone. They are ubiquitously found in all mammalian cells, including adipocytes, as critical structural components of cellular membranes [68]. In addition, several class members have been shown to function as signaling molecules [69]. The importance of SLs in adipose tissue structure and function has emerged in the past decade as our understanding of adipocyte biology has advanced. Adipocytes are unique cells as they must respond rapidly to whole-body energy conditions by expanding or reducing their size to accommodate lipid storage or lipid mobilization, respectively. To achieve this, they must be equipped with essential metabolic pathways to produce and degrade sphingolipids that respond to the energy demands of the body.

Changes in adipocyte size can occur rapidly, which may demand an acute increase in de novo sphingolipid biosynthesis to maintain the proper lipid composition in the plasma membrane and growing LD(s). Several SL species are present in LDs at relatively low concentrations and their specific functions are poorly understood [70]. De novo sphingolipid biosynthesis occurs the ER. It begins with condensation of L-serine and saturated fatty acyl-CoA (most commonly palmitoyl-CoA) to form 3-ketosphinganine. This is the rate-limiting step in the pathway catalyzed by serine palmitoyltransferase (SPT) [71]. Adipose tissue-specific deletion of the catalytic subunit of this enzyme in mice resulted in the normal development of adipose tissue early in life, with progressively increasing cell death and inflammation with age [72]. Adipocyte-specific SPTLC2-deficient (aSPTLC2-KO) mice have markedly reduced adipose tissue mass, resulting in impaired fat distribution, causing systemic insulin resistance and severe lipodystrophy [73]. This unequivocal evidence emphasizes the significance of these lipids in maintaining a healthy adipose tissue.

In the second step, 3-ketosphinganine reductase (KSR) produces sphinganine, which is subsequently acylated to form dihydroceramide by dihydroceramide synthases, often referred to as ceramide synthases (CERS). These are a family of enzymes that differ in their specificity for fatty acyl chain length and saturation. CERS1/4 mainly produces C18-Cer, CERS5/6 results in the formation of C16-ceramides, and CERS2 generates very long-chain ceramides, predominantly C24-cermides. The possible pleiotropic roles of the individual enzymes in adipose tissue are yet to be elucidated; however, the significance of CERS 5 and CERS 6 yielding C-16 ceramide is well illustrated by the consequences of their deletion in animals. Mice deficient in CERS6 exhibit reduced C16:0 ceramide levels in white and brown adipose tissue when placed on a high-fat diet, which in turn results in suppressed adipose inflammation and increased β-oxidative capacity [74]. Similarly, loss of CERS5 is associated with reduced weight gain, reduced white adipose tissue inflammation, and improved glucose levels in animal models of diet-induced obesity [75].

The 4,5-trans-double bond is introduced in the sphingoid backbone of dihydroceramide by the enzyme dihydroceramide desaturase 1 (DES1), yielding ceramide [76]. Although germline ablation of this gene has been shown to result in lethality [77], its removal from adult animals resulted in healthy animals with reduced adipose tissue depots and liver mass, as well as improved lipid and glucose handling [78]. Furthermore, adipose tissue-specific ablation of this gene in obese rodents resulted in improved whole-body glucose metabolism, reduced weight gain by reducing adipose tissue mass, and resolved hepatic steatosis independent of adipose tissue browning.

Once synthesized in the ER, ceramide is transported to the Golgi apparatus, where it is used for the synthesis of all complex sphingolipids, including sphingomyelin (SM) and glycosphingolipids (cerebrosides and gangliosides), as well as other metabolites, such as ceramide-1-phosphate and sphingosine [71]. In the Golgi apparatus, ceramide is converted to SM by sphingomyelin synthase (SMS) 1 (SMS1). Ablation of SMS1 enzyme in animals results in reduced body weight and loss of fat tissue mass, accompanied by a severe lipodystrophic phenotype [79, 80]. Overexpression of this enzyme in a human liver cell line is associated with decreased TAG synthesis and LD formation [81]. However, whether similar phenomena occur in adipocytes remains to be elucidated. A second enzyme, namely SMS2, with the same catalytic activity, exists in mammalian cells, and it is primarily localized to the plasma membrane, where it contributes to lipid remodeling in cell signaling [82]. SMS2-null mice appear to have normal adipose tissue development and function, with no reported severe abnormalities [80].

The addition of glucose to ceramide in the Golgi apparatus is facilitated by glucosylceramide synthase (GCS or UDP-glucose:ceramide glucosyltransferase (UGCG)), leading to the production of the first cerebroside, glucosylceramide, which can be further transformed into more complex glycosphingolipids including gangliosides [83]. Pharmacological targeting of this step in animal models of obesity leads to decreased glycosphingolipid content and has been shown to suppress adipogenesis, reduce the formation of white adipocytes, improve overall adipocyte function, and reduce inflammation in adipose tissue [84, 85]. Thus emphasizes the significance of these lipids in maintaining a healthy adipose tissue.

Ceramide is one of the several key metabolites in this pathway that possesses bioactive properties that mediate cell signaling and regulate various cellular processes. It has been extensively studied in the context of adipose tissue biology, obesity, diabetes, and cardiometabolic diseases [86].

Ceramide can also be produced via the turnover pathway of complex sphingolipids that involve different lysosomal acid hydrolases, and can be recycled by the reacylation of sphingosine. Lysosomal degradation of SM is catalyzed by acid sphingomyelinase (aSMase), which releases the phosphocholine group from the SM-regenerating ceramide. Mice deficient in this enzyme are protected against diet-induced adipose tissue expansion, hyperglycemia, and insulin resistance via liver-mediated metabolic adaptations as well as adipose tissue-specific changes associated with the absence of adipocyte hypertrophy and increased expression of genes associated with brown adipocyte differentiation [87, 88].

Lysosomal degradation of glycosphingolipids is achieved by the action of specific acid hydrolyses that release the respective glyco- moiety to produce ceramide. Acid ceramidase then catalyzes the diacylation of ceramide to release sphingosine. This enzyme has been shown to be activated in mouse adipose tissue overexpressing adiponectin receptors (AdipoRs), resulting in improved overall metabolic health when animals are challenged with a high-fat diet [89, 90].

In the next step, sphingosine can be phosphorylated by either isoform of sphingosine kinase (SPHK1,2) to form sphingosine-1-phosphate (S1P), another crucial bioactive sphingolipid with robust cell-signaling properties. This is the final step in the turnover pathway before the sphingoid backbone is entirely degraded to hexadecenal and phosphoethanolamine, a reaction catalyzed by S1P lyase [91]. A noteworthy characteristic of S1P in adipocytes is its role as a key mediator of the lipolysis metabolic pathway and leptin secretion, supported by the dose-dependent increase in glycerol released by adipocytes treated with S1P in vitro [92]. In addition, depletion of SPHK1 in mice results in increased expression of adipogenesis markers and anti-inflammatory molecules in adipose tissue, whereas it abrogates pro-inflammatory molecules, suggesting that this enzyme is required for lipolysis-driven adipose tissue inflammation [93]. SPHK2 deficiency appears to play a similar role in adipogenesis regulation [94]. S1P lyase deficiency in mice was accompanied by decreased adipose tissue depots. However, the metabolic changes are liver-mediated rather than adipocyte-specific [95].

Although sphingolipids are a minor class of lipids, they are crucial for maintaining metabolically healthy adipose tissue. Therefore, the preservation of structurally and functionally healthy sphingolipid metabolism is an essential determinant of health.

## *Glucose Metabolism in Adipose Tissue*

Although adipose tissue contributes less than 10% to whole-body glucose disposal following a meal, its critical role in systemic glucose homeostasis is well documented. Adipocytes express at least four glucose transporters (GLUTs), GLUT1, GLUT4, GLUT8, and GLUT10 [96]. Of these, the most abundantly expressed is GLUT4, which is also the major insulin-regulated glucose transporter. Several seminal studies elegantly demonstrated that GLUT4-mediated glucose transport into adipocytes regulates whole-body glucose homeostasis, and disruption of this process leads to disease states, such as insulin resistance and diabetes [97]. Moreover, the activity of GLUT4 has been associated with enhanced lipogenesis via activation of the nuclear transcription factor carbohydrate-response element-binding protein (ChREBP), illustrating the existence of an important cross-talk between the two metabolic pathways in adipose tissue [98].

Intracellular glucose is phosphorylated to glucose-6-phosphate (G6P) by the enzyme hexokinase (HK). This irreversible step effectively traps this sugar molecule inside the cell and commits it to further metabolism. Mammalian cells have four isoforms (1–4) of this enzyme, hexokinase, which catalyzes the phosphorylation of glucose to glucose 6-phosphate with broad substrate specificity and the ability to phosphorylate other hexoses. Adipocytes express HK1–3, with HK2 having the highest expression among the three isoforms [99]. Loss of HK2 activity in the adipose tissue of mice resulted not only in reduced glucose disposal by adipocytes but also in increased glucose production in the liver, contributing to the development of glucose intolerance. Similarly to the knockdown of GLUT4, the loss of adipose HK2 was accompanied by decreased lipogenesis and increased FA release in WAT, further emphasizing the role of adipose tissue glucose metabolism in healthy whole-body glucose and lipid homeostasis [100]. G6P can enter one of two major pathways in adipocytes: glycolysis followed by the tricarboxylic acid (TCA) cycle or hexose-monophosphate (HMP) shunt.

### Glucose Metabolism Via Glycolysis and TCA Cycle

Glycolysis is a central metabolic pathway used by all tissues, including adipose tissue, for glucose oxidation to provide energy as ATP and intermediates in other metabolic pathways. It consists of nine steps divided into two phases, in which G6P is metabolized into two pyruvate molecules. The rate-limiting step of glycolysis is the reaction catalyzed by phosphofructokinase (PFK) 1 (PFK-1). This irreversible phosphorylation reaction is the most important control point and is subject to allosteric and hormonal regulation, which determines the overall rate of glycolysis. In mammalian tissues, PFK is a tetramer composed of four subunits encoded by different genes that are arranged in a tissue-specific manner to form homo- and heterotetramers. The tissue-specific arrangement results in three different isoforms of PFK with specific tissue distributions, namely PFK M-type, L-type, and C-type (or PF in humans) [101]. Silencing of the PFK M isoform in mice results in reduced fat depots and lower body weight, most likely because of glycolytic oscillations responsible for enhancing the production of the TAG backbone precursor α-glycerophosphate [102].

PFK-1 in adipose tissue is controlled by the available concentrations of the substrates ATP and fructose 6-phosphate as well as by other regulatory molecules. It is inhibited by ATP and citrate and activated by 5′-AMP, phosphorylation via cyclic AMP-dependent protein kinase, and fructose 2,6-bisphosphate, similar to its regulation in other tissues [103]. This enzyme has been reported to play a role in adipocyte TAG synthesis, lipogenesis, diet-induced systemic insulin resistance, and adipose tissue inflammatory response, which further illustrates the interconnectivity between glucose and lipid metabolism in adipose tissue [104].

The last regulatory step in glycolysis is the conversion of phosphoenolpyruvate to pyruvate, catalyzed by pyruvate kinase (PK). In mammalian tissues, four different PK isoenzymes are expressed in a tissue-specific manner: liver-specific PK (PKL), red blood cells—specific PK (PKR), and two muscle isoforms (PK muscle isoenzymes PKM1 and PKM2), which are generated via alternative splicing of the respective encoding gene and are differentially expressed in various tissues [105, 106]. Adipocytes exhibit high levels of PKM2 expression. PKM2 and PKM1 possess both overlapping and non-overlapping functions with PKM1 and share some similarities with the liver isoform [107]. The expression of both PKM isoforms is increased by thermogenic activation in BAT; however, silencing of PKM2 in mature brown adipocytes leads to augmentation of the expression of selected thermogenic genes, revealing a novel role of the glycolytic pathway in BAT [108, 109].

The fate of the final product of glycolysis, pyruvate, in adipocytes depends on the energetic state of the cell, oxygen availability, and metabolic demands of the body. Pyruvate can be reduced to lactate by lactate dehydrogenase (LDH) to regenerate $NAD^+$ and allow for the continuation of glycolysis. Although counterintuitive, adipocytes actively produce lactate, with as much as 70% of the ingested glucose being converted to this metabolite, a magnitude similar to that observed in skeletal muscles [110, 111]. Moreover, recent studies have demonstrated that lactate production is prioritized in adipocytes, further supporting the idea that glucose metabolism in adipose tissue is essential for whole-body glucose homeostasis [112]. Finally, adipocyte-derived lactate may play roles beyond being an intermediate of glycolysis, such as acting as a sensing and signaling molecule and mediating inflammation [112, 113].

An alternative pathway for further pyruvate metabolism, which occurs in tissues with high oxidative capacity, is oxidative decarboxylation by the pyruvate dehydrogenase complex (PDC). PDC is a sizeable multi-enzyme complex built from multiple copies of the three enzymes. It irreversibly converts pyruvate into acetyl-CoA, a key metabolite that can be further degraded in the TCA cycle to carbon dioxide and water, releasing the energy utilized to synthesize ATP via oxidative phosphorylation. The activity of mammalian PDC is regulated through the phosphorylation–dephosphorylation cycle. Phosphorylation, resulting in inactivation of the complex, is catalyzed by pyruvate dehydrogenase kinase (PDK), and dephosphorylation and the resultant reactivation is catalyzed by pyruvate dehydrogenase phosphatase

(PDP). The most abundant PDK isozyme in adipose tissue is PDK4, which has been implicated in BAT lipogenesis, thermogenesis, and glyceroneogenesis [114].

Several intermediates produced during glucose metabolism in adipocytes serve as substrates for the synthesis of metabolic precursors utilized in the process of lipogenesis, realizing critically important crosstalk between different metabolic pathways in adipose tissue.

## DHAP and Synthesis of Glycerol-3-Phosphate

Dihydroxyacetone phosphate (DHAP) is a glycolytic metabolite produced in a reaction that cleaves fructose-1,6-bisphosphate into two three-carbon molecules. It can be subsequently reduced to yield G3P, which provides a glycerol backbone for TAG synthesis. The reaction is catalyzed by cytosolic glycerol 3-phosphate dehydrogenase (GPD1) and provides an essential link between glucose metabolism and TAG synthesis because adipocytes are unable to re-use glycerol released during lipolysis because they do not express glycerol kinase. Instead, glycerol is transported through the blood to the liver, where it expresses the kinase that phosphorylates glycerol and produces glycerol 3-phosphate. GPD1 utilizes NADH as a coenzyme and is primarily regulated at the transcriptional level by peroxisome proliferator-activated receptor (PPAR)-gamma (PPARγ) in adipose tissues. Its levels have been reported to increase during adipogenesis [115].

G3P is in constant demand in adipose tissue owing to the ongoing TG/FA recycling that exists in adipocytes, especially in situations when recycling is increased. Interestingly, a growing body of evidence suggests that this pathway may not be the major source of G3P in triglyceride synthesis. A biosynthetic pathway for producing G3P from sources other than glucose and glycerol, glyceroneogenesis, is emerging as an important source of carbon for the glycerol backbone of TAG in mammalian cells, including adipocytes [116].

## Glyceroneogenesis, Oxaloacetate, and Pyruvate

Glyceroneogenesis was discovered due to the finding that WAT contains both pyruvate carboxylase and the cytosolic form of phosphoenolpyruvate carboxykinase (PEPCK-C) [117]. Until this discovery, these two enzymes were believed to be involved only in gluconeogenesis. Glyceroneogenesis is initiated by the carboxylation of pyruvate to oxaloacetate and its entry into the TCA cycle, where it is reduced to malate by NAD-dependent malate dehydrogenase (MDH). Malate is transported out of the mitochondria via the dicarboxylate shuttle and re-oxidized back to oxalacetate by the cytosolic form of the former enzyme, which is subsequently decarboxylated to phosphoenolpyruvate by PEPCK-C and converted to 1,3-bisphosphoglycerate via a reversal step of glycolysis. In the next step, this metabolite is oxidized to glyceraldehyde-3-phosphate by glyceraldehyde-3-phosphate dehydrogenase (GAPDH), which exists in equilibrium with DHAP via isomerase. In the final step, DHAP is reduced to 3-glycerol phosphate by 3-glycerol phosphate dehydrogenase (GPDH), as described above [116]. The relative contribution of this pathway to G3P synthesis, compared to glycolysis, is approximately 50% in humans and as much as 80% in rodents. In adipose tissue, it increases during fasting and appears to be the major pathway glyceride-glycerol synthesis in adipose tissue of rats fed a diet supplemented with sucrose for 5 days, accompanied by a significant elevation in the rates of lipogenesis [118, 119]. Taken together, these observations suggest that glucose is most likely not the major precursor of the glycerol backbone for TAG synthesis in adipose tissue, indicating the simultaneous occurrence of both glyceroneogenesis and glycolysis in this tissue type.

#### Acetyl-CoA and De Novo Lipogenesis (DNL)

The conversion of acetyl-CoA to palmitate via FA synthesis *de novo*, followed by palmitate conversion to other FAs and their subsequent incorporation into TAG and storage in LDs, is known as *de novo* lipogenesis (DNL). In humans, FA synthesis occurs primarily in the liver and lactating mammary glands and, and to a lesser extent, in adipose tissue. Approximately 20% of palmitate in TAGs is derived from DNL in the human adipose tissue [1]. *De novo* FA synthesis occurs in the cytosol where acetyl-CoA carboxylase (ACC) and FA synthase (FAS) work together to produce palmitate, the predominant FA produced by FAS [120]. Palmitate can be subsequently elongated and/or desaturated to produce a diverse range of FA species. This process requires energy such as ATP and the availability of the reducing equivalents NADPH to incorporate carbons from acetyl-CoA into the growing FA chain. NADPH is a product of oxidative reactions in the HMP shunt, which provides another link between glucose and lipid metabolism [121]. Numerous studies in animal models and humans have demonstrated that DNL in adipose tissue is metabolically beneficial and decreases insulin resistance, obesity, and high-fat diet [122].

In summary, glucose metabolism in adipose tissue, although a quantitatively insignificant contributor to whole-body disposal of glucose, is qualitatively a critical player in the maintenance of healthy systemic glucose homeostasis.

### *Protein and Amino Acid Metabolism in Adipose Tissue*

A distinct feature of adipose tissue that has been extensively studied and characterized in the last decade is its function as an endocrine organ that actively secretes hundreds of hormones and small peptides acting in endo-, para-, and autocrine fashion to crosstalk with the rest of the cells and tissues in the body. This requires highly active protein synthesizing cellular machinery to ensure adequate control of transcription and translation of newly synthesized proteins and proper folding and post-translational modification to form a final functional protein product. In adipocytes, these processes use pathways and mechanisms that are most likely not different from those in other tissue types. Although the role of adipose tissue in systemic protein and amino acid (AA) metabolism is not well understood, a growing body of research is focusing on deciphering the role of specific AAs.

In the recent decade, a considerable body of evidence has suggested that adipose tissue is capable of metabolizing significant quantities of branch-chain amino acids (BCAAs), leucine, isoleucine, and valine, and it is second only to skeletal muscle in its capacity to catabolize these AAs [123, 124]. More recent studies utilizing selective downregulation of specific BCAA metabolic enzymes in animals have demonstrated that adipose tissue can modulate circulating BCAA levels in vivo [125]. Although the physiological importance of adipose tissue BCAA oxidation remains obscure, these observations support the idea that coordinated downregulation of BCAA oxidative enzymes may dramatically alter adipose tissue BCAA oxidative flux. Consistent with this is the reported downregulation of BCAA oxidative enzyme expression in adipose tissue of obese patients [126]. Studies in BAT using stable isotope tracing showed that in this tissue type, the contribution of BCAA carbons as downstream substrates in the TCA cycle is far less than the respective contributions of glucose, lactate, and FAs [127, 128]. However, their carbon skeleton can be efficiently utilized for de novo lipogenesis in brown adipocytes [129].

Additionally, brown adipocytes use BCAA as a key nitrogen source for synthesizing non-essential amino acids and their downstream products, including Glu, Ala, Asp, N-acetyl-AAs, and glutathione. Many of these molecules are actively secreted by BAT, and when the BCAA-nitrogen flux into brown adipocyte mitochondria is impaired, this results in decreased circulating BCAA-derived metabolites accompanied by increased oxidative stress and decreased hepatic insulin signaling [130]. Taken together, these studies suggest that active BCAA uptake to BAT is critical as a nitrogen donor for metabolite synthesis, rather than being used as carbon fuel for thermogenesis.

Another amino acid of particular importance for adipose tissue function is L-arginine (L-Arg), which acts as a precursor for synthesizing the signaling molecule nitric oxide (NO) via the NOS pathway. In this pathway, L-Arg is converted to NO via one of several NOS enzymes that differ in tissue expression and dependency on $Ca^{2+}$ [131]. Human and rat adipocytes predominantly express endothelial (eNOS) and inducible (iNOS) nitric oxide synthase enzymes [132, 133]. Subsequent studies revealed that eNOS is most likely the major source of NO production in both white and brown adipocytes and plays an essential role in maintaining normal glucose and lipid metabolism, mitochondrial biogenesis, oxidative phosphorylation, reactive oxygen species (ROS) generation, and adipokine secretion [134]. Modulating NO-mediated pathways through dietary supplementation with L-arginine has been suggested as a potential therapeutic strategy against metabolic syndrome and dyslipidemia [135].

## *Thermogenesis in Adipose Tissue*

Thermogenesis is a metabolic process in which the body utilizes energy substrates to produce heat as the primary end-product of metabolism. This process reflects the thermodynamic inefficiency of utilizing substrates for ATP generation in mammals, which results in heat production. Heat generation, in turn, is a vital contributor to the regulation of body temperature. Thermogenesis is used to describe a facultative or adaptive process of heat generation "on-demand" in response to external stimuli, such as cold exposure or nutrient oversupply. In mammals, thermogenesis can be shivering or non-shivering. Shivering thermogenesis is characterized by involuntary irregular muscular activity resulting in heat generation. On the other hand, nonshivering thermogenesis is defined as an increase in metabolic heat production that is not associated with muscle activity. The primary site for non-shivering thermogenesis is the adipose tissue, mainly brown and beige adipocytes.

Non-shivering thermogenesis is achieved via the uncoupling of oxidative phosphorylation from ATP production during electron transfer in the mitochondrial electron transport chain (ETC), releasing extra energy as heat instead of preserving it as ATP. In normal mitochondria, ATP synthesis is coupled with electron transport via the $H^+$ gradient. When the $H^+$ gradient in the inner mitochondrial membrane is disrupted, protons are allowed to flow back into the matrix without flowing through the ATP synthase, resulting in uncoupling and heat generation. This process is mediated by UCP1, which is localized to the inner mitochondrial membrane and facilitates proton leakage [136]. UCP1 is highly expressed in BAT, whose brown color is secondary to the high mitochondrial content underlying its thermogenic potential, and is the best-characterized thermogenic effector in mammals [137]. UCP1-dependent thermogenesis is regulated by both exogenous and endogenous stimuli.

The most prominent activator of BAT thermogenesis is cold exposure, which results in increased metabolic activity and concomitantly increased thermogenesis. This process includes the upregulation of GLUT1 and GLUT4 enhanced expression of the glycolytic enzymes HK, LDH, and PFK, which stimulate glucose metabolism, providing sufficient electron donors to supply electrons to the ETC [138].

Another well-characterized stimulus includes β-adrenergic activators that mediate β-adrenergic signaling via β3-adrenergic receptors (β3-ARs) in rodents and β2-AR in humans [139, 140]. Signaling through these receptors initiates downstream signaling cascades that affect cellular metabolism and initiate the transcription of thermogenic genes. This leads to increased UCP1 protein levels, lower FFA storage in WAT, and higher FFA oxidation and storage in BAT [141, 142]. Thermogenic genes can also be upregulated via the stimulation and activation of thyroid receptors α and β (TRα and β), which are expressed in both WAT and BAT [143]. Other activators of thermogenesis include exercise, and natural extracts are effective in activating adipocytes to gain thermogenic capacity, while simultaneously enhancing metabolic regulation. In addition, numerous mechanisms for activating thermogenesis beyond UCP1 have been identified and characterized to varying extents [9, 144].

Thermogenesis is the key activity that leads to the modulation of systemic metabolism and improvement of energy expenditure, which are two fundamental mechanisms that may be beneficial against obesity and diabetes. Therefore, pharmacological and genetic manipulations of this metabolic process have been explored as promising approaches to treat this disease and may provide future avenues for drug development.

## Regulation of Adipose Tissue Metabolism

Adipose tissue plays a critical role in regulating systemic glucose and lipid metabolism and secretes bioactive molecules that affect almost every cell and tissue in the body. Therefore, well-coordinated and timely regulation of different metabolic pathways in this tissue type is essential for maintaining whole-body metabolic health. Any dysregulation may ultimately result in dysfunctional adipose tissue and initiate pathological changes in other cells and tissues, subsequently leading to disease. The regulation of adipose tissue metabolism is under hormonal and nutritional control achieved at the level of regulating specific metabolic enzymes and pathways to meet the demands of the whole body to maintain a balance in the availability and accessibility of energy substrates to other tissues underlying the essential function of adipose tissue in energy homeostasis.

### *Hormonal Regulation*

The two major hormones that are master regulators of metabolism are insulin and glucagon which will be discussed in this chapter. Other hormones, including steroid hormones, estrogen and androgens, glucocorticoids, and thyroid hormones, employ additional hormonal control of adipose tissue metabolism. The effects of these hormones are less well understood as regulators of normal adipocyte metabolism and are better known for their roles in disease states such as obesity, as discussed elsewhere [145].

#### Insulin

Insulin is produced in pancreatic β-cells and is released in response to increased blood glucose and nutrient availability. It binds to the insulin receptor (IR) expressed on the surface of many tissues, including the adipose tissue. There are two IR isoforms, IR-A and IR-B; however, IR-B is much more specific to insulin and mediates most of its metabolic effects [146]. This isoform is the predominant isoform expressed in adipocytes. Insulin binding to IR triggers downstream signaling cascades that result primarily in altered cellular metabolism and secondary changes to gene expression levels to activate intracellular metabolic pathways and utilize the available substrates.

Adipocytes are also susceptible to insulin exposure. The binding of insulin to adipose tissue insulin receptors results in changes in all metabolic pathways described in this chapter. Insulin affects glucose metabolism in adipose tissue by increasing the rate of glucose transport across the cell membrane, followed by an increase in the rate of glycolysis via the activation of HK and PFK. The effects on lipid metabolism include a decreased rate of lipolysis, lowering of FFAs in the plasma, and stimulation of the lipogenic pathways of fatty acid and triacylglycerol synthesis. High insulin levels favor the dephosphorylated form of HSL, which is inactivated and inhibits lipolysis of stored TAG. Insulin also upregulates lipoprotein lipase (LPL), an extracellular enzyme attached to the endothelial cells of the capillary walls in many tissues, particularly adipose tissue. LPL acts on circulating lipoproteins to release FFAs, which can enter adipocytes and be stored as TAG molecules [147]. Insulin can also affect protein metabolism in adipocytes by increasing the rate of AA transport and protein synthesis [148].

### Glucagon

Glucagon is one of several counterregulatory hormones that counteract the effects of insulin in the human body. The others include the catecholamines epinephrine and norepinephrine, cortisol, and growth hormone. It is produced in pancreatic α-cells from a larger proglucagon peptide and is secreted in response to low blood glucose levels. Encoded by the proglucagon gene, the proglucagon peptide is subject to post-translational processing via specific convertase enzymes expressed in a tissue-specific manner to produce glucagon in pancreatic α-cells and other peptides, namely glucagon-like peptides (GLPs), GLP-1, GLP-2, oxyntomodulin, and glicentin in L-cells of the intestine and brain, respectively [149]. The effects of glucagon in different tissues are mediated by glucagon binding to the glucagon receptor, a member of the class II G protein-coupled receptor (GPCR) superfamily [150]. Glucagon receptors are expressed in adipocytes, but to a lesser extent than in liver and kidney cells [151]. Upon binding to its receptor, glucagon activates a downstream signaling cascade mediated by c-AMP-dependent activation of several signaling proteins, including PKA activation and subsequent phosphorylation of downstream targets aimed at mobilizing TAG stores to release energy substrates for use by other tissues [152]. The primary effect of glucagon on lipid metabolism is inhibition of FA synthesis through phosphorylation and subsequent inactivation of ACC by adenosine monophosphate (AMP)-activated protein kinase (AMPK). This facilitates the transport of FA into the mitochondrial matrix for β-oxidation. In addition, glucagon stimulates HSL in adipocytes, promoting lipolysis independently of sympathetic nervous system innervation, resulting in increased FFAs in the circulation, which bind to albumin and are delivered to the liver and other peripheral tissues [153–155]. However, catecholamines appear to be the primary activators of lipolysis in adipose tissue.

In addition to regulating lipid metabolism in adipose tissue, glucagon also plays a pivotal role in controlling energy expenditure via thermogenesis in BAT. It has been shown to increase energy expenditure by stimulating oxygen consumption and heat production in vitro and in vivo [156–158]. Cold exposure increases circulating glucagon levels and the concomitant stimulation of non-shivering thermogenesis [159]. Acute glucagon administration affects energy expenditure via BAT-dependent and -independent mechanisms, which are poorly understood [145].

The glucagon regulation of different metabolic pathways in adipose tissue is complex and not yet fully understood, most likely involving multi-directional cross-talk between the adipose tissue, the brain, and potentially other vital peripheral organs.

## *Nutritional Regulation*

The regulation of adipose tissue metabolism by the availability of macronutrients for providing energy fuels during the feed-fast cycle in the body is mediated through the fluctuating circulating levels of the abovementioned hormones, insulin and glucagon, in response to food intake. The effect of certain micronutrients, such as vitamins, minerals, and other minor dietary components such as bioactive dietary compounds (or neutraceuticals), on different metabolic pathways within adipocytes is an emerging field of research. However, data on the effects of specific nutrients and the underlying mechanisms, particularly in humans, are unclear. Moreover, most dietary interventional studies have focused on the impact of nutrition on disease states such as insulin resistance and obesity.

Specific vitamins and dietary compounds have been implicated in the regulation of adipose tissue metabolic properties by affecting lipid turnover and proper adipose tissue functional integrity. These molecules act as modulators of cellular signaling at the molecular level and affect gene expression of key regulators of food intake, lipid metabolism, adipogenesis into white, beige and brown adipose tissue, thermogenesis, lipotoxicity, adipo/cytokine synthesis, and inflammation [160]. Only a few of these nutritional regulators that are important for adipose tissue metabolic integrity are summarized here.

### Bioactive Dietary Compounds

Nutraceuticals are a large group of chemically different molecules that do not typically provide energy. Instead, they are minor constituents of our diet, often added as spices such as cinnamon, rosemary, ginger, pepper, saffron, garlic, onion, turmeric, chili, and others. They contain specific biologically active compounds such as curcumin, catechins, capsaicin, gingerol, resveratrol, and berberine, which have been shown to have beneficial health effects. In adipose tissue, nutraceuticals modulate cellular signaling pathways involved in the storage and dissipation of energy and enhance satiety, ultimately leading to weight loss. The underlying mechanisms of these effects include increased thermogenesis and reduced lipid accumulation in adipocytes by regulating the expression of related transcription factors such as CCAAT/enhancer-binding proteins (C/EBPs) and PPARγ as well as modulating the activities of certain enzymes related to lipogenesis, such as ACC, FAS, and GPDH [161]. Each individual bioactive compound has a specific mode of action in the control of metabolic functions in adipose tissues, some of which have been studied and summarized in more detail [160].

Another essential property of these compounds is their anti-inflammatory and antioxidant properties. In vitro studies have shown that they decrease preadipocyte proliferation and differentiation, and viability. These effects are mediated by interference with specific signaling pathways involved in inflammation.

### Vitamins

Fat-soluble vitamins A, D, and E have emerged as regulators of adipose tissue metabolism. Adipose tissue is one of the primary storage depots for these micronutrients, where their intracellular concentration is usually proportional to the LD size.

Vitamin A in the human diet is present in two forms: retinyl esters and β-carotene or other carotenoids like xanthophylls and zeaxanthins [162]. Endogenously, carotenoids are converted to retinol, which is subsequently oxidated to 9-*cis*-retinoic acid (RA) and all-trans RA (ATRA), which are the biologically active metabolites that act as important regulators of gene transcription via retinoid acid receptors (RARs) [163]. RA is involved in the regulation of several metabolic processes that govern adipogenesis and lipogenesis in adipocytes. The mechanisms underlying these effects remain largely speculative [164]. In addition, RA regulates the expression of genes that affect lipid metabolism in adipocytes, such as UCP1 and UCP3, and plays a role in regulating thermogenesis in rodent BAT [165]. However, in humans, these effects are minimal [166].

The active form of vitamin D, calcitriol (1,25$(OH)_2$D), exhibits anti-inflammatory activity in adipocytes, suggesting that it may help decrease adipose tissue inflammation associated with obesity and metabolic dysfunction. Furthermore, a vitamin D-deficient diet results in reduced expression of lipid oxidation-related genes, such as PGC1-α, PPARα, UCP1, and others in the adipose tissue [167, 168]. Although most of these studies have been conducted in animals and data from human studies are yet to be obtained, based on the detrimental effects of vitamin D deficiency on adipose tissue metabolism, it has been suggested that it plays a beneficial role in adipose tissue metabolism and functional integrity.

Vitamin E (α-tocopherol) is the far less studied fat-soluble vitamin. It acts as a fat-soluble antioxidant and signaling molecule that mediates a number of cellular events relevant to fatty acid uptake, metabolism, accumulation, storage, and release, as well as adipokine secretion and release from adipose tissue [169–171].

## Conclusion

Adipose tissue is a complex metabolic organ designed to respond appropriately to changes in nutritional status by undergoing dynamic remodeling, including quantitative and qualitative alterations in its structure. To accommodate these changes, it must be able to quickly adapt to intracellular metabolism and provide a continuous supply of energy donors to maintain whole-body energy homeostasis. This is achieved by the coordinated regulation of the metabolism of lipids, glucose, and certain amino acids highlighted in this chapter. The regulation of adipose tissue metabolism is under the coordinated action of two hormones, the master-metabolic regulators insulin and glucagon. Additional regulation exists via certain nutrients; however, their significance has mainly been studied in animals, and their role in humans is unclear.

The metabolic integrity of adipose tissue is an essential determinant of overall adipocyte health, assuring the proper synthesis and release of hundreds of molecules that act as endocrine mediators in other tissues. These include a variety of peptides, microRNAs, and lipids, all of which are direct products of adipose tissue metabolism.

## References

1. Strawford A, et al. Adipose tissue triglyceride turnover, de novo lipogenesis, and cell proliferation in humans measured with 2H2O. Am J Physiol Endocrinol Metab. 2004;286(4):E577–88.
2. Gilsanz V, et al. Changes in brown adipose tissue in boys and girls during childhood and puberty. J Pediatr. 2012;160(4):604–609.e1.
3. Carpentier AC, et al. Brown adipose tissue energy metabolism in humans. Front Endocrinol (Lausanne). 2018;9:447.
4. Enerbäck S. Human Brown adipose tissue. Cell Metab. 2010;11(4):248–52.
5. Wu J, et al. Beige adipocytes are a distinct type of thermogenic fat cell in mouse and human. Cell. 2012;150(2):366–76.
6. Smorlesi A, et al. The adipose organ: white-brown adipocyte plasticity and metabolic inflammation. Obes Rev. 2012;13(Suppl 2):83–96.
7. Ikeda K, Maretich P, Kajimura S. The common and distinct features of Brown and Beige adipocytes. Trends Endocrinol Metab. 2018;29(3):191–200.
8. Roberts-Toler C, O'Neill BT, Cypess AM. Diet-induced obesity causes insulin resistance in mouse brown adipose tissue. Obesity (Silver Spring). 2015;23(9):1765–70.
9. Cohen P, Kajimura S. The cellular and functional complexity of thermogenic fat. Nat Rev Mol Cell Biol. 2021;22(6):393–409.
10. Kahn CR, Wang G, Lee KY. Altered adipose tissue and adipocyte function in the pathogenesis of metabolic syndrome. J Clin Invest. 2019;129(10):3990–4000.
11. Ghesmati Z, et al. An update on the secretory functions of brown, white, and beige adipose tissue: towards therapeutic applications. Rev Endocr Metab Disord. 2023;25:279.
12. Kennedy EP, Smith SW, Weiss SB. New synthesis of lecithin in an isolated enzyme system. Nature. 1956;178(4533):594–5.
13. Coleman RA, Mashek DG. Mammalian triacylglycerol metabolism: synthesis, lipolysis, and signaling. Chem Rev. 2011;111(10):6359–86.

14. Yu J, et al. Update on glycerol-3-phosphate acyltransferases: the roles in the development of insulin resistance. Nutr Diabetes. 2018;8(1):34.
15. Garcia-Fabiani MB, et al. Methylation of the Gpat2 promoter regulates transient expression during mouse spermatogenesis. Biochem J. 2015;471(2):211–20.
16. Wendel AA, et al. Glycerol-3-phosphate acyltransferase 1 deficiency in Ob/Ob mice diminishes hepatic steatosis but does not protect against insulin resistance or obesity. Diabetes. 2010;59(6):1321–9.
17. Coleman RA, Lee DP. Enzymes of triacylglycerol synthesis and their regulation. Prog Lipid Res. 2004;43(2):134–76.
18. Cao J, et al. Molecular identification of microsomal acyl-CoA:glycerol-3-phosphate acyltransferase, a key enzyme in de novo triacylglycerol synthesis. Proc Natl Acad Sci U S A. 2006;103(52):19695–700.
19. Nagle CA, et al. Identification of a novel sn-glycerol-3-phosphate acyltransferase isoform, GPAT4, as the enzyme deficient in Agpat6−/− mice. J Lipid Res. 2008;49(4):823–31.
20. Beigneux AP, et al. Agpat6—a novel lipid biosynthetic gene required for triacylglycerol production in mammary epithelium. J Lipid Res. 2006;47(4):734–44.
21. Agarwal AK, et al. AGPAT2 is mutated in congenital generalized lipodystrophy linked to chromosome 9q34. Nat Genet. 2002;31(1):21–3.
22. Donkor J, et al. Three mammalian lipins act as phosphatidate phosphatases with distinct tissue expression patterns. J Biol Chem. 2007;282(6):3450–7.
23. Temprano A, et al. Redundant roles of the phosphatidate phosphatase family in triacylglycerol synthesis in human adipocytes. Diabetologia. 2016;59(9):1985–94.
24. Cases S, et al. Identification of a gene encoding an acyl CoA:diacylglycerol acyltransferase, a key enzyme in triacylglycerol synthesis. Proc Natl Acad Sci. 1998;95(22):13018–23.
25. Cases S, et al. Cloning of DGAT2, a second mammalian diacylglycerol acyltransferase, and related family members*. J Biol Chem. 2001;276(42):38870–6.
26. Irshad Z, et al. Diacylglycerol acyltransferase 2 links glucose utilization to fatty acid oxidation in the brown adipocytes. J Lipid Res. 2017;58(1):15–30.
27. Chen G, et al. Acyl-CoA:diacylglycerol acyltransferase: properties, physiological roles, metabolic engineering and intentional control. Prog Lipid Res. 2022;88:101181.
28. Zimmermann R, et al. Fat mobilization in adipose tissue is promoted by adipose triglyceride lipase. Science. 2004;306(5700):1383–6.
29. Jenkins CM, et al. Identification, cloning, expression, and purification of three novel human calcium-independent phospholipase A2 family members possessing triacylglycerol lipase and acylglycerol transacylase activities. J Biol Chem. 2004;279(47):48968–75.
30. Villena JA, et al. Desnutrin, an adipocyte gene encoding a novel patatin domain-containing protein, is induced by fasting and glucocorticoids: ectopic expression of desnutrin increases triglyceride hydrolysis. J Biol Chem. 2004;279(45):47066–75.
31. Haemmerle G, et al. Defective lipolysis and altered energy metabolism in mice lacking adipose triglyceride lipase. Science. 2006;312(5774):734–7.
32. Bezaire V, et al. Contribution of adipose triglyceride lipase and hormone-sensitive lipase to lipolysis in hMADS adipocytes. J Biol Chem. 2009;284(27):18282–91.
33. Fischer J, et al. The gene encoding adipose triglyceride lipase (PNPLA2) is mutated in neutral lipid storage disease with myopathy. Nat Genet. 2007;39(1):28–30.
34. Lass A, et al. Adipose triglyceride lipase-mediated lipolysis of cellular fat stores is activated by CGI-58 and defective in Chanarin-Dorfman syndrome. Cell Metab. 2006;3(5):309–19.
35. Lefèvre C, et al. Mutations in CGI-58, the gene encoding a new protein of the esterase/lipase/thioesterase subfamily, in Chanarin-Dorfman syndrome. Am J Hum Genet. 2001;69(5):1002–12.
36. Vaughan M, Steinberg D. Effect of hormones on lipolysis and esterification of free fatty acids during incubation of adipose tissue in vitro. J Lipid Res. 1963;4:193–9.
37. Holm C, et al. Molecular mechanisms regulating hormone-sensitive lipase and lipolysis. Annu Rev Nutr. 2000;20:365–93.

38. Recazens E, Mouisel E, Langin D. Hormone-sensitive lipase: sixty years later. Prog Lipid Res. 2021;82:101084.
39. Wang H, et al. Activation of hormone-sensitive lipase requires two steps, protein phosphorylation and binding to the PAT-1 domain of lipid droplet coat proteins*. J Biol Chem. 2009;284(46):32116–25.
40. Shen W-J, et al. Vimentin is a functional partner of hormone sensitive lipase and facilitates lipolysis. J Proteome Res. 2010;9(4):1786–94.
41. Jenkins-Kruchten AE, et al. FATTY ACID-binding protein-hormone-sensitive lipase interaction: FATTY ACID DEPENDENCE ON BINDING*. J Biol Chem. 2003;278(48):47636–43.
42. Albert Jessica S, et al. Null mutation in hormone-sensitive lipase gene and risk of type 2 diabetes. N Engl J Med. 2014;370(24):2307–15.
43. Farhan SMK, et al. A novel LIPE nonsense mutation found using exome sequencing in siblings with late-onset familial partial lipodystrophy. Can J Cardiol. 2014;30(12):1649–54.
44. Tornqvist H, Belfrage P. Purification and some properties of a monoacylglycerol-hydrolyzing enzyme of rat adipose tissue. J Biol Chem. 1976;251(3):813–9.
45. Dinh TP, et al. Brain monoglyceride lipase participating in endocannabinoid inactivation. Proc Natl Acad Sci U S A. 2002;99(16):10819–24.
46. Schlosburg JE, et al. Chronic monoacylglycerol lipase blockade causes functional antagonism of the endocannabinoid system. Nat Neurosci. 2010;13(9):1113–9.
47. Hofer P, et al. The lipolysome—a highly complex and dynamic protein network orchestrating cytoplasmic triacylglycerol degradation. Metabolites. 2020;10(4):147.
48. Krause BR, Hartman AD. Adipose tissue and cholesterol metabolism. J Lipid Res. 1984;25(2):97–110.
49. Kovanen PT, Nikkilä EA, Miettinen TA. Regulation of cholesterol synthesis and storage in fat cells. J Lipid Res. 1975;16(3):211–23.
50. Schreibman PH, Dell RB. Human adipocyte cholesterol. Concentration, localization, synthesis, and turnover. J Clin Invest. 1975;55(5):986–93.
51. Uelmen PJ, et al. Tissue-specific expression and cholesterol regulation of acylcoenzyme a:cholesterol acyltransferase (ACAT) in mice. Molecular cloning of mouse ACAT cDNA, chromosomal localization, and regulation of ACAT in vivo and in vitro. J Biol Chem. 1995;270(44):26192–201.
52. Kraemer FB, et al. Low-density lipoprotein receptors in rat adipocytes: regulation with fasting. Am J Physiol. 1994;266(1 Pt 1):E26–32.
53. Desai KS, et al. Very low density lipoprotein binding to adipocytes. Biochim Biophys Acta. 1980;620(3):341–51.
54. Barbaras R, et al. Binding of lipoproteins and regulation of cholesterol synthesis in cultured mouse adipose cells. Biochim Biophys Acta. 1985;845(3):492–501.
55. Fazio S, Linton MF. Unique pathway for cholesterol uptake in fat cells. Arterioscler Thromb Vasc Biol. 2004;24(9):1538–9.
56. Angel A, Farkas J. Regulation of cholesterol storage in adipose tissue. J Lipid Res. 1974;15(5):491–9.
57. Rosenson RS, et al. Cholesterol efflux and atheroprotection: advancing the concept of reverse cholesterol transport. Circulation. 2012;125(15):1905–19.
58. Chung S, et al. Adipose tissue ATP binding cassette transporter A1 contributes to high-density lipoprotein biogenesis in vivo. Circulation. 2011;124(15):1663–72.
59. Zhang Y, et al. Adipocyte modulation of high-density lipoprotein cholesterol. Circulation. 2010;121(11):1347–55.
60. Frisdal E, et al. Adipocyte ATP-binding cassette G1 promotes triglyceride storage, fat mass growth, and human obesity. Diabetes. 2015;64(3):840–55.
61. Buchmann J, et al. Ablation of the cholesterol transporter adenosine triphosphate-binding cassette transporter G1 reduces adipose cell size and protects against diet-induced obesity. Endocrinology. 2007;148(4):1561–73.

62. Jiang H, et al. Adipocyte phospholipid transfer protein and lipoprotein metabolism. Arterioscler Thromb Vasc Biol. 2015;35(2):316–22.
63. Bartz R, et al. Lipidomics reveals that adiposomes store ether lipids and mediate phospholipid traffic. J Lipid Res. 2007;48(4):837–47.
64. Fagone P, Jackowski S. Phosphatidylcholine and the CDP-choline cycle. Biochim Biophys Acta. 2013;1831(3):523–32.
65. Vance JE. Phospholipid synthesis and transport in mammalian cells. Traffic. 2015;16(1):1–18.
66. Vance JE, Tasseva G. Formation and function of phosphatidylserine and phosphatidylethanolamine in mammalian cells. Biochim Biophys Acta. 2013;1831(3):543–54.
67. Aloulou A, et al. Phospholipases: an overview. Methods Mol Biol. 2018;1835:69–105.
68. Pruett ST, et al. Thematic review series: sphingolipids. Biodiversity of sphingoid bases ("sphingosines") and related amino alcohols*. J Lipid Res. 2008;49(8):1621–39.
69. Hannun YA. Functions of ceramide in coordinating cellular responses to stress. Science. 1996;274(5294):1855–9.
70. Deevska GM, Nikolova-Karakashian MN. The expanding role of sphingolipids in lipid droplet biogenesis. Biochim Biophys Acta Mol Cell Biol Lipids. 2017;1862(10 Pt B):1155–65.
71. Merrill AH Jr, Nixon DW, Williams RD. Activities of serine palmitoyltransferase (3-ketosphinganine synthase) in microsomes from different rat tissues. J Lipid Res. 1985;26(5):617–22.
72. Alexaki A, et al. De novo sphingolipid biosynthesis is required for adipocyte survival and metabolic homeostasis*. J Biol Chem. 2017;292(9):3929–39.
73. Lee SY, et al. Adipocyte-specific deficiency of de novo sphingolipid biosynthesis leads to lipodystrophy and insulin resistance. Diabetes. 2017;66(10):2596–609.
74. Turpin SM, et al. Obesity-induced CerS6-dependent C16:0 ceramide production promotes weight gain and glucose intolerance. Cell Metab. 2014;20(4):678–86.
75. Gosejacob D, et al. Ceramide synthase 5 is essential to maintain C16:0-ceramide pools and contributes to the development of diet-induced obesity*. J Biol Chem. 2016;291(13):6989–7003.
76. Michel C, et al. Characterization of ceramide synthesis: a dihydroceramide desaturase introduces the 4,5-trans-double bond of sphingosine at the level of dihydroceramide*. J Biol Chem. 1997;272(36):22432–7.
77. Holland WL, et al. Inhibition of ceramide synthesis ameliorates glucocorticoid-, saturated-fat-, and obesity-induced insulin resistance. Cell Metab. 2007;5(3):167–79.
78. Chaurasia B, et al. Targeting a ceramide double bond improves insulin resistance and hepatic steatosis. Science. 2019;365(6451):386–92.
79. Yano M, et al. Mitochondrial dysfunction and increased reactive oxygen species impair insulin secretion in sphingomyelin synthase 1-null mice*. J Biol Chem. 2011;286(5):3992–4002.
80. Yano M, et al. Increased oxidative stress impairs adipose tissue function in sphingomyelin synthase 1 null mice. PLoS One. 2013;8(4):e61380.
81. Deevska GM, et al. Novel interconnections in lipid metabolism revealed by overexpression of sphingomyelin Synthase-1. J Biol Chem. 2017;292(12):5110–22.
82. Holthuis JC, Luberto C. Tales and mysteries of the enigmatic sphingomyelin synthase family. Adv Exp Med Biol. 2010;688:72–85.
83. Messner MC, Cabot MC. Glucosylceramide in humans. Adv Exp Med Biol. 2010;688:156–64.
84. van Eijk M, et al. Reducing glycosphingolipid content in adipose tissue of obese mice restores insulin sensitivity, adipogenesis and reduces inflammation. PLoS One. 2009;4(3):e4723.
85. Jang HJ, et al. Glucosylceramide synthase regulates adipo-osteogenic differentiation through synergistic activation of PPARγ with GlcCer. FASEB J. 2020;34(1):1270–87.
86. Lambert JM, Anderson AK, Cowart LA. Sphingolipids in adipose tissue: What's tipping the scale? Adv Biol Regul. 2018;70:19–30.
87. Deevska GM, et al. Acid sphingomyelinase deficiency prevents diet-induced hepatic triacylglycerol accumulation and hyperglycemia in mice. J Biol Chem. 2009;284(13):8359–68.
88. Sydor S, et al. Acid sphingomyelinase deficiency in Western diet-fed mice protects against adipocyte hypertrophy and diet-induced liver steatosis. Mol Metab. 2017;6(5):416–27.

89. Holland WL, et al. Inducible overexpression of adiponectin receptors highlight the roles of adiponectin-induced ceramidase signaling in lipid and glucose homeostasis. Mol Metab. 2017;6(3):267–75.
90. Reibe-Pal S, Febbraio MA. Adiponectin serenades ceramidase to improve metabolism. Mol Metab. 2017;6(3):233–5.
91. Serra M, Saba JD. Sphingosine 1-phosphate lyase, a key regulator of sphingosine 1-phosphate signaling and function. Adv Enzyme Regul. 2010;50(1):349–62.
92. Jun D-J, et al. Sphingosine-1-phosphate modulates both lipolysis and leptin production in differentiated rat white adipocytes. Endocrinology. 2006;147(12):5835–44.
93. Wang J, et al. Sphingosine kinase 1 regulates adipose proinflammatory responses and insulin resistance. Am J Physiol Endocrinol Metab. 2014;306(7):E756–68.
94. Mastrandrea LD. Role of sphingosine kinases and sphingosine 1-phosphate in mediating adipogenesis. J Diabetes Mellitus. 2013;3:52.
95. Bektas M, et al. Sphingosine 1-phosphate Lyase deficiency disrupts lipid homeostasis in liver*. J Biol Chem. 2010;285(14):10880–9.
96. Chadt A, Al-Hasani H. Glucose transporters in adipose tissue, liver, and skeletal muscle in metabolic health and disease. Pflugers Arch. 2020;472(9):1273–98.
97. Kahn BB. Adipose tissue, inter-organ communication, and the path to type 2 diabetes: the 2016 banting medal for scientific achievement lecture. Diabetes. 2019;68(1):3–14.
98. Herman MA, et al. A novel ChREBP isoform in adipose tissue regulates systemic glucose metabolism. Nature. 2012;484(7394):333–8.
99. Tan VP, Miyamoto S. HK2/hexokinase-II integrates glycolysis and autophagy to confer cellular protection. Autophagy. 2015;11(6):963–4.
100. Shimobayashi M, et al. Diet-induced loss of adipose hexokinase 2 correlates with hyperglycemia. Elife. 2023;12:12.
101. Dunaway GA, Kasten TP. Nature of the subunits of the 6-phosphofructo-1-kinase isoenzymes from rat tissues. Biochem J. 1987;242(3):667–71.
102. Getty-Kaushik L, et al. Mice deficient in phosphofructokinase-M have greatly decreased fat stores. Obesity. 2010;18(3):434–40.
103. Sale EM, Denton RM. Adipose-tissue phosphofructokinase. Rapid purification and regulation by phosphorylation in vitro. Biochem J. 1985;232(3):897–904.
104. Huo Y, et al. Disruption of inducible 6-phosphofructo-2-kinase ameliorates diet-induced adiposity but exacerbates systemic insulin resistance and adipose tissue inflammatory response. J Biol Chem. 2010;285(6):3713–21.
105. Noguchi T, Inoue H, Tanaka T. The M1- and M2-type isozymes of rat pyruvate kinase are produced from the same gene by alternative RNA splicing. J Biol Chem. 1986;261(29):13807–12.
106. Noguchi T, et al. The L- and R-type isozymes of rat pyruvate kinase are produced from a single gene by use of different promoters. J Biol Chem. 1987;262(29):14366–71.
107. Pogson CI. Adipose-tissue pyruvate kinase. Properties and interconversion of two active forms. Biochem J. 1968;110(1):67–77.
108. Isidor MS, et al. Pyruvate kinase M2 represses thermogenic gene expression in brown adipocytes. FEBS Lett. 2020;594(7):1218–25.
109. Hao Q, et al. Transcriptome profiling of brown adipose tissue during cold exposure reveals extensive regulation of glucose metabolism. Am J Physiol Endocrinol Metab. 2015;308(5):E380–92.
110. Jansson PA, Smith U, Lönnroth P. Evidence for lactate production by human adipose tissue in vivo. Diabetologia. 1990;33(4):253–6.
111. DiGirolamo M, Newby FD, Lovejoy J. Lactate production in adipose tissue: a regulated function with extra-adipose implications. FASEB J. 1992;6(7):2405–12.
112. Krycer JR, et al. Lactate production is a prioritized feature of adipocyte metabolism. J Biol Chem. 2020;295(1):83–98.
113. Feng T, et al. Adipocyte-derived lactate is a signalling metabolite that potentiates adipose macrophage inflammation via targeting PHD2. Nat Commun. 2022;13(1):5208.

114. Klyuyeva A, et al. Tissue-specific kinase expression and activity regulate flux through the pyruvate dehydrogenase complex. J Biol Chem. 2019;294(3):838–51.
115. Oh S, et al. Glycerol 3-phosphate dehydrogenases (1 and 2) in cancer and other diseases. Exp Mol Med. 2024;56:1066.
116. Nye C, et al. Reassessing triglyceride synthesis in adipose tissue. Trends Endocrinol Metab. 2008;19(10):356–61.
117. Ballard FJ, Hanson RW, Leveille GA. Phosphoenolpyruvate carboxykinase and the synthesis of glyceride-glycerol from pyruvate in adipose tissue. J Biol Chem. 1967;242(11):2746–50.
118. Nye CK, Hanson RW, Kalhan SC. Glyceroneogenesis is the dominant pathway for triglyceride glycerol synthesis in vivo in the rat. J Biol Chem. 2008;283(41):27565–74.
119. Reshef L, et al. Glyceroneogenesis and the triglyceride/fatty acid cycle*. J Biol Chem. 2003;278(33):30413–6.
120. Wallace M, et al. Enzyme promiscuity drives branched-chain fatty acid synthesis in adipose tissues. Nat Chem Biol. 2018;14(11):1021–31.
121. Akram M, et al. Hexose monophosphate shunt, the role of its metabolites and associated disorders: a review. J Cell Physiol. 2019;234(9):14473–82.
122. Wallace M, Metallo CM. Tracing insights into de novo lipogenesis in liver and adipose tissues. Semin Cell Dev Biol. 2020;108:65–71.
123. Tischler ME, Goldberg AL. Leucine degradation and release of glutamine and alanine by adipose tissue. J Biol Chem. 1980;255(17):8074–81.
124. Rosenthal J, Angel A, Farkas J. Metabolic fate of leucine: a significant sterol precursor in adipose tissue and muscle. Am J Physiol. 1974;226(2):411–8.
125. Herman MA, et al. Adipose tissue branched chain amino acid (BCAA) metabolism modulates circulating BCAA levels. J Biol Chem. 2010;285(15):11348–56.
126. She P, et al. Obesity-related elevations in plasma leucine are associated with alterations in enzymes involved in branched-chain amino acid metabolism. Am J Physiol Endocrinol Metab. 2007;293(6):E1552–63.
127. Jung SM, et al. In vivo isotope tracing reveals the versatility of glucose as a brown adipose tissue substrate. Cell Rep. 2021;36(4):109459.
128. Bornstein MR, et al. Comprehensive quantification of metabolic flux during acute cold stress in mice. Cell Metab. 2023;35(11):2077–2092.e6.
129. Green CR, et al. Branched-chain amino acid catabolism fuels adipocyte differentiation and lipogenesis. Nat Chem Biol. 2016;12(1):15–21.
130. Verkerke ARP, et al. BCAA-nitrogen flux in brown fat controls metabolic health independent of thermogenesis. Cell. 2024;187(10):2359–2374.e18.
131. Alderton WK, Cooper CE, Knowles RG. Nitric oxide synthases: structure, function and inhibition. Biochem J. 2001;357(3):593–615.
132. Elizalde M, et al. Expression of nitric oxide synthases in subcutaneous adipose tissue of nonobese and obese humans. J Lipid Res. 2000;41(8):1244–51.
133. Ribiere C, et al. White adipose tissue nitric oxide synthase: a potential source for NO production. Biochem Biophys Res Commun. 1996;222(3):706–12.
134. Bahadoran Z, Mirmiran P, Ghasemi A. Adipose organ dysfunction and type 2 diabetes: role of nitric oxide. Biochem Pharmacol. 2024;221:116043.
135. Dai Z, et al. Nitric oxide and energy metabolism in mammals. Biofactors. 2013;39(4):383–91.
136. Jacobsson A, et al. Mitochondrial uncoupling protein from mouse brown fat. Molecular cloning, genetic mapping, and mRNA expression. J Biol Chem. 1985;260(30):16250–4.
137. Chouchani ET, Kazak L, Spiegelman BM. New advances in adaptive thermogenesis: UCP1 and beyond. Cell Metab. 2019;29(1):27–37.
138. Basse AL, et al. Regulation of glycolysis in brown adipocytes by HIF-1α. Sci Rep. 2017;7(1):4052.
139. de Jong JMA, et al. The β(3)-adrenergic receptor is dispensable for browning of adipose tissues. Am J Physiol Endocrinol Metab. 2017;312(6):E508–e518.

140. Blondin DP, et al. Human Brown adipocyte thermogenesis is driven by β2-AR stimulation. Cell Metab. 2020;32(2):287–300.e7.
141. Warner A, et al. Activation of β3-adrenoceptors increases in vivo free fatty acid uptake and utilization in brown but not white fat depots in high-fat-fed rats. Am J Physiol Endocrinol Metab. 2016;311(6):E901–e910.
142. Mottillo EP, et al. Coupling of lipolysis and de novo lipogenesis in brown, beige, and white adipose tissues during chronic β3-adrenergic receptor activation. J Lipid Res. 2014;55(11):2276–86.
143. Ribeiro MO, et al. Thyroid hormone--sympathetic interaction and adaptive thermogenesis are thyroid hormone receptor isoform—specific. J Clin Invest. 2001;108(1):97–105.
144. Van Thi-Tuong N, Van Vu V, Van Pham P. Brown adipocyte and browning thermogenesis: metabolic crosstalk beyond mitochondrial limits and physiological impacts. Adipocytes. 2023;12(1):2237164.
145. Kleinert M, et al. Glucagon regulation of energy expenditure. Int J Mol Sci. 2019;20(21):5407.
146. Belfiore A, et al. Insulin receptor isoforms in physiology and disease: an updated view. Endocr Rev. 2017;38(5):379–431.
147. Robinson DS, Speake BK. Role of insulin and other hormones in the control of lipoprotein lipase activity. Biochem Soc Trans. 1989;17(1):40–2.
148. Dimitriadis G, et al. Insulin effects in muscle and adipose tissue. Diabetes Res Clin Pract. 2011;93:S52–9.
149. Baggio LL, Drucker DJ. Biology of incretins: GLP-1 and GIP. Gastroenterology. 2007;132(6):2131–57.
150. Mayo KE, et al. International Union of Pharmacology. XXXV. The glucagon receptor family. Pharmacol Rev. 2003;55(1):167–94.
151. Watanabe J, Kanai K, Kanamura S. Glucagon receptors in endothelial and Kupffer cells of mouse liver. J Histochem Cytochem. 1988;36(9):1081–9.
152. Jelinek LJ, et al. Expression cloning and signaling properties of the rat glucagon receptor. Science. 1993;259(5101):1614–6.
153. Richter WO, Robl H, Schwandt P. Human glucagon and vasoactive intestinal polypeptide (VIP) stimulate free fatty acid release from human adipose tissue in vitro. Peptides. 1989;10(2):333–5.
154. Lefebvre P, Luyckx A, Bacq ZM. Effects of denervation on the metabolism and the response to glucagon of white adipose tissue of rats. Horm Metab Res. 1973;5(4):245–50.
155. Perea A, et al. Physiological effect of glucagon in human isolated adipocytes. Horm Metab Res. 1995;27(8):372–5.
156. Joel CD. Stimulation of metabolism of rat brown adipose tissue by addition of lipolytic hormones in vitro. J Biol Chem. 1966;241(4):814–21.
157. Kuroshima A, Yahata T. Thermogenic responses of brown adipocytes to noradrenaline and glucagon in heat-acclimated and cold-acclimated rats. Jpn J Physiol. 1979;29(6):683–90.
158. Yahata T, Habara Y, Kuroshima A. Effects of glucagon and noradrenaline on the blood flow through brown adipose tissue in temperature-acclimated rats. Jpn J Physiol. 1983;33(3):367–76.
159. Billington CJ, et al. Glucagon stimulation of brown adipose tissue growth and thermogenesis. Am J Physiol. 1987;252(1 Pt 2):R160–5.
160. Bertoncini-Silva C, et al. Bioactive dietary components-anti-obesity effects related to energy metabolism and inflammation. Biofactors. 2023;49(2):297–321.
161. Lu M, et al. Molecular mechanisms of the anti-obesity effect of bioactive ingredients in common spices: a review. Food Funct. 2018;9(9):4569–81.
162. Maoka T. Carotenoids as natural functional pigments. J Nat Med. 2020;74(1):1–16.
163. Carazo A, et al. Vitamin a update: forms, sources, kinetics, detection, function, deficiency, therapeutic use and toxicity. Nutrients. 2021;13(5):1703.
164. Bonet ML, Ribot J, Palou A. Lipid metabolism in mammalian tissues and its control by retinoic acid. Biochim Biophys Acta. 2012;1821(1):177–89.

165. Alvarez R, et al. A novel regulatory pathway of brown fat thermogenesis. Retinoic acid is a transcriptional activator of the mitochondrial uncoupling protein gene. J Biol Chem. 1995;270(10):5666–73.
166. Murholm M, et al. Retinoic acid has different effects on UCP1 expression in mouse and human adipocytes. BMC Cell Biol. 2013;14:41.
167. Abbas MA. Physiological functions of vitamin D in adipose tissue. J Steroid Biochem Mol Biol. 2017;165(Pt B):369–81.
168. Chang E, Kim Y. Vitamin D insufficiency exacerbates adipose tissue macrophage infiltration and decreases AMPK/SIRT1 activity in obese rats. Nutrients. 2017;9(4):338.
169. Shen XH, et al. Vitamin E regulates adipocytokine expression in a rat model of dietary-induced obesity. Exp Biol Med (Maywood). 2010;235(1):47–51.
170. Zingg JM, Meydani M, Azzi A. alpha-Tocopheryl phosphate—an active lipid mediator? Mol Nutr Food Res. 2010;54(5):679–92.
171. Lirangi M, et al. α-Tocopheryl-phosphate regulation of gene expression in preadipocytes and adipocytes. Biofactors. 2012;38(6):450–7.

# Chapter 4
# Endocrine Functions of the Adipose Tissue

**Mihail I. Mitov**

## Abbreviations

| | |
|---|---|
| ACRP30 | Adipocyte complement-related protein (s. adiponectin) |
| AdipoR | Adiponectin receptor |
| AgRP | Agouti-related protein |
| AHN | Arcuate hypothalamus nucleus |
| AMPK | Adenosine monophosphate-activated protein kinase |
| ANGPTL2 | Angiopoietin-like protein 2 |
| BAT | Brown adipose tissue |
| BMI | Body mass index |
| CAP1 | Adenylyl cyclase-associated protein 1 |
| CCN4 (*s.* WISP1) | Cellular communication network factor 4 |
| CSF | Colony-stimulating factor |
| CTRP | C1q/TNF-related protein |
| CXCL10 | C-X-X motif chemokine ligand 10 |
| DCN | Decorin |
| DHEA | Dehydroepiandrosterone |
| DHEA-S | Dehydroepiandrosterone sulfate |
| EGF | Epidermal growth factor |
| EMT | Epithelial-to-mesenchymal transition |
| EV | Extracellular vehicle |
| FSTL1 | Follistatin-like 1 |
| GABA | Gamma-aminobutyric acid |
| G-CSF | Granulocyte CSF |

M. I. Mitov (✉)
Idaho College of Osteopathic Medicine, Meridian, ID, USA

Sam Houston State University, College of Osteopathic Medicine, Conroe, TX, USA
e-mail: mmitov@icom.edu


D. Avtanski (ed.), *Adipose Tissue*, Contemporary Endocrinology,
https://doi.org/10.1007/978-3-031-72570-8_4

| | |
|---|---|
| GLP-1 | Glucagon-like peptide-1 |
| GLP-1RA | Glucagon-like peptide-1 receptor agoist |
| IL | Interleukin |
| LCN2 | Lipocalin-2 |
| LEP-R | Leptin receptor |
| LH | Lateral hypothalamic (area) |
| LPS | Lipopolysaccharide |
| MC4R | Melanocortin-4 receptor |
| MCP-1 | Monocyte chemoattractant protein-1 |
| M-CSF | Macrophage colony-stimulating factor |
| Metrnl | Meteorin like |
| miRNA | MicroRNA |
| NPY | Neuropeptide Y |
| NRG4 | Neuregulin-4 |
| PAI-1 | Plasminogen activator inhibitor-1 |
| POMC | Proopiomelanocortin |
| PPARα | Peroxisome proliferator-activated receptor-alpha |
| PVN | Paraventricular nucleus |
| RBP4 | Retinol binding protein 4 |
| RELM | Resistin-like molecules |
| ROR1 | Receptor tyrosine kinase-like orphan receptor 1 |
| SFRP5 | Secreted frizzled-related protein 5 |
| STAT3 | Signal transducer and activator of transcription 3 |
| T2DM | Type 2 diabetes mellitus |
| TBI | Traumatic brain injury |
| TLR4 | Toll-like receptor 4 |
| TNBC | Triple-negative breast cancer |
| TNF | Tumor necrosis factor |
| VH | Ventromedial hypothalamic (nucleus) |
| VMH | Ventromedial hypothalamic (nucleus) |
| WAT | White adipose tissue |
| WISP1 (*s.* CCN4) | Wingless-type inducible signaling pathway protein 1 |
| 11β-HSD1 | 11beta-hydroxysteroid dehydrogenase type 1 |
| 17β-HSD | 17beta-hydroxysteroid dehydrogenase |
| α-MSH | α-melanocyte stimulating hormone |

## Introduction

This chapter focuses on how the adipose tissue transmits endocrine signals to other cells, tissues, and organ systems. By definition, "endocrine" is a type of signaling between cells where signaling molecules (hormones) are secreted by specialized endocrine cells and carried through the circulatory system to act on target cells, tissues, and organs at a distance. The term hormone was first used by the British physiologist Ernest Starling about 120 years ago to describe the biochemical messengers that travel from endocrine cells to target cells along the circulatory system and coordinate physiological activities and the growth and development of various parts of the organismal bodies [1, 2].

Historically, the main physiological functions of white adipose tissue (WAT) have predominantly focused on long-term energy storage, thermal insulation, and mechanical protection [3]. In the 1960s and the 1970s, brown adipose tissue (BAT) was experimentally proven to be the site of non-shivering thermogenesis in rodents [4]. The concept that adipose tissues can also have endocrine functions was confirmed after the discovery of the first two most characteristic adipocyte secretory products, leptin [5] and adiponectin [6], in 1994–1995.

## Overview of Adipokines

The modern view and understanding of cell signaling functions of adipose tissue is that the cross-talk between fat cells and their environment is mediated in three main ways: **(1)** nutritional mechanisms (cell signaling pathways that are affected by the availability of nutrients), **(2)** neural pathways (information pathways connecting neurons to other neurons), and **(3)** molecular signal messengers (through autocrine, paracrine, and endocrine routes) collectively termed adipokines or adipocytokines (from Greek *adipo-*, fat; *cytos-*, cell; and *-kinos*, movement) [7]. Currently, adipose tissue is considered to be a dynamic secretory tissue/organ that generates a myriad of adipokine molecules, noncoding RNAs, extracellular vesicles (EVs), transferring signals to multiple other cells, tissues, and organs of the body [8].

The first adipokine (adipsin) was discovered in 1987 when adipose tissue was first reported to be the main site for the metabolism of sex steroids [9], and the production of adipsin as a new endocrine factor was found to be significantly downregulated in rodent obesity [10]. As scientific discoveries related to the endocrine function of adipose tissue progressed in the next couple of decades, the identification and characterization of leptin in 1994 further validated the concept that adipose tissue has an important function as an endocrine organ. In 2012, it was reported that more than 600 adipose tissue secretory proteins had already been identified [11], potentially providing a broad databank of adipokines to identify novel biomarkers or signaling factors associated with metabolic diseases. In the current decade, the number of confirmed adipokines has increased significantly (Table 4.1) [11].

Some cytokines, such as tumor necrosis factor (TNF), were first discovered a decade before the concept of adipokines was developed in the late 1980s when the first adipokines were identified. Since the discovery of the first adipose tissue-derived adipokine (adipsin in 1986 [24]), followed by the discovery of leptin, resistin, and others, the interest in the developmental, functional, and pathophysiological aspects of adipose tissue has grown significantly [8]. This increased research focus and interest in adipose tissue has coincided with the exponential increase in global rates of obesity- (and type 2 diabetes mellitus (T2DM)-) related morbidity and

**Table 4.1** List of some of the more notable currently known adipokines

| Adipokine | Year of discovery | Structure | Function |
|---|---|---|---|
| Leptin | 1994 [5] | 167 AA polypeptide | Influence appetite/satiety |
| Adiponectin | 1995 [6] | 244 AA polypeptide | Modulates glucose regulation and fatty acid oxidation |
| Apelin | 1998 [12] | 77 AA pre-proprotein | G-protein-coupled APJ receptor |
| Chemerin | 1997 [13] | 14 kDa protein | Adipocyte differentiation and stimulation of lipolysis |
| Interleukin-6 (IL-6) | 1986 [14] | 26 kDa glycoprotein | Mediator for obesity-derived chronic inflammation in insulin resistance |
| Monocyte chemoattractant protein-1 (MCP-1) | 1989 [15] | 8 kDa protein glycoprotein | Inhibits insulin-stimulated glucose uptake and alters adipocyte gene expression |
| Visfatin | 2004 [16] | 53 kDa protein | Insulin-mimetic, promotes vascular inflammation and atherosclerosis |
| Retinol binding protein 4 (RBP4) | 1968 [17] | 21 kDa protein | Potentially linked to insulin resistance and metabolic diseases |
| Intelectin-1 (omentin) | 2001 [18] | 34 kDa protein | Regulation of energy metabolism, insulin sensitivity, and reproduction |
| Plasminogen activator inhibitor-1 (PAI-1) | 1983 [19] | 46 kDa glycoprotein | Elevated in insulin resistance, obesity, and metabolic syndrome |
| Serpin A12 (Vaspin) | 2005 [20] | 47 kDa glycoprotein | Insulin sensitizing effect and inhibition on food intake |
| Progranulin | 2003 [21] | 68 kDa glycoprotein | Growth-promoting, neuroprotective and anti-inflammatory |
| C-X-C motif chemokine ligand 10 (CXCL10) | 1985 [22] | 9 kDa protein | Pro-inflammatory, correlates positively with obesity and type 2 diabetes |
| Tumor necrosis factor (TNF) | 1975 [23] | 52 kDa homotrimer protein | Classical, pleiotropic pro-inflammatory cytokine/adipokine |

mortality in recent decades [25, 26]. Adipose tissue-produced signaling molecules are among the major tissue-organs mechanisms for controlling food intake and involve interplay between the gut, brain, muscle, and other organs [27]. Current efforts are devoted to developing novel therapeutic approaches for regulating adipokine and neuropeptide functions to combat morbidity and mortality related to the increased incidence of obesity and T2DM.

## *Leptin*

In 1994, leptin was discovered in the Laboratory of Dr. Jeffery Friedman as a new hormone product of the obese (*ob*) gene after his team worked on characterizing genetically obese (*ob/ob*) mice [5]. This groundbreaking discovery confirmed the earlier hypothesis of Dr. Douglas Coleman, who first proposed decades ago that an undiscovered hormonal system regulates body weight and appetite. For these contributions to science, Dr. Coleman and Friedman received the 2010 Albert Lasker Basic Medical Research Award for discovering the hormone leptin, which regulates appetite and body weight, proving the strong connection between obesity and genetics.

In the last century (circa 1940–1990), pioneering research focused on various regions of the brain through experimental lesions (and/or stimulations) led to the discovery of a "satiety center" in the ventromedial hypothalamic (VMH) nucleus. During the same period, a "hunger center" was identified in the lateral hypothalamic (LH) area of animal brains. These earlier reports showed that VMH and LH nuclei lesions can result in abnormal feeding behavior in animals and humans [28, 29]. The current view on this topic is that the ventromedial hypothalamic (VH) nucleus represents the brain's main satiety center, and when stimulated, the VMH nucleus causes a sensation of fullness (and reduces feeding behavior). Destruction of the VMH nucleus leads to overeating and weight gain. In 1995, it was further discovered that leptin regulates metabolism and appetite by suppressing the synthesis or release of neuropeptide Y (NPY) in the arcuate hypothalamus nucleus (AHN) [30].

The molecular mechanism by which leptin communicates with the brain to signal satiety involves leptin transmembrane receptors (LEP-R), which have molecular structure similar to the class I family of cytokine receptors (including receptors for interleukins (ILs), growth hormone, prolactin, etc.) [31, 32]. Specifically in the context of brain mechanisms to signaling satiety, the leptin receptors isoform b (LEP-Rb) has been most important and found in the ventromedial, arcuate, lateral, and dorsomedial hypothalamic nuclei, which are crucial in regulating energy balance and body mass [32, 33]. Further studies have shown that leptin can inhibit the neural pathways activated by appetite stimulants, also known as orexigenic pathways, resulting in reduced food and energy intake. Leptin can also affect anorexigenic pathways to suppress appetite [34].

The mechanisms of action of leptin on the main neuronal centers that regulate/cause hyperphagia or hypophagia constitute a relatively simple algorithm of interactions in which leptin interacts via leptin receptors in the melanocortin system. In this system, NPY and agouti-related protein (AgRP) are well-known peptides with orexigenic effects in the brain's hypothalamus region, affecting two specific neuronal circuits [35, 36]. The two main neuronal populations in the melanocortin system are orexigenic appetite-increasing NPY/AgRP neurons and anorexigenic appetite-suppressing proopiomelanocortin (POMC) neurons, both located in the AHN. Once stimulated by various circulating hormones such as leptin, insulin, or ghrelin, these neurons can significantly change the appetite, feeding, and metabolic rates. The regulation of these neuronal circuits in the AHN brain area is critical for maintaining the body's energy balance, and lesions or dysfunction in these circuits could produce significant changes in food intake (hyperphagia/obesity or hypophagia/starvation) [37]. In addition, in AHN, the prohormone POMC is post-translationally cleaved into several smaller peptides, one of which results in the formation of α-melanocyte stimulating hormone (α-MSH).

During a "fed state," leptin binds to LEP-Rb receptors expressed at the surface of POMC neurons to signal for the initiation of the POMC cleavage in order to produce the active form of the hormone α-MSH. Next, the anorexigenic effects of the α-MSH neuropeptide are signaled to a population of MC4R second-order neurons into the paraventricular nucleus (PVN) by a mechanism in which α-MSH binds to the melanocortin-4 receptors (MC4Rs) [38, 39]. Leptin could upregulate the frequency of action potentials in anorexigenic POMC neurons by depolarizing through a nonspecific cation channel and suppressing inhibition by local orexigenic NPY/gamma-aminobutyric acid (GABA) neurons [40].

During a "starvation/hunger state," there is an increase in the activity of AgRP- and NPY-expressing neurons. Hunger signals stimulate the release of NPY in AHN, which binds to its G protein-coupled receptors to increase feeding behavior and storage of calories. Increasing levels of NPY and the overall stimulation of this pathway promote feeding behavior. There are five known mammalian NPY receptors, Y1R-Y5R (in primates, including humans, Y3R is non-functional), which have been grouped into the same family because they all bind NPY regardless of their overall low sequence similarity [41]. It would be conflicting for both pathways to operate simultaneously; therefore, the NPY system also releases AgRP, an antagonist of MC4R [40, 42, 43].

An increased level of leptin in the blood could be used as a marker of leptin resistance, which is known to correlate with other inflammatory markers and to predict cardiovascular risk in obese individuals [44]. Leptin resistance could be defined by reduced sensitivity (or failure in response of the brain) to leptin, marked by a decrease in the potency of leptin to suppress appetite (and affect the basal metabolic rates). Leptin resistance results in increased food intake, leading to an increase in body mass index (BMI), causing overweight and obesity, which increases morbidity and mortality from cardiovascular diseases and other metabolic disorders [45].

In summary, leptin regulates food intake, metabolic rate, and body mass through the above-mentioned mechanisms. Furthermore, adipose tissue (through leptin production) also affects reproductive function and system [46] and plays a vital role in fetal growth [47, 48], proinflammatory immune responses [49, 50], cardiovascular system angiogenesis [51, 52], and lipolysis [53–56]. Leptin resistance is a significant challenge in the treatment of obesity. However, recent discoveries in the functional relationships between leptin, glucocorticoids, and glucagon-like peptide 1 (GLP-1) receptor agonists (GLP-1RA) have increased the hope for effective future therapies [57, 58].

## *Adiponectin*

In 1995, the polypeptide adiponectin was discovered to be produced exclusively in adipocytes in the laboratory of Dr. Philipp Scherer [6], who first described it as a new hormone and adipokine, which was later discovered to be also biosynthesized in the placenta [59], muscles [60, 61], and even in the brain [62]. Adiponectin is a 244-amino-acid-long polypeptide that is abundant in plasma at higher levels compared to most other hormones [63]. In the mid-1990s, circulating adiponectin levels were found to be inversely correlated with overall adiposity and fat mass, making it unique than most other adipokines, including leptin [64].

Adiponectin (also known as Acrp30, GBP-28, apM1, and AdipoQ) is traditionally considered a "guardian angel adipocytokine" because it has multiple positive systemic (pleiotropic) effects [65]. Unlike other hormones, this adipokine has been reported to significantly improve metabolic health and homeostasis when secreted by adipose tissue at higher levels. In such situations (abundant adiponectin in circulation), it has been proven to increase insulin sensitivity, improve lipid clearance, and reduce systemic inflammation [66]. Numerous studies have shown that increasing adiponectin levels improves metabolic health by enhancing glucose tolerance [67], decreasing liver steatosis and fibrosis [68], and producing a more metabolically healthy obesity phenotype of obesity in diabetic swine [69] and adiponectin transgenic mouse models [70, 71]. On the opposing end of the spectrum for the adiponectin physiological effects, several studies have shown that decreased adiponectin levels can lead to glucose intolerance and hyperlipidemia [72–74]. Furthermore, previous studies using both type 1 and type 2 diabetic mice have shown that a single injection of purified recombinant 30 kDa (Acrp30) adiponectin results in a 2–3-fold elevation in circulating Acrp30 levels, producing a transient decrease in basal glucose levels and a reduction in hyperglycemia (without increasing blood insulin levels) [75].

It is important to note that there is no consensus in the scientific literature or community regarding the effects of adiponectin, which has been reported to have both pro-inflammatory and anti-inflammatory effects [76]. Numerous previous studies have reported the potential beneficial effects of increased circulating adiponectin toward favoring insulin-sensitizing, anti-atherogenic, and anti-inflammatory effects. Conversely, other studies in human cohorts (mostly in patients with autoimmune diseases) have shown that adiponectin might have a pro-inflammatory function, where the pathophysiology of the autoimmune disease might correlate with elevated adiponectin levels. For example, it has been previously reported that high serum levels of adiponectin could be positively associated with the severity of inflammation and pathological progression in systemic lupus erythematosus [77, 78], rheumatoid arthritis [79], chronic kidney disease [80], inflammatory bowel disease [81], and cystic fibrosis [82].

The most likely reason (and mechanism) for the existence of such conflicting dualism of the pro-inflammatory and anti-inflammatory effects of adiponectin's biological functions is that it circulates in the blood in multiple isoforms. Multiple studies have confirmed that adiponectin in the blood exists in at least three different molecular weight forms: (1) low-molecular-weight trimers, (2) middle-molecular-weight hexamers, and (3) high-molecular-weight multimers [83]. Additionally, after its synthesis, adiponectin undergoes a series of post-translational modifications via glycosylation and hydroxylation [84]. During this process, three major oligomeric isoforms of adiponectin are created: a 60 kDa low-molecular-weight trimer, a 150 kDa middle-molecular weight hexamer, and a high-molecular-weight multimer (420 kDa). In these isoforms, the structural differences are in the multiple lysine residues of adiponectin, where the isoforms differ in the extent of hydroxylation and subsequent glycosylation. Furthermore, in human adiponectin isoforms, the post-translational modification of isoform variants includes several hydroxylated proline residues and an intermolecular disulfide bond formed at the cysteine residue at the N-terminus. With these modifications, adiponectin's low molecular isoform in its active form is a 60 kDa glycoprotein, which is a trimer of adiponectin 28 kDa monomers joined by their C-terminal globular and collagen-like domains. If trimers multimerize further, this leads to the production of 150 kDa hexamers and 420 kDa multimers, each comprising 12–32 monomers [76].

Another potentially important mechanism explaining the dual pro-inflammatory and anti-inflammatory effects of adiponectin is the natural variety of adiponectin receptors in various tissues and organs. Both adiponectin receptors (AdipoR1 and AdipoR2) exhibit structural similarity, with 67% amino acid overlap and similar organization of their intracellular $NH_2$-terminal and extracellular COOH-terminal domains [85]. In addition, both receptors are differentially expressed in tissues and organs, with AdipoR1 being broadly expressed and present in skeletal muscle and the liver, whereas AdipoR2 is predominantly expressed in the liver [86]. Experimental genetic studies have shown that the removal or disruption of AdipoR1 receptors decreases adiponectin-induced 5′ adenosine monophosphate-activated protein kinase (AMPK) activation at the cellular level, and disturbances of AdipoR2 result in decreased peroxisome proliferator-activated receptor-alpha (PPARα) activation [87]. Additionally, T-cadherin (a cell surface protein lacking transmembrane and cytoplasmic domains common to all other cadherins) has also been reported as a receptor for adiponectin. T-cadherin binds and participates in adiponectin signaling of both 150 kDa hexamers and 420 kDa multimers but not for globular or 60 kDa low-molecular isoforms in various muscle tissues [88].

Adiponectin has been extensively studied for its potential as a therapeutic target for drug development, aiming to reduce the morbidity and mortality of T2DM and metabolic syndrome [89]. The lack of sufficient knowledge or understanding of the mechanisms determining the specific pro- or anti-inflammatory properties of adiponectin isoforms, receptors, and post-translational modifications has slowed down the development of potential anti-inflammatory therapeutics against metabolic disorders and autoimmune diseases. More research is needed to decipher the underlying mechanisms controlling adiponectin isoform and receptor levels, and the body's response to their interactions under different metabolic conditions. Another challenge in drug development targeting adiponectin is the technical problems in accurately measuring adiponectin isoforms in biological samples and the absence of universal standards in research laboratories. Finally, dietary interventions can be used as pivotal modulators of adiponectin levels, creating non-pharmacological and non-invasive opportunities for intervention in prediabetes and metabolic syndrome management [90].

## *Resistin*

In the early 2000s, resistin was discovered in the laboratory of Dr. Mitchell A. Lazar from the University of Pennsylvania [91]. The name of this adipokine originates from its ability to interfere with the physiological effects of insulin. Resistin is a 12.5 kDa cysteine-rich polypeptide secreted from white adipocytes [92], and when it was discovered in mice, it was hypothesized to be a major molecular factor linking obesity and diabetes [91]. In obese rodents, plasma levels of adipocyte-produced resistin are significantly increased and are strongly related to insulin resistance [93]. However, in contrast to rodents, human resistin is predominantly expressed and secreted by macrophages, and its levels increase under inflammatory conditions [94]. Studies have demonstrated a correlation between increased resistin levels and insulin resistance [95], microvascular complications in patients with T2DM [95], and cardiovascular disease events in older adults [96].

Resistin induces inflammation in many chronic inflammatory diseases, infectious diseases, and inflammation-associated cancers [97]. More recently, multiple reports have suggested that resistin is part of the host defense system as an innate immunity peptide that exhibits broad-spectrum antimicrobial effects by modulating immunity [98].

In recent years, it has become clear that the physiological role of resistin is more complex than that originally described or anticipated. Resistin in humans and rodents has variations in its molecular structure, and three isoforms (resistin-like molecules (RELM)-α, β, and γ) have been identified and described in the literature [99]. The receptor downstream signaling of resistin remains unknown, and therefore, the molecular mechanisms of resistin action are poorly understood [100]. However, multiple studies have suggested potential receptors for resistin isoforms, such as decorin (DCN) [101], mouse receptor tyrosine kinase-like orphan receptor 1 (ROR1) [102], toll-like receptor 4 (TLR4) [103], and adenylyl cyclase-associated protein 1 (CAP1) [104]. The signaling between RELM-α, −β, and -γ (resistin isoforms) and various possible receptors is complex and adds to the pleiotropic nature of the physiological action of resistin in health and disease.

Furthermore, multiple studies have indicated that resistin expression is not limited only to adipocytes and macrophages but is also present and expressed by various cells from other organs and systems, including the skeletal muscle, stomach, small and large intestine, adrenal gland, brain, and pituitary gland [105, 106]. Resistin can be overexpressed in the brain after traumatic brain injury (TBI), and this increase could lead to the systemic elevation of plasma resistin levels. After acute TBI (<6 h), elevated resistin levels have shown a strong correlation with a worse clinical prognosis for patients [107]. However, under normal healthy conditions, the plasma range of resistin in humans has been identified to be between 7 and 22 ng/ml [108, 109], and some proinflammatory cytokines, such as TNF [110], IL-6 [111], and lipopolysaccharide (LPS) [112], can upregulate resistin gene expression and increase the serum levels of resistin [113–115].

Owing to the inherent complexity of resistin, the research focus in this field has shifted from viewing it as a molecular link between obesity and insulin resistance to its newly emerging role as a biomarker of cancer promotion and progression in obese subjects [116–118]. In addition, resistin has been proposed as an inflammatory marker in human atherosclerosis [119] and rheumatoid arthritis [120] because its pro-inflammatory properties are superior to those of its insulin resistance-inducing effects.

## Cytokines and Steroid Hormones

As previously described in this chapter, many adipokines have pro- or anti-inflammatory properties. Some initially discovered cytokines molecules, such as TNF, have been later proven to be adipokine (molecules that are made by fat cells in the body).

In 1975, Dr. Lloyd J. Old from the Memorial Sloan Kettering Cancer Center, New York, discovered TNF as a new cytotoxic factor produced by macrophages [23]. More than two decades later, in 1997, it was discovered that adipocytes under obese conditions also produce TNF mRNA and elevated TNF protein levels [121].

Currently, TNF is a well-studied cytokine with a pleiotropic function that can regulate immune function, cell differentiation, proliferation, apoptosis, and energy metabolism [122]. Generally, cytokines within adipose tissue can be produced by any of the main adipose tissue cell types (adipocytes, preadipocytes, and other cell types). Previous studies utilizing mRNA expression approaches to study TNF have shown that adipocytes can synthesize TNF along several interleukins (more specifically, IL-1β [123] and IL-6 [124].

Other adipocyte-produced signaling molecules with immunological actions include two categories of factors related to families of complement system products and macrophage colony-stimulating factors (CSFs) (M-CSFs).

As an example of a complement system product, a 30 kDa adipocyte complement-related protein (ACRP30) was reported 25 years ago as a secreted protein with an unknown function at the time, which was exclusively expressed in differentiated adipocytes [125]. After numerous studies have been completed in recent years, the protein is now known as adiponectin, which is one of the best-known adipocytokines with well-documented pro- and anti-inflammatory functions [126].

An example of adipocyte-produced CSFs is granulocyte CSF (G-CSF). G-CSF is a 174-amino acid polypeptide with a molecular weight of 45–50 kDa and is highly expressed primary preadipocytes isolated from mammary fat pads of human breast cancer patients [127]. Treatment approaches using recombinant human G-CSF protein in cell lines of triple-negative breast cancer (TNBC) can enhance the epithelial-to-mesenchymal transition (EMT), migration, and invasion of cancer cells by activating the signal transducer and activator of transcription 3 (STAT3).

Some additional, relatively less studied adipokines associated with the pathogenesis of obesity and T2DM include the following pro- and anti-inflammatory factors [128]: follistatin-like 1 (FSTL1) (a glycoprotein with potential proinflammatory functions that contribute to angiogenesis, immune response regulation, and cell proliferation [129]), wingless-type inducible signaling pathway protein 1 (WISP1, or CCN4) (known to be upregulated in obesity and induce a pro-inflammatory response in macrophages *in vitro* [130]), asprosin (found in 2016 as a novel adipokine cleavage product of profibrillin, utilizing a cAMP second messenger system to modulate hepatic glucose production and appetite stimulation [131]), secreted frizzled-related protein 5 (SFRP5) (discovered in 2010 to have inhibitory effects on the Wnt signaling pathway; known effects include anti-inflammatory and insulin-sensitizing action and positively associated with insulin sensitivity [132, 133].), meteorin-like (metrnl, or subfatin) (identified as a novel adipokine in 2014; has some positive metabolic effects in obesity, T2DM, and dyslipidemia), neuregulin-4 (NRG4) (an epidermal growth factor (EGF)-like signaling molecule; reduced levels may be causally linked to obesity-related impaired glucose metabolism [134]), TAFA5 (also known as FAM19A5) is a 14 kDa type I transmembrane protein with known functions linked to inflammation, differentiation, and cognition [135]; its silencing in adipocytes significantly promotes vascular smooth muscle cell activation; downregulation of TAFA5 during obesity may trigger cardiometabolic diseases [136]).

Adipose tissue functions as an endocrine organ that secretes numerous peptide hormones and other bioactive molecules (such as steroids, which can be biosynthesized de novo from cholesterol in WAT [137]) that can act in auto-, para-, and endocrine ways to regulate systemic metabolism. The higher mass of dysfunctional adipose tissue in obesity may cause or exacerbate metabolic abnormalities that lead to T2DM [138], cardiovascular diseases [139], and multiple types of cancer [140].

Although the adrenal gland is the primary source of circulating corticosteroids, and gonads are the primary source of sex hormones, adipose tissue expresses an almost full range of enzymes needed to produce, convert, and activate steroid hormones [141]. In postmenopausal individuals, estrogen production, is shifted to other body sites, including adipose tissue [142]. After menopause, adipose tissue plays a leading role in estrogen production, and circulating C19 steroid precursors androstenedione, dehydroepiandrosterone (DHEA), and DHEA sulfate (DHEA-S) serve as reservoirs for extragonadal estrogen synthesis [143].

Adipose tissue can also actively synthesize androgens with the help of the 17beta-hydroxysteroid dehydrogenase (17β-HSD), an essential enzyme in sex steroids conversion. A literature review can bring multiple studies that have reported that 17β-HSD expression in obesity is systemically increased, which suggests that it can be an important factor contributing to circulating androgen excess in obesity [144]. Animal studies have reported that the whole steroidogenic enzymatic set and testosterone biosynthesis capacity were found in the fat depots of female rats [145].

These findings support the hypothesis that adipose tissue may contribute substantially to hyperandrogenism in obese women.

Furthermore, adipocytes can process and activate existing circulating androgens, and, by converting less potent androgens into more potent hormones, can further modulate the production of testosterone and androstenedione through the actions of various adipocytokines [146]. Androgens have been reported in the literature to contribute to adipocyte proliferation and differentiation. They stimulate adipocytokine production and promote visceral adipose tissue accumulation, adipogenesis, lipid storage, and lipolysis. Androgens can also modulate body fat distribution patterns in humans through spot and depot-specific effects on the cells of each fat compartment [147].

Cortisol can also be produced and secreted locally by adipose tissue to some extent. Obesity, along with chronic exposure to stress and inadequate sleep, has been reported to negatively affect the hypothalamic-pituitary-adrenal axis and elevate systemic cortisol levels. This results in upregulated activity of the 11β-HSD1 enzyme, which locally increases cortisol synthesis in adipose tissue under obesity and stress conditions [148].

Adipose tissue produces cortisol by the activity of the 11β-HSD1 enzyme in adipocytes [149], although 11β-HSD1 mostly increases the local glucocorticoid concentrations within adipose tissue and does not seem to contribute significantly to systemic cortisol concentrations [150]. Increased long-term exposure to elevated cortisol levels is known to be strongly related to abdominal obesity [151] and increased blood cholesterol, triglycerides, blood glucose, blood pressure, muscle weakness, and mass [152–154]. Cortisol has been validated in many studies as a potent steroid hormone that affects adipose tissue in many direct and indirect ways, influencing insulin sensitivity, fatty acid metabolism, adipocyte differentiation, adipokine expression, and body fat distribution [149, 155].

## Regulation of Adipokine Secretion

Regulation of adipokine secretion is an important topic because mechanisms and factors regulating adipokines levels in circulation could be used to develop new pharmacological and lifestyle treatments for metabolic syndrome, type 1 and 2 diabetes, cancer, autoimmune conditions, and other more or less common diseases [156–158]. Unfortunately, due to adipokines signaling complexity and their pleiotropic systemic effects, only a few clinical trials have shown limited success for adipokine-based treatments in improving cardiometabolic health [158].

In the previous sections of this chapter, detailed information was provided on individual adipokine secretion and how it might be influenced by various pathological conditions, such as obesity, inflammation, and diabetes. For example, the inflammatory regulator IL-6 has emerged in recent decades as one of the potential mediators of obesity-derived chronic inflammation with insulin resistance [159]. In recent years, more details emerged regarding the molecular mechanisms regulating the expression and circulating levels of IL-6. For example, it has been experimentally demonstrated that IL-1β and TNF exhibit cooperativity in regulating IL-6 expression in adipocytes depending on CREB binding and H3K14 acetylation [123].

Additionally, microRNAs (miRNAs) have been discovered as regulators controlling diverse biological processes (including adipokine secretion) at the level of post-transcriptional repression. For example, miR-146b is known as a factor that can downregulate cytokine signaling (suppress the expression of inflammatory genes) in the inflammatory process, and it has been shown in mature adipocytes that miR-146b can be highly expressed in response to an increase in levels of TNF and IL-6 [160]. This proves miRNAs can be important regulators of adipokine production in obesity complications.

The mechanism of IL-6 signaling starts with binding to its cellular receptor (IL-6Rα) as a dimeric complex to form a trimeric receptor complex, further activating the intracellular JAK/STAT3 pathway. Recruiting JAKs to the membrane leads to their activation and triggers the phosphorylation of tyrosine on a cytoplasmic tail subunit of the receptor at docking sites for STAT proteins. As a final step, the phosphorylated STAT3 translocates into the nucleus to activate target transcription [161].

Other categories of specific molecular factors that have been identified to regulate adipokine secretion include sphingolipids, namely ceramides, which accumulate at high concentrations in the liver and muscle of obese and diabetic rats [162]. Ceramides are part of the sphingolipids which are containing the organic aliphatic amino alcohol sphingosine. Their general molecular structure consists of a long-chain sphingoid base (sphingosine, sphinganine, or phytosphingosine) and an additional fatty acid linked to the sphingoid base to form the ceramide backbone [163]. In early studies of obese and T2DM patients, plasma ceramide levels have been found to be strongly correlated with diminished insulin sensitivity [164]. In 2007, endogenously increased ceramide synthesis came into the spotlight as a potentially major molecular factor leading to increased insulin resistance in humans in vivo [165]. Sphingolipid metabolism studies in relation to the regulation of specific adipokines have shown that adiponectin potently stimulates ceramidase activity through its two receptors (AdipoR1 and AdipoR2). As a result of this interaction, ceramidase activity initiates pleiotropic actions of adiponectin [165]. Meanwhile, other molecular regulators that inhibit adipocyte differentiation, such as interferon, retinoic acid, IL-1, and TNF, have also been found to interfere with adiponectin secretion [166].

In recent years, it has been established that multiple circulating factors, including insulin, growth hormone, glucocorticoids, sex steroids, and thyroid hormones, significantly affect adipokine secretion and regulation [167]. In the last decade, a strong association between resistin or adiponectin levels and free thyroxine was discovered in newly diagnosed hypo- and hyperthyroidism patients compared with healthy control populations [168]. This establishes and reaffirms the hypothesis of a complex interplay between the adipose tissue and endocrine system.

## Conclusion

This book chapter provides details from early historical and recent studies illustrating the complexity of adipose tissue as an active endocrine organ with a significant impact on metabolic and hormonal regulation. The bioactive molecular secretions of adipocytes (known as adipokines) can affect other tissues and organs locally and distantly via autocrine, paracrine, and endocrine routes. When these effects impact the immune system and have systemic inflammatory effects, adipokines are referred to as pro-inflammatory adipocytokines to emphasize their significant immunological systemic effects. Adipokines can be classified into three major categories: **(1)** Pro-inflammatory (including TNF, leptin, IL-6, resistin, RBP4, lipocalin-2 (LCN2), IL-18, angiopoietin-like protein 2 (ANGPTL2)), that are positively related to body weight, fat mass, insulin resistance, and general pro-inflammatory properties [169]; **(2)** Anti-inflammatory adipokines, (such as adiponectin, some C1q/TNF-related proteins (CTRPs), progranulin, omentin-1, and other new emerging adipokines) that may protect against the complications of obesity by exerting anti-inflammatory or antioxidant effects [170]); and **(3)** Mixed type or neutral adipokines, which include adipokines that have not been well studied or have shown both pro- and anti-inflammatory effects depending on the context (such as adiponectinin certain autoimmune or genetic conditions, as well as relatively new adipokines such as TAFA5/FAM19A5, NRG4, etc.) whose effects on inflammation and the immune system are not yet fully understood.

The endocrine functions of WAT and BAT play a significant role in adipose tissue dysfunction, with possible clinical implications and opportunities for the development of new biomarkers or targets for the pharmacological treatment of diseases such as metabolic syndrome, type 1 and 2 diabetes, cancer, autoimmune conditions, and other common diseases that disturb the homeostasis of adipose tissue. However, it is important to note that adipokines are not the only factors involved in these conditions, and other factors such as insulin resistance, inflammation, and genetic predisposition also play significant roles.

## References

1. Starling, E.H. The Croonian Lectures on the chemical correlation of the functions of the body: delivered before the Royal College of Physics of London: on June 20th, 22nd, 27th & 29th, 1905. 1905: Women's Printing Society.
2. Tata JR. One hundred years of hormones. EMBO Rep. 2005;6(6):490–6.
3. Saely CH, Geiger K, Drexel H. Brown versus white adipose tissue: a mini-review. Gerontology. 2012;58(1):15–23.
4. Trayhurn P. Origins and early development of the concept that brown adipose tissue thermogenesis is linked to energy balance and obesity. Biochimie. 2017;134:62–70.
5. Zhang Y, et al. Positional cloning of the mouse obese gene and its human homologue. Nature. 1994;372(6505):425–32.

6. Scherer PE, et al. A novel serum protein similar to C1q, produced exclusively in adipocytes. J Biol Chem. 1995;270(45):26746–9.
7. Rosen ED, Spiegelman BM. What we talk about when we talk about fat. Cell. 2014;156(1–2):20–44.
8. Funcke JB, Scherer PE. Beyond adiponectin and leptin: adipose tissue-derived mediators of inter-organ communication. J Lipid Res. 2019;60(10):1648–84.
9. Siiteri PK. Adipose tissue as a source of hormones. Am J Clin Nutr. 1987;45(1 Suppl):277–82.
10. Flier JS, et al. Severely impaired adipsin expression in genetic and acquired obesity. Science. 1987;237(4813):405–8.
11. Lehr S, Hartwig S, Sell H. Adipokines: a treasure trove for the discovery of biomarkers for metabolic disorders. Proteomics Clin Appl. 2012;6(1–2):91–101.
12. Tatemoto K, et al. Isolation and characterization of a novel endogenous peptide ligand for the human APJ receptor. Biochem Biophys Res Commun. 1998;251(2):471–6.
13. Nagpal S, et al. Tazarotene-induced gene 2 (TIG2), a novel retinoid-responsive gene in skin. J Invest Dermatol. 1997;109(1):91–5.
14. Hirano T, et al. Complementary DNA for a novel human interleukin (BSF-2) that induces B lymphocytes to produce immunoglobulin. Nature. 1986;324(6092):73–6.
15. Yoshimura T, et al. Purification and amino acid analysis of two human monocyte chemoattractants produced by phytohemagglutinin-stimulated human blood mononuclear leukocytes. J Immunol. 1989;142(6):1956–62.
16. Fukuhara A, et al. Visfatin: a protein secreted by visceral fat that mimics the effects of insulin. Science. 2005;307(5708):426–30.
17. Kanai M, Raz A, Goodman DS. Retinol-binding protein: the transport protein for vitamin A in human plasma. J Clin Invest. 1968;47(9):2025–44.
18. Lee JK, et al. Human homologs of the Xenopus oocyte cortical granule lectin XL35. Glycobiology. 2001;11(1):65–73.
19. Loskutoff DJ, et al. Detection of an unusually stable fibrinolytic inhibitor produced by bovine endothelial cells. Proc Natl Acad Sci. 1983;80(10):2956–60.
20. Hida K, et al. Visceral adipose tissue-derived serine protease inhibitor: a unique insulin-sensitizing adipocytokine in obesity. Proc Natl Acad Sci USA. 2005;102(30):10610–5.
21. He Z, Bateman A. Progranulin (granulin-epithelin precursor, PC-cell-derived growth factor, acrogranin) mediates tissue repair and tumorigenesis. J Mol Med (Berl). 2003;81(10):600–12.
22. Luster AD, Unkeless JC, Ravetch JV. Gamma-interferon transcriptionally regulates an early-response gene containing homology to platelet proteins. Nature. 1985;315(6021):672–6.
23. Carswell EA, et al. An endotoxin-induced serum factor that causes necrosis of tumors. Proc Natl Acad Sci USA. 1975;72(9):3666–70.
24. Cook KS, et al. Adipsin: a circulating serine protease homolog secreted by adipose tissue and sciatic nerve. Science. 1987;237(4813):402–5.
25. Oh KJ, et al. Metabolic adaptation in obesity and type II diabetes: myokines, adipokines and hepatokines. Int J Mol Sci. 2016;18(1):8. https://doi.org/10.3390/ijms18010008.
26. Guariguata L, et al. Global estimates of diabetes prevalence for 2013 and projections for 2035. Diabetes Res Clin Pract. 2014;103(2):137–49.
27. Sobrino Crespo C, et al. Peptides and food intake. Front Endocrinol (Lausanne). 2014;5:58.
28. Brooks CM, Lockwood RA, Wiggins ML. A study of the effect of hypothalamic lesions on the eating habits of the albino rat. Am J Phys. 1946;147(4):735–41.
29. Bertozzi, S. [Abnormal hunger due to brain lesion: pituitary tumor and paroxystic bulimia]. Riv Patol Nerv Ment. 1950;71(2):343–66.
30. Stephens TW, et al. The role of neuropeptide Y in the antiobesity action of the obese gene product. Nature. 1995;377(6549):530–2.
31. Frühbeck G. Intracellular signalling pathways activated by leptin. Biochem J. 2006;393(Pt 1):7–20.
32. Wada N, et al. Leptin and its receptors. J Chem Neuroanat. 2014;61-62:191–9.

33. Fei H, et al. Anatomic localization of alternatively spliced leptin receptors (Ob-R) in mouse brain and other tissues. Proc Natl Acad Sci USA. 1997;94(13):7001–5.
34. Ahima RS. Revisiting leptin's role in obesity and weight loss. J Clin Invest. 2008;118(7):2380–3.
35. Marcos P, Coveñas R. Regulation of homeostasis by neuropeptide Y: involvement in food intake. Curr Med Chem. 2022;29(23):4026–49.
36. Gouveia A, de Oliveira Beleza R, Steculorum SM. AgRP neuronal activity across feeding-related behaviours. Eur J Neurosci. 2021;54(10):7458–75.
37. Sohn JW. Network of hypothalamic neurons that control appetite. BMB Rep. 2015;48(4):229–33.
38. Li MD. Leptin and beyond: an odyssey to the central control of body weight. Yale J Biol Med. 2011;84(1):1–7.
39. Diano S. New aspects of melanocortin signaling: a role for PRCP in α-MSH degradation. Front Neuroendocrinol. 2011;32(1):70–83.
40. Cowley MA, et al. Leptin activates anorexigenic POMC neurons through a neural network in the arcuate nucleus. Nature. 2001;411(6836):480–4.
41. Brothers SP, Wahlestedt C. Therapeutic potential of neuropeptide Y (NPY) receptor ligands. EMBO Mol Med. 2010;2(11):429–39.
42. Fu L-Y, van den Pol AN. Agouti-related peptide and MC3/4 receptor agonists both inhibit excitatory hypothalamic ventromedial nucleus neurons. J Neurosci. 2008;28(21):5433–49.
43. Elias CF, et al. Leptin differentially regulates NPY and POMC neurons projecting to the lateral hypothalamic area. Neuron. 1999;23(4):775–86.
44. Martin SS, Qasim A, Reilly MP. Leptin resistance: a possible interface of inflammation and metabolism in obesity-related cardiovascular disease. J Am Coll Cardiol. 2008;52(15):1201–10.
45. Liu J, et al. The leptin resistance. Adv Exp Med Biol. 2018;1090:145–63.
46. Childs GV, et al. The importance of leptin to reproduction. Endocrinology. 2021;162(2):bqaa204. https://doi.org/10.1210/endocr/bqaa204.
47. de Knegt VE, et al. The role of leptin in fetal growth during pre-eclampsia. Int J Mol Sci. 2021;22(9):4569.
48. Grisaru-Granovsky S, Samueloff A, Elstein D. The role of leptin in fetal growth: a short review from conception to delivery. Eur J Obstet Gynecol Reprod Biol. 2008;136(2):146–50.
49. Carlton ED, Demas GE, French SS. Leptin, a neuroendocrine mediator of immune responses, inflammation, and sickness behaviors. Horm Behav. 2012;62(3):272–9.
50. Kiernan K, MacIver NJ. The role of the adipokine leptin in immune cell function in health and disease. Front Immunol. 2020;11:622468.
51. Zhang F, et al. Leptin: structure, function and biology. Vitam Horm. 2005;71:345–72.
52. Tahergorabi Z, Khazaei M. Leptin and its cardiovascular effects: focus on angiogenesis. Adv Biomed Res. 2015;4:79.
53. Farr OM, Gavrieli A, Mantzoros CS. Leptin applications in 2015: what have we learned about leptin and obesity? Curr Opin Endocrinol Diabetes Obes. 2015;22(5):353–9.
54. De Vos P, et al. Induction of ob gene expression by corticosteroids is accompanied by body weight loss and reduced food intake. J Biol Chem. 1995;270(27):15958–61.
55. Saladin R, et al. Transient increase in obese gene expression after food intake or insulin administration. Nature. 1995;377(6549):527–9.
56. Obradovic M, et al. Leptin and obesity: role and clinical implication. Front Endocrinol (Lausanne). 2021;12:585887.
57. Freire RH, Alvarez-Leite JI. Appetite control: hormones or diet strategies? Curr Opin Clin Nutr Metab Care. 2020;23(5):328–35.
58. Brandfon S, et al. Advances in anti-obesity pharmacotherapy: current treatments, emerging therapies, and challenges. Cureus. 2023;15(10):e46623.
59. Chen J, et al. Secretion of adiponectin by human placenta: differential modulation of adiponectin and its receptors by cytokines. Diabetologia. 2006;49(6):1292–302.

60. da Silva Rosa SC, Liu M, Sweeney G. Adiponectin synthesis, secretion and extravasation from circulation to interstitial space. Physiology (Bethesda). 2021;36(3):134–49.
61. Martinez-Huenchullan SF, et al. Skeletal muscle adiponectin induction in obesity and exercise. Metabolism. 2020;102:154008.
62. Psilopanagioti A, et al. Expression of adiponectin and adiponectin receptors in human pituitary gland and brain. Neuroendocrinology. 2009;89(1):38–47.
63. Shapiro L, Scherer PE. The crystal structure of a complement-1q family protein suggests an evolutionary link to tumor necrosis factor. Curr Biol. 1998;8(6):335–8.
64. Hu E, Liang P, Spiegelman BM. AdipoQ is a novel adipose-specific gene dysregulated in obesity. J Biol Chem. 1996;271(18):10697–703.
65. Parida S, Siddharth S, Sharma D. Adiponectin, obesity, and cancer: clash of the bigwigs in health and disease. Int J Mol Sci. 2019;20(10):2519.
66. Scherer PE. Adipose tissue: from lipid storage compartment to endocrine organ. Diabetes. 2006;55(6):1537–45.
67. Wolf G. Adiponectin: a regulator of energy homeostasis. Nutr Rev. 2003;61(8):290–2.
68. Moschen AR, Wieser V, Tilg H. Adiponectin: key player in the adipose tissue-liver crosstalk. Curr Med Chem. 2012;19(32):5467–73.
69. Hu X, et al. Adiponectin decreases plasma glucose and improves insulin sensitivity in diabetic Swine. Acta Biochim Biophys Sin Shanghai. 2007;39(2):131–6.
70. Combs TP, et al. A transgenic mouse with a deletion in the collagenous domain of adiponectin displays elevated circulating adiponectin and improved insulin sensitivity. Endocrinology. 2004;145(1):367–83.
71. Kim JY, et al. Obesity-associated improvements in metabolic profile through expansion of adipose tissue. J Clin Invest. 2007;117(9):2621–37.
72. Nawrocki AR, et al. Mice lacking adiponectin show decreased hepatic insulin sensitivity and reduced responsiveness to peroxisome proliferator-activated receptor gamma agonists. J Biol Chem. 2006;281(5):2654–60.
73. Xia JY, et al. Acute loss of adipose tissue-derived adiponectin triggers immediate metabolic deterioration in mice. Diabetologia. 2018;61(4):932–41.
74. Frankenberg ADV, Reis AF, Gerchman F. Relationships between adiponectin levels, the metabolic syndrome, and type 2 diabetes: a literature review. Arch Endocrinol Metab. 2017;61(6):614–22.
75. Berg AH, et al. The adipocyte-secreted protein Acrp30 enhances hepatic insulin action. Nat Med. 2001;7(8):947–53.
76. Choi HM, Doss HM, Kim KS. Multifaceted physiological roles of adiponectin in inflammation and diseases. Int J Mol Sci. 2020;21(4):1219.
77. Sada KE, et al. Altered levels of adipocytokines in association with insulin resistance in patients with systemic lupus erythematosus. J Rheumatol. 2006;33(8):1545–52.
78. Rovin BH, et al. Plasma, urine, and renal expression of adiponectin in human systemic lupus erythematosus. Kidney Int. 2005;68(4):1825–33.
79. Choi H-M, et al. Adiponectin may contribute to synovitis and joint destruction in rheumatoid arthritis by stimulating vascular endothelial growth factor, matrix metalloproteinase-1, and matrix metalloproteinase-13 expression in fibroblast-like synoviocytes more than proinflammatory mediators. Arthritis Res Ther. 2009;11(6):R161.
80. Przybyciński J, et al. Adiponectin in chronic kidney disease. Int J Mol Sci. 2020;21(24):9375.
81. Karmiris K, et al. Circulating levels of leptin, adiponectin, resistin, and ghrelin in inflammatory bowel disease. Inflamm Bowel Dis. 2006;12(2):100–5.
82. Moriconi N, et al. Body composition and adiponectin serum concentrations in adult patients with cystic fibrosis. J Clin Endocrinol Metab. 2006;91(4):1586–90.
83. Van Berendoncks AM, et al. Disturbed adiponectin—AMPK system in skeletal muscle of patients with metabolic syndrome. Eur J Prev Cardiol. 2015;22(2):203–5.
84. Magkos F, Sidossis LS. Recent advances in the measurement of adiponectin isoform distribution. Curr Opin Clin Nutr Metab Care. 2007;10(5):571–5.

85. Wang ZV, Scherer PE. Adiponectin, the past two decades. J Mol Cell Biol. 2016;8(2):93–100.
86. Yamauchi T, et al. Cloning of adiponectin receptors that mediate antidiabetic metabolic effects. Nature. 2003;423(6941):762–9.
87. Yamauchi T, et al. Targeted disruption of AdipoR1 and AdipoR2 causes abrogation of adiponectin binding and metabolic actions. Nat Med. 2007;13(3):332–9.
88. Fukuda S, et al. The unique prodomain of T-cadherin plays a key role in adiponectin binding with the essential extracellular cadherin repeats 1 and 2. J Biol Chem. 2017;292(19):7840–9.
89. Ghadge AA, Khaire AA, Kuvalekar AA. Adiponectin: a potential therapeutic target for metabolic syndrome. Cytokine Growth Factor Rev. 2018;39:151–8.
90. Abdalla MMI. Therapeutic potential of adiponectin in prediabetes: strategies, challenges, and future directions. Ther Adv Endocrinol Metab. 2024;15:20420188231222371.
91. Steppan CM, et al. The hormone resistin links obesity to diabetes. Nature. 2001;409(6818):307–12.
92. McTernan PG, Kusminski CM, Kumar S. Resistin. Curr Opin Lipidol. 2006;17(2):170–5.
93. Park HK, Ahima RS. Resistin in rodents and humans. Diabetes Metab J. 2013;37(6):404–14.
94. Codoñer-Franch P, Alonso-Iglesias E. Resistin: insulin resistance to malignancy. Clin Chim Acta. 2015;438:46–54.
95. Siddiqui K, Scaria Joy S, George TP. Circulating resistin levels in relation with insulin resistance, inflammatory and endothelial dysfunction markers in patients with type 2 diabetes and impaired fasting glucose. Endocrine Metab Sci. 2020;1(3):100059.
96. Gencer B, et al. Association between resistin levels and cardiovascular disease events in older adults: the health, aging and body composition study. Atherosclerosis. 2016;245:181–6.
97. Hsieh Y-Y, et al. Resistin-induced stromal cell-derived factor-1 expression through Toll-like receptor 4 and activation of p38 MAPK/ NFκB signaling pathway in gastric cancer cells. J Biomed Sci. 2014;21(1):59.
98. Li Y, et al. Resistin, a novel host defense peptide of innate immunity. Front Immunol. 2021;12:699807.
99. Pine GM, Batugedara HM, Nair MG. Here, there and everywhere: resistin-like molecules in infection, inflammation, and metabolic disorders. Cytokine. 2018;110:442–51.
100. Rak A, et al. Adiponectin and resistin: potential metabolic signals affecting hypothalamo-pituitary gonadal axis in females and males of different species. Reproduction. 2017;153(6):R215–r226.
101. Daquinag A, et al. An isoform of decorin is a resistin receptor on the surface of adipose progenitor cells. Cell Stem Cell. 2011;9:74–86.
102. Sanchez-Solana B, Laborda J, Baladron V. Mouse resistin modulates adipogenesis and glucose uptake in 3T3-L1 preadipocytes through the ROR1 receptor. Mol Endocrinol. 2012;26:110–27.
103. Benomar Y, et al. Central resistin overexposure induces insulin resistance through toll-like receptor 4. Diabetes. 2013;62:102–14.
104. Lee S, et al. Adenylyl cyclase-associated protein 1 is a receptor for human resistin and mediates inflammatory actions of human monocytes. Cell Metab. 2014;19:484–97.
105. Nogueiras R, et al. Resistin is expressed in different rat tissues and is regulated in a tissue- and gender-specific manner. FEBS Lett. 2003;548(1–3):21–7.
106. Wilkinson M, et al. Adipokine gene expression in brain and pituitary gland. Neuroendocrinology. 2007;86(3):191–209.
107. Dong XQ, et al. Resistin is associated with mortality in patients with traumatic brain injury. Crit Care. 2010;14(5):R190.
108. Patel L, et al. Resistin is expressed in human macrophages and directly regulated by PPAR gamma activators. Biochem Biophys Res Commun. 2003;300(2):472–6.
109. Fain JN, et al. Resistin release by human adipose tissue explants in primary culture. Biochem Biophys Res Commun. 2003;300(3):674–8.

110. Fasshauer M, et al. Tumor necrosis factor alpha is a negative regulator of resistin gene expression and secretion in 3T3-L1 adipocytes. Biochem Biophys Res Commun. 2001;288(4):1027–31.
111. Kaser S, et al. Resistin messenger-RNA expression is increased by proinflammatory cytokines in vitro. Biochem Biophys Res Commun. 2003;309(2):286–90.
112. Lu SC, et al. Lipopolysaccharide increases resistin gene expression in vivo and in vitro. FEBS Lett. 2002;530(1–3):158–62.
113. Bokarewa M, et al. Resistin, an adipokine with potent proinflammatory properties. J Immunol. 2005;174(9):5789–95.
114. Silswal N, et al. Human resistin stimulates the pro-inflammatory cytokines TNF-alpha and IL-12 in macrophages by NF-kappaB-dependent pathway. Biochem Biophys Res Commun. 2005;334(4):1092–101.
115. Pang SS, Le YY. Role of resistin in inflammation and inflammation-related diseases. Cell Mol Immunol. 2006;3(1):29–34.
116. Deb A, et al. Resistin: a journey from metabolism to cancer. Transl Oncol. 2021;14(10):101178.
117. Hou WK, et al. Adipocytokines and breast cancer risk. Chin Med J. 2007;120(18):1592–6.
118. Dalamaga M. Resistin as a biomarker linking obesity and inflammation to cancer: potential clinical perspectives. Biomark Med. 2014;8(1):107–18.
119. Reilly MP, et al. Resistin is an inflammatory marker of atherosclerosis in humans. Circulation. 2005;111(7):932–9.
120. Senolt L, et al. Resistin in rheumatoid arthritis synovial tissue, synovial fluid and serum. Ann Rheum Dis. 2007;66(4):458–63.
121. Coppack SW. Pro-inflammatory cytokines and adipose tissue. Proc Nutr Soc. 2001;60(3):349–56.
122. Cawthorn WP, Sethi JK. TNF-alpha and adipocyte biology. FEBS Lett. 2008;582(1):117–31.
123. Al-Roub A, et al. IL-1β and TNFα cooperativity in regulating IL-6 expression in adipocytes depends on CREB binding and H3K14 acetylation. Cells. 2021;10(11):3228.
124. Luan D, et al. Adipocyte-secreted IL-6 sensitizes macrophages to IL-4 signaling. Diabetes. 2023;72(3):367–74.
125. Fruebis J, et al. Proteolytic cleavage product of 30-kDa adipocyte complement-related protein increases fatty acid oxidation in muscle and causes weight loss in mice. Proc Natl Acad Sci USA. 2001;98(4):2005–10.
126. Gao M, Cui D, Xie J. The role of adiponectin for immune cell function in metabolic diseases. Diabetes Obes Metab. 2023;25(9):2427–38.
127. Liu L, et al. Cancer-associated adipocyte-derived G-CSF promotes breast cancer malignancy via Stat3 signaling. J Mol Cell Biol. 2020;12(9):723–37.
128. Cheng JX, Yu K. New discovered adipokines associated with the pathogenesis of obesity and type 2 diabetes. Diabetes Metab Syndr Obes. 2022;15:2381–9.
129. Xu X, et al. Follistatin-like 1 as a novel adipomyokine related to insulin resistance and physical activity. J Clin Endocrinol Metab. 2020;105(12):dgaa629.
130. Murahovschi V, et al. WISP1 is a novel adipokine linked to inflammation in obesity. Diabetes. 2015;64(3):856–66.
131. Romere C, et al. Asprosin, a fasting-induced glucogenic protein hormone. Cell. 2016;165(3):566–79.
132. Rydzewska M, et al. Serum secreted frizzled-related protein 5 in relation to insulin sensitivity and its regulation by insulin and free fatty acids. Endocrine. 2021;74(2):300–7.
133. Xu Q, et al. Plasma Sfrp5 levels correlate with determinants of the metabolic syndrome in Chinese adults. Diabetes Metab Res Rev. 2017;33(6) https://doi.org/10.1002/dmrr.2896.
134. Guo D, et al. Adiposity measurements and metabolic syndrome are linked through circulating neuregulin 4 and adipsin levels in obese adults. Front Physiol. 2021;12:667330.
135. Yari FA, et al. Circulating levels of FAM19A5 are inversely associated with subclinical atherosclerosis in non-alcoholic fatty liver disease. BMC Endocr Disord. 2021;21(1):153.

136. Wang Y, et al. Novel adipokine, FAM19A5, inhibits neointima formation after injury through sphingosine-1-phosphate receptor 2. Circulation. 2018;138(1):48–63.
137. Li J, Papadopoulos V, Vihma V. Steroid biosynthesis in adipose tissue. Steroids. 2015;103:89–104.
138. Klein S, et al. Why does obesity cause diabetes? Cell Metab. 2022;34(1):11–20.
139. Powell-Wiley TM, et al. Obesity and cardiovascular disease: a scientific statement from the American Heart Association. Circulation. 2021;143(21):e984–e1010.
140. Pati S, et al. Obesity and cancer: a current overview of epidemiology, pathogenesis, outcomes, and management. Cancers (Basel). 2023;15(2):485.
141. Kiani AK, et al. Steroid-converting enzymes in human adipose tissues and fat deposition with a focus on AKR1C enzymes. Eur Rev Med Pharmacol Sci. 2021;25(1 Suppl):23–32.
142. Mair KM, Gaw R, MacLean MR. Obesity, estrogens and adipose tissue dysfunction—implications for pulmonary arterial hypertension. Pulm Circ. 2020;10(3):2045894020952019.
143. Bhardwaj P, et al. Estrogens and breast cancer: mechanisms involved in obesity-related development, growth and progression. J Steroid Biochem Mol Biol. 2019;189:161–70.
144. Quinkler M, et al. Androgen generation in adipose tissue in women with simple obesity—a site-specific role for 17beta-hydroxysteroid dehydrogenase type 5. J Endocrinol. 2004;183(2):331–42.
145. Wagner IV, et al. Adipose tissue is a potential source of hyperandrogenism in obese female rats. Obesity (Silver Spring). 2018;26(7):1161–7.
146. Shabbir S, et al. The interplay between androgens and the immune response in polycystic ovary syndrome. J Transl Med. 2023;21(1):259.
147. Zerradi M, et al. Androgens, body fat distribution and adipogenesis. Curr Obes Rep. 2014;3(4):396–403.
148. Lucassen EA, Cizza G. The hypothalamic-pituitary-adrenal axis, obesity, and chronic stress exposure: sleep and the HPA axis in obesity. Curr Obes Rep. 2012;1(4):208–15.
149. Lee MJ, et al. Deconstructing the roles of glucocorticoids in adipose tissue biology and the development of central obesity. Biochim Biophys Acta. 2014;1842(3):473–81.
150. Westerbacka J, et al. Body fat distribution and cortisol metabolism in healthy men: enhanced 5beta-reductase and lower cortisol/cortisone metabolite ratios in men with fatty liver. J Clin Endocrinol Metab. 2003;88(10):4924–31.
151. van der Valk ES, Savas M, van Rossum EFC. Stress and obesity: are there more susceptible individuals? Curr Obes Rep. 2018;7(2):193–203.
152. Jones C, Gwenin C. Cortisol level dysregulation and its prevalence-is it nature's alarm clock? Physiol Rep. 2021;8(24):e14644.
153. Pruszkowska-Przybylska P, et al. Cortisol concentration affects fat and muscle mass among Polish children aged 6–13 years. BMC Pediatr. 2021;21(1):365.
154. Braun TP, Marks DL. The regulation of muscle mass by endogenous glucocorticoids. Front Physiol. 2015;6:12.
155. Macfarlane DP, Forbes S, Walker BR. Glucocorticoids and fatty acid metabolism in humans: fuelling fat redistribution in the metabolic syndrome. J Endocrinol. 2008;197(2):189–204.
156. Clemente-Suárez VJ, et al. The role of adipokines in health and disease. Biomedicines. 2023;11(5):1290.
157. Andrade-Oliveira V, Câmara NOS, Moraes-Vieira PM. Adipokines as drug targets in diabetes and underlying disturbances. J Diabetes Res. 2015;2015:681612.
158. Metz S, Huang LO, Kilpeläinen TO. Genetic variation, adipokines, and cardiometabolic disease. Curr Opin Pharmacol. 2020;52:33–9.
159. Kim JH, Bachmann RA, Chen J. Interleukin-6 and insulin resistance. Vitam Horm. 2009;80:613–33.
160. Shi C, et al. IL-6 and TNF-α induced obesity-related inflammatory response through transcriptional regulation of miR-146b. J Interf Cytokine Res. 2014;34(5):342–8.
161. Lin W, et al. Functional role of skeletal muscle-derived interleukin-6 and its effects on lipid metabolism. Front Physiol. 2023;14:1110926.

162. Turinsky J, O'Sullivan DM, Bayly BP. 1,2-Diacylglycerol and ceramide levels in insulin-resistant tissues of the rat in vivo. J Biol Chem. 1990;265(28):16880–5.
163. Hammerschmidt P, Brüning JC. Contribution of specific ceramides to obesity-associated metabolic diseases. Cell Mol Life Sci. 2022;79(8):395.
164. Haus JM, et al. Plasma ceramides are elevated in obese subjects with type 2 diabetes and correlate with the severity of insulin resistance. Diabetes. 2009;58(2):337–43.
165. Holland WL, et al. Inhibition of ceramide synthesis ameliorates glucocorticoid-, saturated-fat-, and obesity-induced insulin resistance. Cell Metab. 2007;5(3):167–79.
166. Spiegelman BM. Regulation of gene expression in the adipocyte: implications for obesity and proto-oncogene function. Trends Genet. 1988;4(7):203–7.
167. Deng Y, Scherer PE. Adipokines as novel biomarkers and regulators of the metabolic syndrome. Ann N Y Acad Sci. 2010;1212:E1–e19.
168. Chen Y, et al. Changes in profile of lipids and adipokines in patients with newly diagnosed hypothyroidism and hyperthyroidism. Sci Rep. 2016;6(1):26174.
169. Ambroszkiewicz J, et al. Anti-inflammatory and pro-inflammatory adipokine profiles in children on vegetarian and omnivorous diets. Nutrients. 2018;10(9):1241.
170. Ohashi K, et al. Role of anti-inflammatory adipokines in obesity-related diseases. Trends Endocrinol Metab. 2014;25(7):348–55.

# Chapter 5
# Adipose Tissue Dysfunction: Clinical Implications and Complications

**Just Dronkers, Michelle Lobeek, Laura M. G. Meems, and Thomas M. Gorter**

## Abbreviations

| | |
|---|---|
| AF | Atrial fibrillation |
| AGL | Acquired generalized lipodystrophy |
| APL | Acquired partial lipodystrophy |
| AT | Adipose tissue |
| BMI | Body mass index |
| BP | Blood pressure |
| CAD | Coronary artery disease |
| CGL | Congenital generalized lipodystrophy |
| COPD | Chronic Obstructive Pulmonary Disease |
| CV | Cardiovascular |
| CVD | Cardiovascular diseases |
| DAG | Diacylglycerol |
| EAT | Epicardial adipose tissue |
| EMA | European Medicines Agency |
| ER | Endoplasmic reticulum |
| FA | Fatty acid |
| FDA | Food and Drug Administration |
| FFA | Free fatty acid |
| FPLD | Familial partial lipodystrophy |
| GIP | Glucose-dependent insulinotropic polypeptide |
| GLP-1 | Glucagon-like peptide-1 |
| GLUT4 | Glucose transporter type 4 |
| HAART | Highly active antiretroviral therapy |

J. Dronkers · M. Lobeek · L. M. G. Meems · T. M. Gorter (✉)
Department of Cardiology, University Medical Center Groningen, University of Groningen, Groningen, The Netherlands
e-mail: j.dronkers@umcg.nl; m.lobeek@umcg.nl; l.m.g.meems@umcg.nl; tm.gorter@umcg.nl


D. Avtanski (ed.), *Adipose Tissue*, Contemporary Endocrinology,
https://doi.org/10.1007/978-3-031-72570-8_5

| | |
|---|---|
| HALS | HIV-associated lipodystrophy syndrome |
| $HbA_{1c}$ | Hemoglobin $A_{1C}$ |
| HDL-C | High-density lipoprotein cholesterol |
| HF | Heart failure |
| HFpEF | HF with preserved ejection fraction |
| HFrEF | HF with reduced ejection fraction |
| IL-6 | Interleukin-6 |
| INSR | Insulin receptor |
| MetS | Metabolic syndrome |
| MHO | Metabolically healthy obese |
| MUO | Metabolically unhealthy obese |
| PC | Pyruvate carboxylase |
| PKC | Protein kinase C |
| RAAS | Renin-angiotensin-aldosterone system |
| SAT | Subcutaneous adipose tissue |
| SGLT-2 | Sodium-glucose cotransporter 2 |
| T2DM | Type 2 diabetes mellitus |
| TNF | Tumor necrosis factor |
| VAT | Visceral adipose tissue |
| VTE | Venous thromboembolism |
| WC | Waist circumference |

## Introduction

The worldwide prevalence of obesity has increased to epidemic proportions and in 2022, more than one billion people in the world were obese [1]. Obesity, a metabolic disease typically characterized by the accumulation of visceral and ectopic fat, is a major contributor to the development of cardiovascular (CV) risk factors, such as hypertension, type 2 diabetes mellitus (T2DM), dyslipidemia, and CV diseases (CVD), and therefore greatly contributes to increased morbidity and mortality rates. Beyond overall obesity, the onset and progression of CVD are driven by adipose tissue (AT) dysfunction. However, the spectrum of obesity consists of individuals with metabolic healthy and unhealthy obesity. Furthermore, the pathophysiological transformation from healthy to dysfunctional AT is complex and has been the focus of research in recent years. The treatment of obesity has greatly evolved over the years and consists mainly of lifestyle behaviors, such as caloric restriction and physical activity, and surgical and/or pharmacological weight loss interventions. This chapter discusses the pathophysiology and clinical implications of obesity and AT dysfunction, reviews the pathophysiological role of AT dysfunction in the development of CV risk factors and disease, presents several real-world cases of adiposity-related CVD, and summarizes the available non-pharmacological and pharmacological treatments for obesity and AT dysfunction.

## Adipose Tissue Dysfunction

### *Pathophysiology of Obesity*

Adipose tissue (AT) has pleiotropic functions under physiological conditions. It serves as protection for organs and body parts, exerts effects on reproduction, and is most importantly involved in energy homeostasis. AT serves as storage of triglycerides in periods of excess energy that may be released during energy deprivation [2]. However, if this energy balance is disrupted with prolonged periods of excess energy, adipocytes can no longer adequately store this energy, resulting in AT dysfunction. In response to the altered energy balance, three important consecutive changes occur in dysfunctional AT. First, AT expands by increasing adipocyte size (hypertrophy) and/or by proliferation and differentiation of pre-adipocytes (hyperplasia) [2]. The extent of hypertrophy and hyperplasia is limited, especially in the visceral AT (VAT) [3]. As a result, intracellular pathways are activated leading to reactive oxygen species production and endoplasmic reticulum (ER) stress (Fig. 5.1) [4]. Second, expansion of AT increases the demand for nutrients and oxygen that are accommodated by increased angiogenesis, i.e., the sprouting of new blood vessels [5]. This angiogenic capacity is, however, reduced in AT of obese individuals, contributing to enhanced adipocyte stress [6, 7]. Third, adipocyte stress stimulates the secretion of pro-inflammatory cytokines, resulting in the infiltration of immune cells (Fig. 5.1) [4, 8]. Severe AT inflammation is characterized by the accumulation of immune cells surrounding dying adipocytes and this histological pattern is known as a crown-like structure [9]. Infiltrated immune cells contribute to a local pro-inflammatory environment by secreting inflammatory factors in the circulation that ultimately leading to a systemic low-grade inflammation.

In the late 1960s, it was observed that obese individuals had a higher free fatty acid (FFA) turnover rate than lean individuals, suggesting that AT in obese individuals releases more fatty acids (FAs) [10]. This process of lipolysis is induced by tumor necrosis factor (TNF) and interleukin-6 (IL-6), both pro-inflammatory cytokines (Fig. 5.1) [11, 12]. The enhanced lipolysis by inflammatory signals was also confirmed in humans as infusion of IL-6 in healthy men [13] or release of IL-6 by exercise increased FFA levels [14]. The enhanced FFA release subsequently induces ER stress that further aggravates adipocyte inflammatory signaling [15, 16]. This results in a detrimental feedback loop that maintains the pro-inflammatory environment in AT and continues to increase FFA release. Elevated circulatory FFA levels are particularly important for the development of T2DM and metabolic syndrome (MetS).

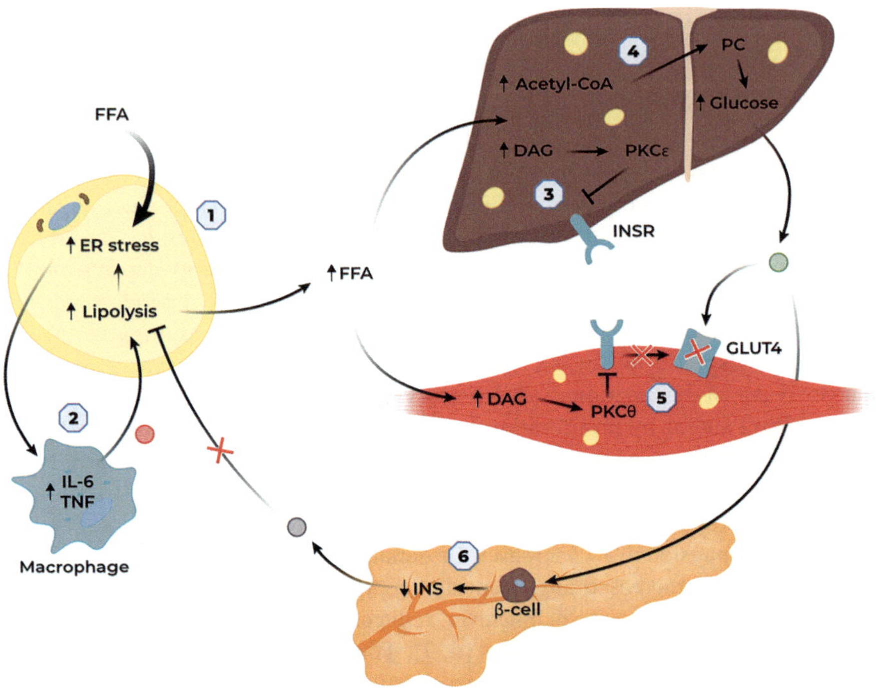


**Fig. 5.1** Molecular mechanism of dysfunctional adipose tissue leading to insulin resistance. (*1*) If the storage capacity of FFA in adipocytes is reached, ER stress increases, which subsequently enhances lipolysis and FFA release. (*2*) ER stress results in the secretion of chemoattractants and infiltration of immune cells. These cells produce pro-inflammatory cytokines that further increase lipolysis and ER stress. (*3*) Excess circulating FFAs are stored in the liver, which increases DAG. This activates PKCε, which inhibits INSR and leads to hepatic insulin resistance. (*4*) Acetyl-CoA levels increase with elevating FFA levels, which activates PC and increases glucose production and secretion. (*5*) Similar to the liver, DAG-PKCθ-INSR prevents the translocation of GLUT4 to the cellular membrane and the uptake of glucose, leading to glycemia. (*6*) Prolonged glycemia hampers insulin production in pancreatic β-cells, eventually leading to a decline in circulating insulin and loss of the anti-lipolytic effect of insulin. *FFA* free fatty acid, *ER* endoplasmic reticulum, *DAG* diacylglycerol, *PKC* protein kinase C, *INSR* insulin receptor, *Acetyl-CoA* Acetyl coenzyme A, *GLUT4* glucose transporter type 4, *IL-6* interleukin-6, *TNF* tumor necrosis factor, *PC* pyruvate carboxylase

## Clinical Implications of Obesity

In modern well-developed societies, food is abundantly available and physical activity is limited, resulting in a positive energy balance (e.g., higher intake than usage). The body resolves this state of excess energy by accumulating AT, which leads to an increased body weight. Individuals become overweight, classified as a body mass index (BMI) ≥ 25 kg/m$^2$, and if the excess energy state persists, a BMI ≥ 30 kg/m$^2$ may be observed categorizing individuals as obese [17]. However, AT distribution may vary greatly between individuals with similar BMI. As illustrated in Fig. 5.2, women accumulate fat predominantly in subcutaneous AT (SAT),

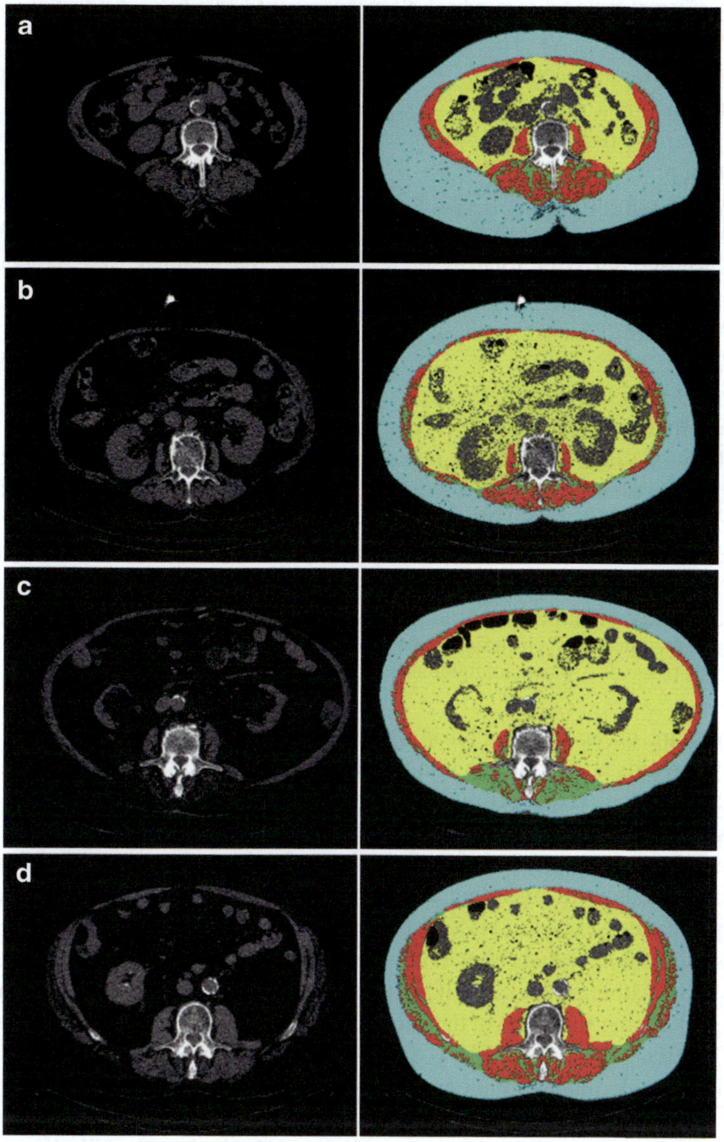


**Fig. 5.2** Varying adipose tissue distributions among four individuals with obesity. (**a**) A 73-year-old woman with a BMI of 34 kg/m$^2$, a high amount of SAT, and a moderate amount of VAT. She was diagnosed with hypertension and HF with preserved ejection fraction. (**b**) A 67-year-old woman with a BMI of 34 kg/m$^2$, high amount of SAT, and high VAT. She was diagnosed with atrial fibrillation with no history of T2DM, hypertension, CAD, dyslipidemia, or stroke. (**c**) An 87-year-old man with a BMI of 34 kg/m$^2$, a moderate amount of SAT, and a high amount of VAT. He was diagnosed with HF with a preserved ejection fraction, atrial fibrillation, and chronic kidney disease. (**d**) A 63-year-old man with a BMI of 35 kg/m$^2$ and a high amount of SAT and VAT. He was diagnosed with CAD and underwent coronary artery bypass grafting. Cyan and yellow illustrate the measured SAT and VAT volume, respectively. *BMI* body mass index, *SAT* subcutaneous adipose tissue, *VAT* visceral adipose tissue, *HF* heart failure, *T2DM* type 2 diabetes mellitus, *CAD* coronary artery disease

**Table 5.1** Characteristics of four individuals with various adipose tissue distributions despite the same BMI

| Individual | A | B | C | D |
|---|---|---|---|---|
| Age, years | 73 | 67 | 87 | 63 |
| Gender | Woman | Woman | Man | Man |
| BMI, kg/m$^2$ | 34 | 34 | 34 | 35 |
| SAT, cm$^2$ | 323 | 349 | 244 | 303 |
| VAT, cm$^2$ | 185 | 344 | 527 | 389 |
| WC, cm | 115 | 128 | 129 | 128 |
| EAT, ml | 155 | 218 | 369 | 321 |
| *Presence of risk factors* | | | | |
| Hypertension | Yes | No | Yes | Yes |
| Type 2 diabetes | No | No | No | No |
| Atrial fibrillation | No | Yes | Yes | No |
| Heart failure | Yes | Yes | Yes | Yes |
| Dyslipidemia | No | No | No | Yes |
| Myocardial infarction | No | No | No | Yes |
| Renal disease | No | No | Yes | No |
| History of stroke | No | No | No | No |
| COPD | Yes | No | No | No |

*BMI* body mass index, *EAT* epicardial adipose tissue, *SAT* subcutaneous adipose tissue, *VAT* visceral adipose tissue, *WC* waist circumference, *COPD* chronic obstructive pulmonary disease.

while men are prone to store fat in VAT. However, even within the same sex, the tendency towards accumulation in a particular depot differs. Notably, each individual has a variable CV risk profile despite their similar BMI (Table 5.1).

Obese individuals who are relatively healthy (i.e., absence of other cardiometabolic disorders) are classified as metabolically healthy obese (MHO). Conversely, obese individuals with prevalent cardiometabolic diseases are classified as metabolically unhealthy obese (MUO). However, different classification criteria exist with distinct cutoff values that complicate the identification of particular MHO patients [18]. Although these MHO patients are viewed as rather healthy, the risk of CVD is still greater than that in their lean counterparts [19]. In addition, almost 50% of MHO patients progress to MUO over a follow-up of 12 years [20]. This indicates that early lifestyle management and weight loss therapies should be recommended for all obese individuals, irrespective of their MHO or MUO classifications.

Therefore, obesity is not a benign condition and may impair daily life (Fig. 5.3). Obese individuals score 4.9 points lower than lean individuals on a health-related quality of life questionnaire that encompasses five levels: mobility, self-care, activities, pain, and anxiety [21]. Higher BMI is associated with anxiety and depression, especially in children, and self-esteem may be affected contributing to a worse mental health status [22]. Furthermore, obesity has a profound impact on the body posture and structure. Greater abdominal fat leads to a shift of the center of gravity to the front of the feet, reducing the control of posture and overall stability. This may increase the risk of osteoarthritis, as obesity and poor posture increase knee and hip

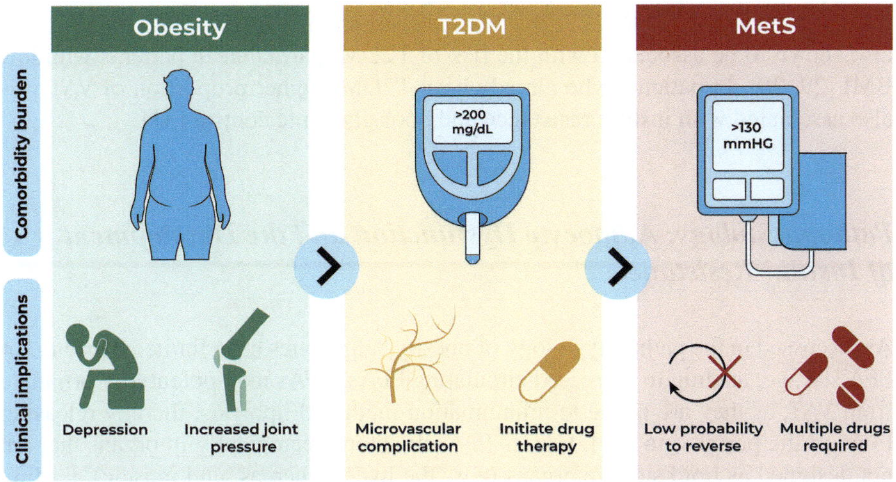


**Fig. 5.3** Clinical manifestations of obesity, type 2 diabetes, and metabolic syndrome. The development of obesity, T2DM, and MetS is associated with multiple clinical manifestations. The manifestations become more severe with the progression from obesity to MetS, and they may accumulate; individuals with T2DM and MetS will also carry the burden of obesity-related health issues. *T2DM* type 2 diabetes mellitus, *MetS* metabolic syndrome

joint pressures. The combination of altered posture and structure modifies the gait pattern, which impairs daily activities including walking or climbing stairs [23]. As a result, both depressive thoughts (lack of motivation) and reduced mobility will make it harder to reduce weight by exercise or diet.

## Insulin Resistance and Diabetes

### *Obesity and Risk of Type 2 Diabetes*

T2DM is a common comorbidity associated with obesity, and the most common initial signs and symptoms of diabetes are increased urination, thirst, and tiredness. Individuals may also experience impaired wound healing and frequent infections [24]. Diabetes is diagnosed if one of the following criteria applies: 8-h fasting plasma glucose ≥126 mg/dl, 2-h plasma glucose ≥200 mg/dl during a 75 g oral glucose tolerance test, or hemoglobin $A_{1C}$ ($HbA_{1c}$) ≥ 6.5% [25]. The elevated glucose levels are a result of insulin insensitivity; organs are not able to respond to insulin signaling to increase the uptake of glucose.

Indeed, obesity substantially increases the risk of T2DM. In a large cohort study in 1,647,826 adolescent men, the risk of developing T2DM within the next 40 years was nearly 40% in men with a BMI ≥35 kg/m$^2$ [26]. Especially the gain in body weight early in life, as well as the total duration of obesity, is important regarding the development of T2DM [27, 28]. Waist circumference, as surrogate for VAT, was also shown to be associated with the risk of T2DM, particular in patients with low BMI [29, 30]. In patients who already have T2DM, higher proportion of VAT was also associated with insulin resistance and poor glycemic control [31].

## *Pathophysiology: Adipocyte Dysfunction and the Development of Insulin Resistance*

As discussed in the pathophysiology of obesity, adipocytes have limited capacity for FFA storage, leading to increased circulating FFAs. FFAs may potentially originate from VAT, as they are prone to inflammation-mediated lipolysis, thereby releasing FFAs in the portal vein [11]. Excess FFAs are stored ectopically in organs that are not designed as lipid storage organs (e.g., the liver, pancreas, and muscle), leading to organ lipotoxicity and impaired organ function [32]. In stable isotope experiments, it was shown that around 60% of triglycerides in the liver of nonalcoholic fatty liver disease arose from circulating FFAs [33].

FFAs play a pivotal role in insulin sensitivity in the liver and muscle. Accumulation of FFAs in the liver results in high intracellular levels of diacylglycerol, an intermediate product of triglyceride esterification (Fig. 5.1). This activates protein kinase Cε (PKCε), which inhibits downstream insulin receptor signaling [34, 35]. This is supported by clinical evidence as hepatic diacylglycerol content is the best predictor of insulin resistance and highly correlates with activation of hepatic PKCε [36]. Skeletal muscles are affected by FFAs in a similar fashion to the liver. Lipid accumulation increases diacylglycerol in the muscle and activates PKCθ, leading to impaired insulin signaling [12, 37].

Additionally, FFAs derived from AT enhance hepatic acetyl co-enzyme A which allosterically stimulates pyruvate carboxylase, the first enzyme in the gluconeogenesis pathway, thereby increasing hepatic glucose production [38]. Glucose is released from the liver, stimulating β-cells in the pancreas to produce insulin to clear circulating glucose. However, at this stage, many organs are insulin insensitive; aberrant insulin signal transduction impairs glucose transporter type 4 (GLUT4) translocation to the cell membrane, which is the main route of insulin-stimulated glucose uptake in cells [39]. The skeletal muscles play a central role, as over 80% of glucose is stored in this system [40]. Therefore, skeletal muscle insulin insensitivity is the main driver of hyperglycemia. Following prolonged hyperglycemia, pancreatic β-cells function eventually declines [41]. As insulin is a potent inhibitor of adipocyte lipolysis and gluconeogenesis, loss of insulin signaling to the adipocytes further stimulates FFA release and enhances glucose release from the liver (Fig. 5.1) [38]. At this stage, an insulin insensitivity individual is likely to be diagnosed with T2DM.

### *Clinical Implications of Insulin Resistance and Type 2 Diabetes*

The diagnosis of T2DM has many health implications (Fig. 5.3), and individuals with diabetes are often stigmatized [42]. This may exacerbate mental health issues that are already observed in those with obesity. Additionally, hyperglycemia results in physiological complications that are proportional to its magnitude and duration [43]. Traditional complications are related to microvascular dysfunctions affecting dense microcapillary network (retinopathy, peripheral neuropathy, and nephropathy). Larger vessels are also affected by diabetes, which leads to macrovascular dysfunction (coronary heart disease, stroke, and peripheral vascular disease) [44]. These macrovascular complications will be discussed in the *Cardiovascular disease* section of this chapter.

It is estimated that 26% of T2DM patients in the USA show signs of retinopathy, defined as retinal microvascular aberrations, and in around 5% of these patients, vision is severely threatened [45]. The retina is the first microvascular system that is affect by hyperglycemia and retinopathy may be observed in people even before the diagnosis of T2DM [46]. Diabetic neuropathy, affecting around 30% of T2DM patients, is defined as peripheral nerve dysfunction that may lead to pain, numbness, and the loss of mechanoreception in the extremities [47]. Finally, diabetic nephropathy is defined as proteinuria >500 mg in 24 h, with an estimated prevalence between 20% and 50% of diabetic patients [48]. Persistent and untreated nephropathy may progress to chronic kidney disease which is a risk factor for many CVD [49].

To prevent complications, it is imperative that glucose levels remain stable at low levels. Adequate dietary interventions and exercise could facilitate the normalization of glucose levels and contribute to T2DM remission [50]. However, many T2DM patients are not able to maintain these lifestyle interventions. Metformin is used as first line drug intervention to control their glucose levels [51]. In severe diabetic patients with an $HbA_{1c} > 9\%$ or glucose levels >300 mg/dl supplementation with insulin may be considered [52]. Newly developed weight and glucose lowering drugs, sodium-glucose cotransporter 2 (SGLT-2) inhibitors and glucagon-like peptide-1 (GLP-1) agonists, may be used to resolve glycemia and are especially recommended in patients with high CV risk of or established CVD. GLP-1 agonists are advised for patients who benefit from a dual weight and glucose-lowering regimen and are preferred to insulin in T2DM patients when possible [17].

## Metabolic Syndrome

### *Obesity and Metabolic Syndrome*

MetS is a condition in which an individual experiences multiple metabolic complications. It is defined as having ≥3 of the following metabolic impairments [53]: (1) elevated waist circumference (population-specific cut-off), (2) triglycerides

≥150 mg/dL, (3) high-density lipoprotein cholesterol (HDL-C) <40 mg/dL in men and <50 mg/dL in women, (4) hypertension defined as systolic or diastolic blood pressure (BP) ≥130 mmHg or 85 mmHg, respectively, or (5) fasting glucose levels ≥100 mg/dL.

Waist circumference is a measure that correlates with the amount of VAT irrespective of BMI, indicating that excess VAT is particularly abundant in individuals with MetS [54]. This is corroborated by the observation that VAT volume (as measured by computed tomography) is more strongly associated with components of MetS compared to SAT volume [55, 56]. Obesity is also related to dyslipidemia, a disturbance in lipid balance reflected by increased circulating triglycerides and cholesterol or decreased HDL-C levels [57]. In addition, a strong linear association of BMI with systolic and diastolic BP is observed, and obesity increases the risk of developing hypertension [58, 59]. Each combination of the metabolic impairments could define MetS. However, hypertension was the most prevalent comorbidity in individuals with MetS, with a prevalence of 85% in MUO and 35% in MHO. Notably, the prevalence of dyslipidemia increases five-fold to 45%, and hyperglycemia increases six-fold to 60% in MUO compared to MHO [60].

## *Obesity and Hypertension*

In addition to the release of FFAs and secretion of pro-inflammatory cytokines, adipocytes can secrete factors that play a role in BP regulation. Components of the renin-angiotensin-aldosterone system (RAAS), the primary system for BP regulation, are expressed in AT. In obesity, the RAAS becomes dysregulated, leading to its overactivation, thereby contributing to elevated BP in obese individuals [61–63]. In particular, perivascular AT surrounding blood vessels in the mesenterial area expresses high levels of RAAS components, underscoring the detrimental role of AT [64].

Additionally, it has been observed that rodents and humans with leptin deficiency have a lower BP despite a higher fat mass. Leptin may act via neuronal circuitry to influence BP, as inhibition of leptin signaling in neurons reduces BP in obese mice [65]. The reduction in circulating adiponectin, an important protective adipokine in obesity, also correlates with higher BP levels [66]. Adiponectin-knockout mice, which were fed a high-salt diet to induce hypertension, had more severe hypertension than their wild-type counterparts on the same diet. The increase in BP was ameliorated by adiponectin therapy, suggesting a protective role for adiponectin in obesity-related hypertension development [67].

Obesity, particularly VAT adiposity, is a strong predictor of increased aortic stiffness, which contributes to enhanced BP [68, 69]. Female mice with a deletion of the mineralocorticoid receptor are protected to western diet-induced vascular stiffness. This is potentially mediated by aldosterone, as it may bind to the mineralocorticoid receptor and subsequently reduce endothelial nitric oxide synthase activity, which is a key component in endothelial relaxation signaling [70]. In addition, VAT increases pressure in the abdominal cavity [71], and increased renal fat is independently associated with hypertension [72, 73]. Excess VAT accumulation within the rigid kidney capsule may therefore compress the kidneys and increase intrarenal congestion, leading to a higher BP [74].

## *Clinical Implications*

The additional burden of multiple metabolic impairments further deteriorates the lives of patients with MetS (Fig. 5.3). There is a bidirectional association of MetS and depression; the risk of depression is increased in MetS individuals and vice versa [75]. The overall quality of life appears unaffected in MetS patients but may be reduced in MetS patients with a high BMI [76]. Compared to more obvious signs and symptoms in obesity and diabetes, individuals with either dyslipidemia or hypertension are asymptomatic that may explain the unaltered quality of life. However, both dyslipidemia and hypertension increase the risk of incident CVD and CV mortality [77–79]. Screening for metabolic complication in obese individuals is therefore key in preventing CVD at a later stage in life.

If these risk factors are not adequately addressed and an individual progresses to MetS, lifestyle interventions are first used to resolve or control MetS. This consists of changing dietary intake, both quality and quantity, and increasing physical activity, which is beneficial for all MetS components [50]. Nevertheless, persistent or severe metabolic impairments will additionally benefit from the use of specific drugs. Dyslipidemia may be treated with statins that lower lipid levels, primarily by inhibition of the enzyme that catalyzes cholesterol biosynthesis [80]. The use of statins is also recommended to prevent CVD especially in individuals with a high CV risk [81]. Hypertension can be treated with multiple agents, including diuretics, calcium channel blockers, angiotensin-converting enzyme inhibitors, and angiotensin receptor blockers. Although stage I hypertension (BP ≥130–139/80–85 mmHg) is used as cutoff criteria for MetS, anti-hypertensive drugs are only prescribed in individuals with a CVD risk score greater than 10% or stage II hypertension (BP ≥140/90 mmHg) [82]. Considering that many MetS components are included in these risk prediction scores, it is likely that patients with elevated BP require antihypertensive therapy irrespective of their hypertension stage.

Patients with MetS may require more than one intervention to alleviate their symptoms and reduce the risk of CVD. Although lifestyle interventions are central to MetS therapy, the majority of MetS patients require supplementation with metformin, statins, or antihypertensive drugs. The prescription of these drugs is complicated by individual dose and response effects, and requires titration to assess the optimal dose that is monitored throughout life. Daily lifestyle modifications and drug use are additional burdens for patients with MetS. Despite the efficacy of the intervention, self-reported adherence to therapy is low. Around 50% is nonadherent to medication and even 90% does not adhere to lifestyle modifications [83]. MetS is therefore a difficult metabolic disorder to reverse and should be prevented already in early stages of obesity.

## Lipodystrophy

Lipodystrophy encompasses several metabolic disorders. It is characterized by partial or complete loss of major AT in the abdominal or gluteal area, concomitant with, in many cases, accumulation of fat in other regions of the body [84]. These conditions are classified by etiology into two main groups: acquired and inherited (congenital). Each of these groups can be further stratified into generalized (whole-body) or partial (specific regions) lipodystrophy (Table 5.2).

### *Types of Lipodystrophies*

1. **Congenital generalized lipodystrophy (CGL)** is a rare autosomal recessive disorder characterized by significant reduction in SAT throughout the body [85]. Approximately 250 cases are described and the estimated prevalence ranges from 0.1 to 1 in a million [86]. The development of CGL is attributed, although not exclusively, to mutations in the *AGPAT2* or *BSCL2* gene. Both genes are involved in adequate storage of FAs in lipid droplets of adipocytes [84].

**Table 5.2** Characteristics of lipodystrophy

| | Congenital lipodystrohy | | Acquired lipodystrophy | | |
|---|---|---|---|---|---|
| | Generalized (CGL) | Partial (FPLD) | Generalized (AGL) | Partial (APL) | HALS |
| Prevalence | 0.9 per million | 1.5 per million | 0.1 per million | 0.2 per million | 40-70% in patients on HAART |
| Fat distribution | Loss of subcutenous fat | Loss of subcutenous fat extremities and abdomen. Increased face, neck, and visceral cavity | Subcutenous fat and pads extremities reduced | Loss starts in the face and spreads to abdomen | Lipodystrophy and/or hypertrophy Face, arms, and femoral-gluteal region |
| Etiology | *AGPAT2* or *BSCL2* mutation | *LMNA* mutation | Inflammatory disorders | Inflammation | HAART in HIV patients |

*CGL* Congenital generalized lipodystrophy, *FPLD* Familial partial lipodystrophy, *AGL* Acquired generalized lipodystrophy, *APL* Acquired partial lipodystrophy, *HALS* HIV-associated lipodystrophy syndrome, *HAART* highly active antiretroviral therapy.

2. **Inherited familial partial lipodystrophy (FPLD)** is an autosomal dominantly inherited condition with loss of specific AT depots and prevalence is roughly estimated at 1.5 cases per million individuals [86]. The most common type of FPLD is type 2, also known as Dunnigan-type lipodystrophy, caused by genetic mutations in the *LMNA* gene. Mutations in or lack of *LMNA* are implicated in the inhibition of adipocyte differentiation and increased lipolysis [87, 88]. FPLD develops in puberty and leads to atrophy of SAT in the extremities and the abdomen. In addition, AT accumulates in the face, neck, and visceral cavity [85]. Notably, genetic screening for mutations in *LMNA* showed that FPLD prevalence is potentially higher than expected with estimates of 1 in 7000. Although many patients with this mutation lack the diagnosis of lipodystrophy, they have a high burden of metabolic disorders [89].
3. **Acquired generalized lipodystrophy (AGL)** has a similar clinical picture as CGL, which includes greatly reduced SAT volumes, but may additionally exhibit a decrease in intra-abdominal fat or fat pads in the extremities [90]. AGL is a rare condition with less than 0.1 cases per million that is predominantly diagnosed in childhood or early adolescence [86]. Although around 50% of AGL are of idiopathic origin many cases are attributed to inflammatory disorders, either excessive SAT inflammation or autoimmunity [90].
4. **Acquired partial lipodystrophy (APL)** is characterized by the cranial to caudal progression of fat loss. It starts in the face and eventually spreads to the abdomen, affecting all AT in between. Lipoatrophy is not observed in the lower extremities or gluteal region, but these regions accumulate excess fat. The prevalence of APL is also low, with estimates around 0.2 cases per million [86]. Similar to AGL, inflammation is associated with the development of APL, but metabolic complications are scarcely observed in patients [90]. This may be related to the preserved storage function of AT around the hips, thereby limiting circulating FFA levels and preventing metabolic disorder development.

Most cases of acquired lipodystrophy occur in HIV-infected patients treated with highly active antiretroviral therapy (HAART). Survival of HIV patients is greatly reduced by HAART, but approximately 40–70% of patients exhibit HIV-associated lipodystrophy syndrome (HALS) concomitant with metabolic perturbations as an adverse effect of therapy [91]. Although both lipoatrophy and AT hypertrophy can be observed in these patients, AT loss occurs commonly in the face, arms, and femoral-gluteal region [92]. The exact pathophysiology of HALS is unknown, but certain HAARTs may impair adipocyte differentiation, induce reactive oxygen species, and enhance adipocyte lipolysis [84, 93].

### *Metabolic Complications and Implications of Lipodystrophy*

Similar to obesity, lipodystrophy results in increased FFA levels, primarily due to limited or inadequate lipid storage in adipocytes. Although the exact mechanisms linking lipodystrophies to metabolic complications, such as diabetes, ectopic lipid accumulation, and dyslipidemia remain elusive, it is likely that pathways similar to obesity-related metabolic complications are involved. Therefore, patients with lipodystrophy may potentially benefit from the treatment strategies utilized in obesity. In most cases, it is recommended that patients eat healthy and exercise regularly [84], except for patients with severe lipoatrophy and low body weight, as exercise may lead to further AT loss [94]. T2DM in lipodystrophy patients can be managed by standard of care drugs: metformin or insulin [95, 96]. However, these drugs are solely prescribed if the patient is diagnosed with diabetes, as these drugs are not approved by the Food and Drug Administration (FDA) for lipodystrophy [84]. To treat dyslipidemia, statins may be prescribed but use is warranted in HALS patients as statins interact with antiretroviral drugs [97].

Additionally, the general appearance of patients with lipodystrophy may be affected by the excessive loss or accumulation of fat in specific regions, which can impact their mental state. In particular patients with HALS exhibit loss of fat in their face, which can be reconstructed with cosmetic surgery via various implantation techniques, either exogenous silicone or endogenous fat [98, 99]. Conversely, FPLD patients with excess fat in the neck may benefit from repeated AT removal via lipectomy or liposuction [85].

## Cardiovascular Disease

### *Obesity and Risk of Cardiovascular Disease*

As described earlier, obesity and visceral adiposity are strongly related to CV risk factors, such as insulin resistance, dyslipidemia, hypertension, sympathetic nervous system, RAAS activation, and systemic inflammation [55, 100]. These obesity-related risk factors are causative for the development of CVD, such as atherosclerosis, heart failure (HF), stroke, and atrial fibrillation [101]. While increased BMI is associated with a higher incidence of CVDs, large inter-individual variations exist in the amount and distribution of VAT, which alter CV risk. Overweight and obese individuals with a low amount of AT often have a better CV risk profile than individuals with a higher amount of AT [102–105]. Additionally, obese individuals with high amount of upper body AT localization (e.g., intra-abdominal AT and increased triglyceride content in specific organs such as the liver and heart) are at higher risk compared with obese individuals with predominantly gluteal-femoral fat [106]. Basically, individuals with an excess amount of VAT have an increased risk for the development of CVD, independent of their BMI [107, 108].

## *Obesity Paradox*

The partially dissociated relationship between BMI and VAT is important to interpret the obesity paradox. This paradox states that obesity is associated with an increased risk of CVD, especially HF. However, among patients with HF, those with a higher BMI have a better prognosis than those with a lower BMI. For a large part, this observation results from methodological flaws, such as confounding, selection bias, reverse causality, and because BMI and VAT are not always mutually related. Therefore, other measures that more specifically estimate the burden of visceral adiposity, for example waist circumference, are a better reflection of individual risk [109]. Interestingly, in patients with HF, these alternative anthropometric measurements showed no evidence of an obesity paradox [110].

## *Atherosclerosis and Thrombo-Embolic Disease*

Obesity and visceral adiposity drive insulin resistance, atherogenic dyslipidemia, hypertension and the release of pro-inflammatory and pro-thrombotic factors, and these factors are closely related to the development of atherosclerosis and thromboembolic disease [111]. Obesity was previously shown to be associated with incident peripheral artery disease [112], stroke [113], and coronary artery disease (CAD) [114, 115]. However, increased BMI and its associated traditional risk factors, especially the metabolic consequences of visceral adiposity, seem most important in the risk of these atherosclerotic diseases [116–118], and the ancillary effects of AT dysfunction may explain these findings. For instance, it was shown that VAT volume was associated with arterial inflammation, independent of established risk factors [119].

A specific VAT depot of interest in relation to CAD is epicardial adipose tissue (EAT) which is the VAT situated between the outer wall of the myocardium and the visceral layer of the pericardium. EAT is mainly located in the atrioventricular and interventricular grooves and has an unobstructed relation with the myocardium and the coronary arteries [120]. Similar to VAT, EAT can transform from healthy AT into a pro-inflammatory and pro-atherogenic AT depot. EAT dysfunction is associated with CAD through inflammation, oxidative stress, endothelial damage, and lipid accumulation [120].

Venous thromboembolism (VTE), including deep vein thrombosis and pulmonary embolism, is highly common in the general population and even more in patients with obesity and contributes to increased morbidity and mortality. Obesity and visceral adiposity are associated with higher incidence for VTE [121, 122]. Although immobilization or hospitalization for trauma or critical illness exert a direct effect on VTE risk, the number of VTE cases is predominantly attributed to individuals with obesity who are hospitalized or had undergone knee or hip surgery [123–125]. As such, the presence of obesity is included in risk prediction models for

VTE, for instance to guide clinicians on prophylactic treatment with anticoagulation in hospitalized or post-surgical patients [126].

## *Heart Failure*

Obesity is independently associated with the development of HF. The increased risk for HF is 5 and 7% per 1-unit increase in BMI for men and women respectively [127]. Obesity seems more strongly associated with HF with preserved ejection fraction (HFpEF) as compared to HF with reduced ejection fraction (HFrEF) [128]. In large clinical trials of patients with HF, the average BMI of patients with HFpEF was ~30 kg/m$^2$, whereas in HFrEF populations, the average BMI was ~28 kg/m$^2$. Beyond BMI, especially the amount of pericardial AT, located on the outer layer of the pericardial sac, was independently associated with the incidence of HFpEF [129]. In HFpEF, the amount of EAT was higher than matched controls without HFpEF [130].

Obesity leads to neurohormonal activation, sodium retention, and plasma volume expansion, thereby increasing metabolic rate and stress of the heart [131]. Furthermore, several ancillary mechanisms beyond obesity, such as intra-myocardial lipo-toxicity, systemic inflammation and endothelial dysfunction, may lead to adverse myocardial remodeling, hypertrophy, and diastolic dysfunction ultimately resulting in HF [131]. These mechanisms may be more pronounced in patients with HFpEF than HFrEF, particular in women with HFpEF [128].

In patients with HF, the presence of obesity is associated with lower quality of life, more HF symptom burden and impaired exercise tolerance. While obesity and associated plasma volume expansion are related to increased cardiac filling pressures [132], it seems that especially an abundance of VAT and EAT is associated with impaired hemodynamics and exercise intolerance [133–135].

## *Atrial Fibrillation*

Obesity is independently associated with the development of atrial fibrillation (AF) [136–138]. In particular, the risk for AF increases when other anthropometric indices of obesity, such as waist circumference, waist-hip ratio, and fat mass, were used [139, 140]. Besides the onset of AF, obesity and visceral and epicardial adiposity are also associated with AF progression [141, 142], and with higher rates of recurrent AF after ablation [143]. The association between the development of AF in the setting of obesity is mediated through several mechanisms, including (1) hemodynamic and functional impairments associated with neuro-hormonal and RAAS activation and sodium retention leading to hypertension, plasma volume expansion, increased cardiac filling pressures, and ultimately left atrial stretch and enlargement [137, 138, 144, 145] and (2) left atrial electrical remodelling and conduction delay due to peri-atrial AT expansion and fibro-fatty infiltration concomitant with local inflammation [146–150].

## Obesity Treatment

Obesity is a treatable disease, and the cornerstone of obesity treatment is a combined lifestyle behavioral intervention focused on diet, caloric restriction, physical activity, and/or psychological interventions. Individuals who experience insufficient weight loss with lifestyle modifications alone may be offered pharmacological or surgical weight loss interventions. Dietary interventions may lead to modest weight loss in the short term but maintaining a durable weight loss may be quite difficult [151]. On the contrary, surgical weight loss interventions (e.g., Roux-Y gastric bypass) lead to a weight reduction ranging from 25% to 30% that sustains for over 5 years [152]. The field of pharmacological weight loss intervention has greatly evolved over the years, especially with the advent of GLP-1 receptor-modifying drugs, such as semaglutide, which may lead to sustained weight reduction in the range that was previously only seen after bariatric surgery.

### *Non-pharmacological Treatment of Obesity*

Most randomized controlled trials investigating non-pharmacological interventions for obesity have used a lifestyle intervention including caloric restriction, typically with a 500–750 kcal/day energy deficit, combined with increased physical activity. An average weight reduction of 5—10% can be achieved with such intervention, but maintaining a longstanding weight reduction is difficult. A standard, moderate-intensity aerobic exercise program alone is associated with a modest weight loss of 1.5–3.5 kg and an average fat loss of 1.3–2.6 kg [153]. The Diabetes Prevention Program randomized individuals with a BMI ≥24 kg/m$^2$ and prediabetes to lifestyle interventions or metformin. Lifestyle intervention was associated with a reduced risk for T2DM, and with a greater reduction in body weight and $HbA_{1C}$ as compared to metformin [154]. The Look AHEAD trial randomized 5145 overweight and obese patients with T2DM to either an intensive lifestyle intervention or control group. While lifestyle intervention led to an 8.6% reduction in body weight after 1 year, compared to 0.7% reduction in the control group, lifestyle intervention was not associated with lower risk of CV events [155]. Weight loss achieved at the end of the study was 6.0 and 3.5% in the intervention and control groups, respectively. Although a subgroup of individuals who lost >10% body weight had a 21% lower risk of CV events [156]. In a post-hoc analysis of the Look AHEAD trial, it was shown that at 1 year, the intervention group lost 5.6 kg fat and 2.3 kg lean mass, but regained 100% fat between years 1 and 8 [157].

The most recent guideline from the American Society of Metabolic and Bariatric Surgery states that bariatric surgery is recommended for individuals with a BMI ≥35 kg/m$^2$, regardless of the presence, absence, or severity of co-morbidities [158]. Furthermore, bariatric surgery should be considered for individuals with metabolic disease and a BMI of 30–34.9 kg/m$^2$ [158]. Bariatric surgery has sustained beneficial effects, not only on CV risk factors and metabolic control but also on CV mortality and CVD [159].

## *Pharmacological Treatment of Obesity*

Anti-obesity medications have a long history of use. Early drugs used for the treatment of obesity in the 1930s and the 1940s, such as 2.4-dinitrophenol, had serious side effects and were withdrawn from the market. More recent drugs, such as sibutramine and rimonabant, were also withdrawn because of serious CV and psychiatric side effects, respectively. Currently, six drugs have been approved by the FDA and the European Medicines Agency (EMA) for the treatment of obesity: orlistat, naltrexone/bupropion, liraglutide, semaglutide, and setmelanotide. The latter is prescribed to treat rare obesity-related monogenetic deficiencies. Tirzepatide is a dual glucose-dependent insulinotropic polypeptide (GIP) and GLP-1 agonist. Orlistat inhibits the breakdown of fat in ingested food. It leads to modest weight loss but is also associated with common gastrointestinal side effects, such as oily rectal leakage, fecal urgency, and fecal incontinence. Naltrexone/bupropion increases satiety and reduces food cravings and is associated with modest weight loss of 5–6% [160]. Liraglutide is the first GLP1 agonist to be used as a specific treatment for obesity and was investigated in the SCALE program. Liraglutide 3.0 mg *o.i.d.* was associated with a weight reduction ranging from 6% to 8% [161, 162]. In the subsequent STEP trials, semaglutide 2.4 mg once weekly was associated with an average weight reduction ranging from 10% to 15% [163–165]. Semaglutide was the first anti-obesity drug that also demonstrated a reduction in CV events. In the SELECT trial, semaglutide was associated with lower incidence of death from cardiovascular causes, nonfatal myocardial infarction, or nonfatal stroke [166]. Semaglutide was also tested in patients with obesity-related HFpEF in the STEP-HFpEF trial. Semaglutide was associated with improved symptoms and physical capacity, greater improvements in exercise function, and greater weight loss as compared to placebo [167]. Interestingly, weight reduction with semaglutide in HFpEF was also associated with a reduction in C-reactive protein levels. This suggests that beyond weight loss, there may also be a reduction in systemic inflammation possibly due to lower burden of AT dysfunction [168]. The dual GIP and GLP-1 agonist tirzepatide has been tested in the SURMOUNT trials and tirzepatide 15 mg once weekly was associated with weight reduction averaging from 15% to 21% within 72 weeks [169–171].

## *Treatment of Adipose Tissue Dysfunction*

While most studies on anti-obesity treatment have focused on weight loss, reduction in CV risk factors and CV events, relatively limited attention has been paid to the potentially beneficial effects on the underlying AT dysfunction. Such insights are important to understand the extent and mechanisms through which these anti-obesity treatments also improve CV risk beyond general weight reduction.

In an ancillary analysis from the Look AHEAD trial in 100 patients, mean VAT decreased by 8.0% in the intervention group, but this decrease was not associated with improvement in CV risk [172]. In another, smaller cross-sectional study, strict adherence to dietary guidelines was associated with less visceral and liver fat, adjusted for total body fat [173].

Bariatric surgery leads to an important reduction in VAT, especially in patients with T2DM [174]. Beyond a significant reduction in fat mass, bariatric surgery improves AT dysfunction, shifting it to more healthy tissue with improved lipolysis regulation, beneficial adipokine secretome, and decreased local and systemic inflammation, and several of these effects are weight loss-independent [175]. As such, bariatric surgery is increasingly being referred to as metabolic surgery.

In addition, several drugs have been shown to specifically reduce visceral and ectopic fat depots and partly resolve AT dysfunction. For instance, liraglutide has been shown to induce the browning of white AT in mice, demonstrated by upregulated lipolytic activity, enhanced expression of brown AT, and mitochondrial-specific genes [176]. Furthermore, in patients with non-alcoholic steatohepatitis, liraglutide was associated with a reduction of metabolic dysfunction, AT insulin resistance, and AT inflammation [177]. Also, semaglutide seems to have beneficial effects on AT dysfunction since proteins related to the regulation of lipid uptake, lipid storage, and lipolysis were downregulated in white AT of mice [178]. While anti-obesity treatments lead to various grades of body weight and VAT reduction, all anti-obesity strategies described may lead to a reduction in EAT volume and potentially to a change in its proinflammatory state [179]. Especially SGLT-2 inhibitors and GLP-1 agonists may have a beneficial effect on EAT by reducing its volume and lowering its inflammatory response [180–182].

## Conclusion

Obesity is becoming one of the most important treatable risk factors for CVD. Moreover, in addition to being a traditional CV risk factor, unhealthy obesity with AT dysfunction is also a disease of its own. Knowledge of the transition of physiological, healthy AT to pathophysiological, unhealthy AT dysfunction, as well as the interplay between AT dysfunction and the development of CVD, has greatly evolved over the years. Major steps have been set in the treatment of obesity, especially with the development of new pharmacological agents that achieve weight loss that approaches the amount of weight loss following bariatric surgery. However, apart from general weight loss strategies, some important knowledge gaps remain regarding the treatment of AT dysfunction. Future studies are urgently needed to better prevent and treat AT dysfunction to reduce the societal burden of CVD.

## References

1. Phelps NH, Singleton RK, Zhou B, et al. Worldwide trends in underweight and obesity from 1990 to 2022: a pooled analysis of 3663 population-representative studies with 222 million children, adolescents, and adults. Lancet. 2024;403:1027–50. https://doi.org/10.1016/S0140-6736(23)02750-2.
2. Rosen ED, Spiegelman BM. What we talk about when we talk about fat. Cell. 2014;156:20–44.
3. Estève D, Boulet N, Belles C, et al. Lobular architecture of human adipose tissue defines the niche and fate of progenitor cells. Nat Commun. 2019;10(1):2549. https://doi.org/10.1038/s41467-019-09992-3.
4. Shoelson SE, Lee J, Goldfine AB. Inflammation and insulin resistance. J Clin Invest. 2006;116:1793–801.
5. AlZaim I, de Rooij LPMH, Sheikh BN, et al. The evolving functions of the vasculature in regulating adipose tissue biology in health and obesity. Nat Rev Endocrinol. 2023;19:691–707.
6. Gealekman O, Guseva N, Hartigan C, et al. Depot-specific differences and insufficient subcutaneous adipose tissue angiogenesis in human obesity. Circulation. 2011;123:186–94. https://doi.org/10.1161/CIRCULATIONAHA.110.970145.
7. Ngo DTM, Farb MG, Kikuchi R, et al. Antiangiogenic actions of vascular endothelial growth factor-A165b, an inhibitory isoform of vascular endothelial growth factor-A, in human obesity. Circulation. 2014;130:1072–80. https://doi.org/10.1161/CIRCULATIONAHA.113.008171.
8. Jacks RD, Lumeng CN. Macrophage and T cell networks in adipose tissue. Nat Rev Endocrinol. 2024;20:50–61.
9. Muir LA, Cho KW, Geletka LM, et al. Human CD206+ macrophages associate with diabetes and adipose tissue lymphoid clusters. JCI Insight. 2022;7(3):e146563. https://doi.org/10.1172/jci.insight.146563.
10. Björntorp P, Bergman H, Varnauskas E. Plasma free fatty acid turnover rate in obesity. Acta Med Scand. 1969;185:351–6. https://doi.org/10.1111/j.0954-6820.1969.tb07347.x.
11. Wueest S, Item F, Lucchini FC, et al. Mesenteric fat lipolysis mediates obesity-associated hepatic steatosis and insulin resistance. Diabetes. 2016;65:140–8. https://doi.org/10.2337/db15-0941.
12. Yu C, Chen Y, Cline GW, et al. Mechanism by which fatty acids inhibit insulin activation of insulin receptor substrate-1 (IRS-1)-associated phosphatidylinositol 3-kinase activity in muscle. J Biol Chem. 2002;277:50230–6. https://doi.org/10.1074/jbc.M200958200.
13. Van Hall G, Steensberg A, Sacchetti M, et al. Interleukin-6 stimulates lipolysis and fat oxidation in humans. J Clin Endocrinol Metab. 2003;88:3005–10. https://doi.org/10.1210/jc.2002-021687.
14. Wedell-Neergaard AS, Lang Lehrskov L, Christensen RH, et al. Exercise-induced changes in visceral adipose tissue mass are regulated by IL-6 signaling: a randomized controlled trial. Cell Metab. 2019;29:844–855.e3. https://doi.org/10.1016/j.cmet.2018.12.007.
15. Kawasaki N, Asada R, Saito A, et al. Obesity-induced endoplasmic reticulum stress causes chronic inflammation in adipose tissue. Sci Rep. 2012;2:799. https://doi.org/10.1038/srep00799.
16. Zcan UO, Cao Q, Yilmaz E, et al. Endoplasmic reticulum stress links obesity, insulin action, and type 2 diabetes. Science. 2004;306(5695):457–61.

17. Elsayed NA, Aleppo G, Aroda VR, et al. 8. Obesity and weight management for the prevention and treatment of type 2 diabetes: standards of care in diabetes—2023. Diabetes Care. 2023;46:S128–39. https://doi.org/10.2337/dc23-S008.
18. Blüher M. Metabolically healthy obesity. Endocr Rev. 2020;41:405–20.
19. Caleyachetty R, Thomas GN, Toulis KA, et al. Metabolically healthy obese and incident cardiovascular disease events among 3.5 million men and women. JACC. 2017;70(12):1429–37.
20. Mongraw-Chaffin M, Foster MC, Anderson CAM, et al. Metabolically healthy obesity, transition to metabolic syndrome, and cardiovascular risk. J Am Coll Cardiol. 2018;71:1857–65. https://doi.org/10.1016/j.jacc.2018.02.055.
21. Stephenson J, Smith CM, Kearns B, et al. The association between obesity and quality of life: a retrospective analysis of a large-scale population-based cohort study. BMC Public Health. 2021;21(1):1990. https://doi.org/10.1186/s12889-021-12009-8.
22. Robinson E, Roberts C, Vainik U, Jones A. The psychology of obesity: an umbrella review and evidence-based map of the psychological correlates of heavier body weight. Neurosci Biobehav Rev. 2020;119:468–80.
23. Forhan M, Gill SV. Obesity, functional mobility and quality of life. Best Pract Res Clin Endocrinol Metab. 2013;27:129–37.
24. American Diabetes Association. Good to know: diabetes symptoms and tests. Clin Diabetes. 2020;38(1):108. https://doi.org/10.2337/cd20-pe01.
25. American Diabetes Association Professional Practice Committee. 2. Classification and diagnosis of diabetes: standards of medical care in diabetes—2022. Diabetes Care. 2022;45:S17–38. https://doi.org/10.2337/dc22-S002.
26. Andréasson K, Edqvist J, Adiels M, et al. Body mass index in adolescence, risk of type 2 diabetes and associated complications: a nationwide cohort study of men. EClinicalMedicine. 2022;46:101356.
27. Abdullah A, Stoelwinder J, Shortreed S, et al. The duration of obesity and the risk of type 2 diabetes. Public Health Nutr. 2011;14:119–26. https://doi.org/10.1017/S1368980010001813.
28. Kodama S, Horikawa C, Fujihara K, et al. Quantitative relationship between body weight gain in adulthood and incident type 2 diabetes: a meta-analysis. Obes Rev. 2014;15:202–14.
29. Feller S, Boeing H, Pischon T. Body-mass-index, taillenumfang und risiko für diabetes mellitus typ 2: Konsequenzen für den medizinischen alltag. Dtsch Arztebl. 2010;107:470–6. https://doi.org/10.3238/arztebl.2010.0470.
30. Jayedi A, Janbozorgi N, Djafarian K, et al. Dietary networks identified by Gaussian graphical model and general and abdominal obesity in adults. Nutr J. 2021;20:1–12. https://doi.org/10.1186/s12937-021-00746-w.
31. Gastaldelli A, Miyazaki Y, Pettiti M, et al. Metabolic effects of visceral fat accumulation in type 2 diabetes. J Clin Endocrinol Metab. 2002;87:5098–103. https://doi.org/10.1210/jc.2002-020696.
32. Cusi K. Role of obesity and lipotoxicity in the development of nonalcoholic steatohepatitis: pathophysiology and clinical implications. Gastroenterology. 2012;142(4):711–725.e6. https://doi.org/10.1053/j.gastro.2012.02.003.
33. Donnelly KL, Smith CI, Schwarzenberg SJ, et al. Sources of fatty acids stored in liver and secreted via lipoproteins in patients with nonalcoholic fatty liver disease. J Clin Invest. 2005;115:1343–51. https://doi.org/10.1172/JCI23621.
34. Lyu K, Zhang Y, Zhang D, et al. A membrane-bound diacylglycerol species induces PKCє-mediated hepatic insulin resistance. Cell Metab. 2020;32:654–664.e5. https://doi.org/10.1016/j.cmet.2020.08.001.
35. Samuel VT, Liu ZX, Wang A, et al. Inhibition of protein kinase Cε prevents hepatic insulin resistance in nonalcoholic fatty liver disease. J Clin Invest. 2007;117:739–45. https://doi.org/10.1172/JCI30400.
36. Kumashiro N, Erion DM, Zhang D, et al. Cellular mechanism of insulin resistance in nonalcoholic fatty liver disease. Proc Natl Acad Sci USA. 2011;108:16381–5. https://doi.org/10.1073/pnas.1113359108.

37. Szendroedi J, Yoshimura T, Phielix E, et al. Role of diacylglycerol activation of PKCθ in lipid-induced muscle insulin resistance in humans. Proc Natl Acad Sci USA. 2014;111:9597–602. https://doi.org/10.1073/pnas.1409229111.
38. Perry RJ, Camporez JPG, Kursawe R, et al. Hepatic acetyl CoA links adipose tissue inflammation to hepatic insulin resistance and type 2 diabetes. Cell. 2015;160:745–58. https://doi.org/10.1016/j.cell.2015.01.012.
39. Sylow L, Tokarz VL, Richter EA, Klip A. The many actions of insulin in skeletal muscle, the paramount tissue determining glycemia. Cell Metab. 2021;33:758–80.
40. Defronzo RA, Gunnarsson R, Bjorkman O, et al. Effects of insulin on peripheral and splanchnic glucose metabolism in noninsulin-dependent (type 11) diabetes mellitus. J Clin Invest. 1985;76(1):149–55.
41. Tabák AG, Jokela M, Akbaraly TN, et al. Trajectories of glycaemia, insulin sensitivity, and insulin secretion before diagnosis of type 2 diabetes: an analysis from the Whitehall II study. Lancet. 2009;373(9682):2215–21. https://doi.org/10.1016/S0140-6736(09)60619-X.
42. Speight J, Holmes-Truscott E, Garza M, et al. Bringing an end to diabetes stigma and discrimination: an international consensus statement on evidence and recommendations. Lancet Diabetes Endocrinol. 2024;12(1):61–82. https://doi.org/10.1016/S2213-8587(23)00347-9.
43. Fowler GC, Vasudevan DA. Type 2 diabetes mellitus: managing hemoglobin A 1c and beyond. South Med J. 2010;103(9):911–6.
44. Tomic D, Shaw JE, Magliano DJ. The burden and risks of emerging complications of diabetes mellitus. Nat Rev Endocrinol. 2022;18:525–39.
45. Lundeen EA, Burke-Conte Z, Rein DB, et al. Prevalence of diabetic retinopathy in the US in 2021. JAMA Ophthalmol. 2023;141:747–54. https://doi.org/10.1001/jamaophthalmol.2023.2289.
46. Fong DS, Aiello LP, Ferris FL, Klein R. Diabetic retinopathy. Diabetes Care. 2004;27(10):2540–53.
47. Elafros MA, Andersen H, Bennett DL, et al. Towards prevention of diabetic peripheral neuropathy: clinical presentation, pathogenesis, and new treatments. Lancet Neurol. 2022;21:922–36.
48. Selby NM, Taal MW. An updated overview of diabetic nephropathy: diagnosis, prognosis, treatment goals and latest guidelines. Diabetes Obes Metab. 2020;22:3–15.
49. Matsushita K, Coresh J, Sang Y, et al. Estimated glomerular filtration rate and albuminuria for prediction of cardiovascular outcomes: a collaborative meta-analysis of individual participant data. Lancet Diabetes Endocrinol. 2015;3:514–25. https://doi.org/10.1016/S2213-8587(15)00040-6.
50. Magkos F, Yannakoulia M, Chan JL, Mantzoros CS. Management of the metabolic syndrome and type 2 diabetes through lifestyle modification. Annu Rev Nutr. 2009;29:223–56.
51. Schernthaner G, Schernthaner GH. The right place for metformin today. Diabetes Res Clin Pract. 2020;159:107946.
52. Hirsch IB, Gaudiani LM. Using insulin to treat poorly controlled type 2 diabetes in 2020. JAMA. 2020;323:2419–20.
53. Alberti KGMM, Eckel RH, Grundy SM, et al. Harmonizing the metabolic syndrome: a joint interim statement of the International Diabetes Federation Task Force on Epidemiology and Prevention; National Heart, Lung, and Blood Institute; American Heart Association; World Heart Federation; International Atherosclerosis Society; and International Association for the Study of Obesity. Circulation. 2009;120:1640–5.
54. Ross R, Neeland IJ, Yamashita S, et al. Waist circumference as a vital sign in clinical practice: a consensus statement from the IAS and ICCR working group on visceral obesity. Nat Rev Endocrinol. 2020;16:177–89. https://doi.org/10.1038/s41574-019-0310-7.
55. Liu J, Fox CS, Hickson DMA, et al. Impact of abdominal visceral and subcutaneous adipose tissue on cardiometabolic risk factors: the Jackson Heart Study. J Clin Endocrinol Metab. 2010;95:5419–26. https://doi.org/10.1210/jc.2010-1378.

56. Shah RV, Murthy VL, Abbasi SA, et al. Visceral adiposity and the risk of metabolic syndrome across body mass index the MESA study. JACC Cardiovasc Imaging. 2014;7(12):1221–35.
57. Denke MA, Sempos CT, Grundy SM. Excess body weight an underrecognized contributor to high blood cholesterol levels in White American men. Arch Intern Med. 1993;153(9):1093–103.
58. Hall JE, Do Carmo JM, Da Silva AA, et al. Obesity-induced hypertension: interaction of neurohumoral and renal mechanisms. Circ Res. 2015;116:991–1006. https://doi.org/10.1161/CIRCRESAHA.116.305697.
59. Shihab HM, Meoni LA, Chu AY, et al. Body mass index and risk of incident hypertension over the life course: the Johns Hopkins precursors study. Circulation. 2012;126:2983–9. https://doi.org/10.1161/CIRCULATIONAHA.112.117333.
60. Stefan N, Schulze MB. Metabolic health and cardiometabolic risk clusters: implications for prediction, prevention, and treatment. Lancet Diabetes Endocrinol. 2023;11:426–40.
61. Engeli S, Gorzelniak K, Kreutz R, et al. Co-expression of renin-angiotensin system genes in human adipose tissue. J Hypertens. 1999;17(4):555–60. https://doi.org/10.1097/00004872-199917040-00014.
62. Giacchetti G, Faloia E, Mariniello B, et al. Overexpression of the renin-angiotensin system in human visceral adipose tissue in normal and overweight subjects. Am J Hypertens. 2002;15:381–8.
63. Harte A, McTernan P, Chetty R, et al. Insulin-mediated upregulation of the renin angiotensin system in human subcutaneous adipocytes is reduced by rosiglitazone. Circulation. 2005;111:1954–61. https://doi.org/10.1161/01.CIR.0000161954.17870.5D.
64. Gálvez-Prieto B, Bolbrinker J, Stucchi P, et al. Comparative expression analysis of the renin—angiotensin system components between white and brown perivascular adipose tissue. J Endocrinol. 2008;197:55–64. https://doi.org/10.1677/JOE-07-0284.
65. Simonds SE, Pryor JT, Ravussin E, et al. Leptin mediates the increase in blood pressure associated with obesity. Cell. 2014;159:1404–16. https://doi.org/10.1016/j.cell.2014.10.058.
66. Imatoh T, Miyazaki M, Momose Y, et al. Adiponectin levels associated with the development of hypertension: a prospective study. Hypertens Res. 2008;31(2):229–33. https://doi.org/10.1291/hypres.31.229.
67. Ohashi K, Kihara S, Ouchi N, et al. Adiponectin replenishment ameliorates obesity-related hypertension. Hypertension. 2006;47:1108–16. https://doi.org/10.1161/01.HYP.0000222368.43759.a1.
68. Sutton-Tyrrell K, Newman A, Simonsick EM, et al. Aortic stiffness is associated with visceral adiposity in older adults enrolled in the study of health, aging, and body composition. Hypertension. 2001;38(3):429–33.
69. Wildman RP, Mackey RH, Bostom A, et al. Measures of obesity are associated with vascular stiffness in young and older adults. Hypertension. 2003;42:468–73. https://doi.org/10.1161/01.HYP.0000090360.78539.CD.
70. Jia G, Habibi J, Aroor AR, et al. Endothelial mineralocorticoid receptor mediates diet-induced aortic stiffness in females. Circ Res. 2016;118:935–43. https://doi.org/10.1161/CIRCRESAHA.115.308269.
71. Sugerman H, Windsor A, Bessos M, Wolfe L. Intra-abdominal pressure, sagittal abdominal diameter and obesity comorbidity. J Intern Med. 1997;241:71–9. https://doi.org/10.1046/j.1365-2796.1997.89104000.x.
72. Chughtai HL, Morgan TM, Rocco M, et al. Renal sinus fat and poor blood pressure control in middle-aged and elderly individuals at risk for cardiovascular events. Hypertension. 2010;56:901–6. https://doi.org/10.1161/HYPERTENSIONAHA.110.157370.
73. Foster MC, Hwang SJ, Porter SA, et al. Fatty kidney, hypertension, and chronic kidney disease: the Framingham heart study. Hypertension. 2011;58:784–90. https://doi.org/10.1161/HYPERTENSIONAHA.111.175315.
74. Boorsma EM, ter Maaten JM, Voors AA, van Veldhuisen DJ. Renal compression in heart failure: the renal tamponade hypothesis. JACC Heart Fail. 2022;10:175–83.

75. Pan A, Keum N, Okereke OI, et al. Bidirectional association between depression and metabolic syndrome: a systematic review and meta-analysis of epidemiological studies. Diabetes Care. 2012;35:1171–80. https://doi.org/10.2337/dc11-2055.
76. Vetter ML, Wadden TA, Lavenberg J, et al. Relation of health-related quality of life to metabolic syndrome, obesity, depression and comorbid illnesses. Int J Obes. 2011;35:1087–94. https://doi.org/10.1038/ijo.2010.230.
77. Lewington S, Clarke R, Qizilbash N, et al. Age-specific relevance of usual blood pressure to vascular mortality: a meta-analysis of individual data for one million adults in 61 prospective studies. Lancet. 2002;360:1903–13. https://doi.org/10.1016/S0140-6736(02)11911-8.
78. Nordestgaard BG, Varbo A. Lipids and cardiovascular disease 3 triglycerides and cardiovascular disease. Lancet. 2014;384(9943):626–35.
79. Rader DJ, Kees Hovingh G. Lipids and cardiovascular disease 2 HDL and cardiovascular disease. Lancet. 2014;384(9943):618–25.
80. Maron DJ, Fazio S, Linton MF. Current perspectives on statins. Circulation. 2000;101(2):207–13. https://doi.org/10.1161/01.cir.101.2.207.
81. Mangione CM, Barry MJ, Nicholson WK, et al. Statin use for the primary prevention of cardiovascular disease in adults: US Preventive Services Task Force recommendation statement. JAMA. 2022;328:746–53.
82. Whelton PK, Carey RM, Mancia G, et al. Harmonization of the American College of Cardiology/American Heart Association and European Society of Cardiology/European Society of Hypertension Blood Pressure/Hypertension guidelines: comparisons, reflections, and recommendations. Circulation. 2022;146:868–77.
83. Alefishat EA, Abu Farha RK, Al-Debei MM. Self-reported adherence among individuals at high risk of metabolic syndrome: effect of knowledge and attitude. Med Princ Pract. 2017;26:157–63. https://doi.org/10.1159/000453037.
84. Fiorenza CG, Chou SH, Mantzoros CS. Lipodystrophy: pathophysiology and advances in treatment. Nat Rev Endocrinol. 2011;7:137–50.
85. Garg A. Acquired and inherited lipodystrophies. N Engl J Med. 2004;12:1220–54.
86. Chiquette E, Oral EA, Garg A, et al. Estimating the prevalence of generalized and partial lipodystrophy: findings and challenges. Diabetes Metab Syndr Obes. 2017;10:375–83. https://doi.org/10.2147/dmso.s130810.
87. Boguslavsky RL, Stewart CL, Worman HJ. Nuclear lamin A inhibits adipocyte differentiation: implications for Dunnigan-type familial partial lipodystrophy. Hum Mol Genet. 2006;15:653–63. https://doi.org/10.1093/hmg/ddi480.
88. Corsa CAS, Walsh CM, Bagchi DP, et al. Adipocyte-specific deletion of lamin A/C largely models human familial partial lipodystrophy type 2. Diabetes. 2021;70:1970–84. https://doi.org/10.2337/DB20-1001.
89. Gonzaga-Jauregui C, Ge W, Staples J, et al. Clinical and molecular prevalence of lipodystrophy in an unascertained large clinical care cohort. Diabetes. 2020;69:249–58. https://doi.org/10.2337/db19-0447.
90. Misra A, Peethambaram A, Garg A. Clinical features and metabolic and autoimmune derangements in acquired partial lipodystrophy: report of 35 cases and review of the literature. Medicine. 2004;83:18–34.
91. Mallewa JE, Wilkins E, Vilar J, et al. HIV-associated lipodystrophy: a review of underlying mechanisms and therapeutic options. J Antimicrob Chemother. 2008;62:648–60.
92. Shlay JC, Sharma S, Peng G, et al. The effect of individual antiretroviral drugs on body composition in HIV-infected persons initiating highly active antiretroviral therapy for the Terry Beirn Community Programs for Clinical Research on AIDS (CPCRA) and the International Network for Strategic Initiatives in Global HIV Trials (INSIGHT). J Acquir Immune Defic Syndr. 2009;51(3):298–304.
93. Hadigan C, Borgonha S, Rabe J, et al. Increased rates of lipolysis among human immunodeficiency virus-infected men receiving highly active antiretroviral therapy. Metabolism. 2002;51:1143–7. https://doi.org/10.1053/meta.2002.34704.

94. Leyes P, Martínez E, De Talló Forga M. Use of diet, nutritional supplements and exercise in HIV-infected patients receiving combination antiretroviral therapies: a systematic review. Antivir Ther. 2008;3(2):149–59.
95. Hadigan C, Corcoran C, Basgoz N, et al. Metformin in the treatment of HIV lipodystrophy syndrome a randomized controlled trial. JAMA. 2000;284(4):472–7.
96. Moreau F, Boullu-Sanchis S, Vigouroux C, et al. Efficacy of pioglitazone in familial partial lipodystrophy of the Dunnigan type: a case report. Diabetes Metab. 2007;33:385–9. https://doi.org/10.1016/j.diabet.2007.04.005.
97. Fichtenbaum CJ, Gerber JG, Rosenkranz SL, et al. Pharmacokinetic interactions between protease inhibitors and statins in HIV seronegative volunteers: ACTG study A5047. AIDS. 2000;16(4):569–77.
98. James J, Carruthers A, Carruthers J, et al. HIV-associated facial lipoatrophy. Dermatologic Surg. 2002;28(11):979–86.
99. Jones DH, Carruthers A, Brody HJ, et al. Ten-year and beyond follow-up after treatment with highly purified liquid-injectable silicone for HIV-associated facial lipoatrophy: a report of 164 patients. Dermatologic Surg. 2019;45:941–8. https://doi.org/10.1097/DSS.0000000000001889.
100. Mathieu P, Poirier P, Pibarot P, et al. Visceral obesity the link among inflammation, hypertension, and cardiovascular disease. Hypertension. 2009;53:577–84.
101. Van Gaal LF, Mertens IL, De Block CE. Mechanisms linking obesity with cardiovascular disease. Nature. 2006;444(7121):875–80. https://doi.org/10.1038/nature05487.
102. Després J-P, Lemieux I. Abdominal obesity and metabolic syndrome. Nature. 2006;444(7121):881–7. https://doi.org/10.1038/nature05488.
103. Ohlson L-O, Larsson B, Svardsudd K, et al. The influence of body fat distribution on the incidence of diabetes mel I it us 13.5 years of follow-up of the participants in the study of men born in 1913. Diabetes. 1985;34(10):1055–8.
104. Pouliot M-C, Desprk J-P, Lemieux S, et al. Waist circumference and abdominal sagittal diameter: best simple anthropometric indexes of abdominal visceral adipose tissue accumulation and related cardiovascular risk in men and women. Am J Cardiol. 1994;73(7):460–8.
105. Wildman RP, Muntner P, Reynolds K, et al. The obese without cardiometabolic risk factor clustering and the normal weight with cardiometabolic risk factor clustering. Arch Intern Med. 2008;168(15):1617–24.
106. Manolopoulos KN, Karpe F, Frayn KN. Gluteofemoral body fat as a determinant of metabolic health. Int J Obes. 2010;34:949–59.
107. Balkau B, Deanfield JE, Després JP, et al. International day for the evaluation of abdominal obesity (IDEA): a study of waist circumference, cardiovascular disease, and diabetes mellitus in 168 000 primary care patients in 63 countries. Circulation. 2007;116:1942–51. https://doi.org/10.1161/CIRCULATIONAHA.106.676379.
108. Larsson B, Svärdsudd K, Welin L, et al. Abdominal adipose tissue distribution, obesity, and risk of cardiovascular disease and death: 13 year follow up of participants in the study of men born in 1913. Clin Res. 1984;288:1401.
109. Chrysant SG, Chrysant GS. New insights into the true nature of the obesity paradox and the lower cardiovascular risk. J Am Soc Hypertens. 2013;7:85–94.
110. Butt JH, Petrie MC, Jhund PS, et al. Anthropometric measures and adverse outcomes in heart failure with reduced ejection fraction: revisiting the obesity paradox. Eur Heart J. 2023;44:1136–53. https://doi.org/10.1093/eurheartj/ehad083.
111. Powell-Wiley TM, Poirier P, Burke LE, et al. Obesity and cardiovascular disease a scientific statement from the American Heart Association. Circulation. 2021;143:E984–E1010.
112. Hicks CW, Yang C, Ndumele CE, et al. Associations of obesity with incident hospitalization related to peripheral artery disease and critical limb ischemia in the aric study. J Am Heart Assoc. 2018;7(16):e008644. https://doi.org/10.1161/JAHA.118.008644.

113. Guo Y, Yue XJ, Li HH, Song ZX, Yan HQ, Zhang P, Gui YK, Chang L, Li T. Overweight and obesity in young adulthood and the risk of stroke: a meta-analysis. J Stroke Cerebrovasc Dis. 2016;25:2995–3004. https://doi.org/10.1016/j.jstrokecerebrovasdis.2016.08.018.
114. Bogers RP, Bemelmans WJ, Hoogenveen RT, Boshuizen HC, Woodward M, Knekt P, van Dam RM, Hu FB, Visscher TL, Menotti A, Thorpe RJ Jr, Jamrozik K, Calling S, Strand BH, Shipley MJ, BMI-CHD Collaboration Investigators. Association of overweight with increased risk of coronary heart disease partly independent of blood pressure and cholesterol levels: a meta-analysis of 21 cohort studies including more than 300 000 persons. Arch Intern Med. 2007;167(16):1720–8.
115. Hubert HB, Mcnamara PM, Castelli WP. Obesity as an independent risk factor for cardiovascular disease: a 26-year follow-up of participants in the Framingham heart study. Circulation. 1983;67(5):968–77.
116. Horn JW, Feng T, Morkedal B, et al. Obesity and risk for first ischemic stroke depends on metabolic syndrome: the HUNT study. Stroke. 2021;52:3555–61. https://doi.org/10.1161/STROKEAHA.120.033016.
117. Jaakonmäki N, Zedde M, Sarkanen T, et al. Obesity and the risk of cryptogenic ischemic stroke in young adults. J Stroke Cerebrovasc Dis. 2022;31(5):106380. https://doi.org/10.1016/j.jstrokecerebrovasdis.2022.106380.
118. Mørkedal B, Vatten LJ, Romundstad PR, et al. Risk of myocardial infarction and heart failure among metabolically healthy but obese individuals: HUNT (Nord-Trøndelag Health Study), Norway. J Am Coll Cardiol. 2014;63:1071–8. https://doi.org/10.1016/j.jacc.2013.11.035.
119. Reijrink M, de Boer SA, Spoor DS, et al. Visceral adipose tissue volume is associated with premature atherosclerosis in early type 2 diabetes mellitus independent of traditional risk factors. Atherosclerosis. 2019;290:87–93. https://doi.org/10.1016/j.atherosclerosis.2019.09.016.
120. Iacobellis G. Epicardial adipose tissue in contemporary cardiology. Nat Rev Cardiol. 2022;19:593–606.
121. Ageno W, Becattini C, Brighton T, et al. Cardiovascular risk factors and venous thromboembolism: a meta-analysis. Circulation. 2008;117:93–102. https://doi.org/10.1161/CIRCULATIONAHA.107.709204.
122. Borch KH, Brækkan SK, Mathiesen EB, et al. Abdominal obesity is essential for the risk of venous thromboembolism in the metabolic syndrome: the Tromsø study. J Thromb Haemost. 2009;7:739–45. https://doi.org/10.1111/j.1538-7836.2008.03234.x.
123. Parkin L, Sweetland S, Balkwill A, et al. Body mass index, surgery, and risk of venous thromboembolism in middle-aged women a cohort study. Circulation. 2012;125(15):1897–904. https://doi.org/10.1161/CIRCULATIONAHA.
124. White RH, Gettner S, Newman JM, et al. Predictors of rehospitalization for symptomatic venous thromboembolism after total hip arthroplasty. NEJM. 2000;343(24):1758–64.
125. Yang G, De Staercke C, Hooper WC. The effects of obesity on venous thromboembolism: a review. Open J Prev Med. 2012;02:499–509. https://doi.org/10.4236/ojpm.2012.24069.
126. Barbar S, Noventa F, Rossetto V, et al. A risk assessment model for the identification of hospitalized medical patients at risk for venous thromboembolism: the Padua Prediction Score. J Thromb Haemost. 2010;8:2450–7. https://doi.org/10.1111/j.1538-7836.2010.04044.x.
127. Kenchaiah S, Evans JC, Levy D, et al. Obesity and the risk of heart failure. Engl J. 2002;347:305–13.
128. Savji N, Meijers WC, Bartz TM, et al. The Association of Obesity and Cardiometabolic Traits with Incident HFpEF and HFrEF. JACC Heart Fail. 2018;6:701–9. https://doi.org/10.1016/j.jchf.2018.05.018.
129. Kenchaiah S, Ding J, Carr JJ, et al. Pericardial fat and the risk of heart failure. J Am Coll Cardiol. 2021;77:2638–52. https://doi.org/10.1016/j.jacc.2021.04.003.
130. van Woerden G, Gorter TM, Westenbrink BD, et al. Epicardial fat in heart failure patients with mid-range and preserved ejection fraction. Eur J Heart Fail. 2018;20:1559–66. https://doi.org/10.1002/ejhf.1283.

131. Borlaug BA, Jensen MD, Kitzman DW, et al. Obesity and heart failure with preserved ejection fraction: new insights and pathophysiological targets. Cardiovasc Res. 2022;118:3434–50.
132. Obokata M, Reddy YNV, Pislaru SV, et al. Evidence supporting the existence of a distinct obese phenotype of heart failure with preserved ejection fraction. Circulation. 2017;136:6–19. https://doi.org/10.1161/CIRCULATIONAHA.116.026807.
133. Gorter TM, van Woerden G, Rienstra M, et al. Epicardial adipose tissue and invasive hemodynamics in heart failure with preserved ejection fraction. JACC Heart Fail. 2020;8:667–76. https://doi.org/10.1016/j.jchf.2020.06.003.
134. Koepp KE, Obokata M, Reddy YNV, et al. Hemodynamic and functional impact of epicardial adipose tissue in heart failure with preserved ejection fraction. JACC Heart Fail. 2020;8:657–66. https://doi.org/10.1016/j.jchf.2020.04.016.
135. Sorimachi H, Obokata M, Takahashi N, et al. Pathophysiologic importance of visceral adipose tissue in women with heart failure and preserved ejection fraction. Eur Heart J. 2021;42:1595–605.
136. Chatterjee NA, Giulianini F, Geelhoed B, et al. Genetic obesity and the risk of atrial fibrillation: causal estimates from Mendelian randomization. Circulation. 2017;135:741–54. https://doi.org/10.1161/CIRCULATIONAHA.116.024921.
137. Nalliah CJ, Sanders P, Kottkamp H, Kalman JM. The role of obesity in atrial fibrillation. Eur Heart J. 2016;37:1565–72. https://doi.org/10.1093/eurheartj/ehv486.
138. Wang TJ, Parise H, Levy D, et al. Obesity and the risk of new-onset atrial fibrillation. JAMA. 2004;292(20):2471–7.
139. Camm CF, Lacey B, Massa MS, et al. Independent effects of adiposity measures on risk of atrial fibrillation in men and women: a study of 0.5 million individuals. Int J Epidemiol. 2022;51:984–95. https://doi.org/10.1093/ije/dyab184.
140. Frost L, Benjamin EJ, Fenger-Grøn M, et al. Body fat, body fat distribution, lean body mass and atrial fibrillation and flutter. A Danish cohort study. Obesity. 2014;22:1546–52. https://doi.org/10.1002/oby.20706.
141. Sandhu RK, Conen D, Tedrow UB, et al. Predisposing factors associated with development of persistent compared with paroxysmal atrial fibrillation. J Am Heart Assoc. 2014;3(3):e000916. https://doi.org/10.1161/JAHA.114.000916.
142. Tsang TSM, Barnes ME, Miyasaka Y, et al. Obesity as a risk factor for the progression of paroxysmal to permanent atrial fibrillation: a longitudinal cohort study of 21 years. Eur Heart J. 2008;29:2227–33. https://doi.org/10.1093/eurheartj/ehn324.
143. Tabaja C, Younis A, Santageli P, et al. Impact of obesity on catheter ablation of atrial fibrillation: patient characteristics, procedural complications, outcomes, and quality of life. J Cardiovasc Electrophysiol. 2023;34:1648–57. https://doi.org/10.1111/jce.15987.
144. Bekwelem W, Misialek JR, Konety S, et al. Echocardiographic measures of cardiac structure and function are associated with risk of atrial fibrillation in blacks: the Atherosclerosis Risk in Communities (ARIC) study. PLoS One. 2014;9(10):e110111. https://doi.org/10.1371/journal.pone.0110111.
145. Stritzke J, Markus MRP, Duderstadt S, et al. The aging process of the heart: obesity is the main risk factor for left atrial enlargement during aging. The MONICA/KORA (monitoring of trends and determinations in cardiovascular disease/cooperative research in the region of Augsburg) study. J Am Coll Cardiol. 2009;54:1982–9. https://doi.org/10.1016/j.jacc.2009.07.034.
146. Batal O, Schoenhagen P, Shao M, et al. Left atrial epicardial adiposity and atrial fibrillation. Circ Arrhythm Electrophysiol. 2010;3:230–6. https://doi.org/10.1161/CIRCEP.110.957241.
147. Friedman DJ, Wang N, Meigs JB, et al. Pericardial fat is associated with atrial conduction: the Framingham heart study. J Am Heart Assoc. 2014;3(2):e000477. https://doi.org/10.1161/JAHA.113.000477.
148. Gawałko M, Saljic A, Li N, et al. Adiposity-associated atrial fibrillation: molecular determinants, mechanisms, and clinical significance. Cardiovasc Res. 2023;119:614–30.

149. Heckbert SR, Wiggins KL, Blackshear C, et al. Pericardial fat volume and incident atrial fibrillation in the Multi-Ethnic Study of Atherosclerosis and Jackson Heart Study. Obesity. 2017;25:1115–21. https://doi.org/10.1002/oby.21835.
150. Thanassoulis G, Massaro JM, O'Donnell CJ, et al. Pericardial fat is associated with prevalent atrial fibrillation: the Framingham heart study. Circ Arrhythm Electrophysiol. 2010;3:345–50. https://doi.org/10.1161/CIRCEP.109.912055.
151. Ge L, Sadeghirad B, Ball GDC, et al. Comparison of dietary macronutrient patterns of 14 popular named dietary programmes for weight and cardiovascular risk factor reduction in adults: systematic review and network meta-analysis of randomised trials. BMJ. 2020;369:m696. https://doi.org/10.1136/bmj.m696.
152. Sharples AJ, Mahawar K. Systematic review and meta-analysis of randomised controlled trials comparing long-term outcomes of Roux-En-Y gastric bypass and sleeve gastrectomy. Obes Surg. 2020;30:664–72. https://doi.org/10.1007/s11695-019-04235-2.
153. Bellicha A, van Baak MA, Battista F, et al. Effect of exercise training on weight loss, body composition changes, and weight maintenance in adults with overweight or obesity: an overview of 12 systematic reviews and 149 studies. Obes Rev. 2021;22 Suppl 4(Suppl 4):e13256. https://doi.org/10.1111/obr.13256.
154. Knowler WC, Barrett-Connor E, Fowler SE, Hamman RF, Lachin JM, Walker EA, Nathan DM, Diabetes Prevention Program Research Group. Reduction in the incidence of type 2 diabetes with lifestyle intervention or metformin. N Engl J Med. 2002;346(6):393–403. https://doi.org/10.1056/NEJMoa012512.
155. Look AHEAD Research Group. Cardiovascular effects of intensive lifestyle intervention in type 2 diabetes. N Engl J Med. 2013;369:145–54. https://doi.org/10.1056/NEJMoa1212914.
156. The Look AHEAD Research Group. Association of the magnitude of weight loss and changes in physical fitness with long-term cardiovascular disease outcomes in overweight or obese people with type 2 diabetes: a post-hoc analysis of the Look AHEAD randomised clinical trial. Lancet Diabetes Endocrinol. 2016;4:913–21. https://doi.org/10.1016/S2213-8587(16)30162-0.
157. Pownall HJ, Bray GA, Wagenknecht LE, et al. Changes in body composition over 8 years in a randomized trial of a lifestyle intervention: the Look AHEAD study. Obesity. 2015;23:565–72. https://doi.org/10.1002/oby.21005.
158. Eisenberg D, Shikora SA, Aarts E, et al. 2022 American Society for Metabolic and Bariatric Surgery (ASMBS) and International Federation for the Surgery of Obesity and Metabolic Disorders (IFSO): indications for metabolic and bariatric surgery. Surg Obes Relat Dis. 2022;18:1345–56. https://doi.org/10.1016/j.soard.2022.08.013.
159. van Veldhuisen SL, Gorter TM, van Woerden G, et al. Bariatric surgery and cardiovascular disease: a systematic review and meta-analysis. Eur Heart J. 2022;43:1955–69. https://doi.org/10.1093/eurheartj/ehac069.
160. Onakpoya IJ, Lee JJ, Mahtani KR, et al. Naltrexone–bupropion (Mysimba) in management of obesity: a systematic review and meta-analysis of unpublished clinical study reports. Br J Clin Pharmacol. 2020;86:646–67.
161. Le Roux CW, Astrup AV, Fujioka K, et al. 3 years of liraglutide versus placebo for type 2 diabetes risk reduction and weight management in individuals with prediabetes: a randomised, double-blind trial. Lancet. 2017;389:1399–409. https://doi.org/10.1016/S0140-6736(17)30069-7.
162. Pi-Sunyer X, Astrup A, Fujioka K, et al. A randomized, controlled trial of 3.0 mg of liraglutide in weight management. N Engl J Med. 2015;373:11–22. https://doi.org/10.1056/nejmoa1411892.
163. Davies M, Færch L, Jeppesen OK, et al. Semaglutide 2·4 mg once a week in adults with overweight or obesity, and type 2 diabetes (STEP 2): a randomised, double-blind, double-dummy, placebo-controlled, phase 3 trial. Lancet. 2021;397:971–84. https://doi.org/10.1016/S0140-6736(21)00213-0.
164. Garvey WT, Batterham RL, Bhatta M, et al. Two-year effects of semaglutide in adults with overweight or obesity: the STEP 5 trial. Nat Med. 2022;28:2083–91. https://doi.org/10.1038/s41591-022-02026-4.

165. Wilding JPH, Batterham RL, Calanna S, et al. Once-weekly semaglutide in adults with overweight or obesity. N Engl J Med. 2021;384:989–1002. https://doi.org/10.1056/nejmoa2032183.
166. Lincoff AM, Brown-Frandsen K, Colhoun HM, et al. Semaglutide and cardiovascular outcomes in obesity without diabetes. N Engl J Med. 2023;389:2221–32. https://doi.org/10.1056/nejmoa2307563.
167. Kosiborod MN, Abildstrøm SZ, Borlaug BA, et al. Semaglutide in patients with heart failure with preserved ejection fraction and obesity. N Engl J Med. 2023;389:1069–84. https://doi.org/10.1056/nejmoa2306963.
168. Borlaug BA, Kitzman DW, Davies MJ, et al. Semaglutide in HFpEF across obesity class and by body weight reduction: a prespecified analysis of the STEP-HFpEF trial. Nat Med. 2023;29:2358–65. https://doi.org/10.1038/s41591-023-02526-x.
169. Garvey WT, Frias JP, Jastreboff AM, et al. Tirzepatide once weekly for the treatment of obesity in people with type 2 diabetes (SURMOUNT-2): a double-blind, randomised, multicentre, placebo-controlled, phase 3 trial. Lancet. 2023;402:613–26. https://doi.org/10.1016/S0140-6736(23)01200-X.
170. Jastreboff AM, Aronne LJ, Ahmad NN, et al. Tirzepatide once weekly for the treatment of obesity. N Engl J Med. 2022;387:205–16. https://doi.org/10.1056/nejmoa2206038.
171. Wadden TA, Chao AM, Machineni S, et al. Tirzepatide after intensive lifestyle intervention in adults with overweight or obesity: the SURMOUNT-3 phase 3 trial. Nat Med. 2023;29:2909–18. https://doi.org/10.1038/s41591-023-02597-w.
172. Sanguankeo A, Lazo M, Upala S, et al. Effects of visceral adipose tissue reduction on CVD risk factors independent of weight loss: the Look AHEAD study. Endocr Res. 2017;42:86–95. https://doi.org/10.1080/07435800.2016.1194856.
173. van Eekelen E, Geelen A, Alssema M, et al. Adherence to dietary guidelines in relation to visceral fat and liver fat in middle-aged men and women: the NEO study. Int J Obes. 2020;44:297–306. https://doi.org/10.1038/s41366-019-0441-x.
174. Favre L, Marino L, Roth A, et al. The reduction of visceral adipose tissue after Roux-en-Y gastric bypass is more pronounced in patients with impaired glucose metabolism. Obes Surg. 2018;28:4006–13. https://doi.org/10.1007/s11695-018-3455-x.
175. Frikke-Schmidt H, O'Rourke RW, Lumeng CN, et al. Does bariatric surgery improve adipose tissue function? Obes Rev. 2016;17:795–809.
176. Zhu E, Yang Y, Zhang J, et al. Liraglutide suppresses obesity and induces brown fat-like phenotype via soluble guanylyl cyclase mediated pathway in vivo and in vitro. Oncotarget. 2016;7:81077–89.
177. Armstrong MJ, Hull D, Guo K, et al. Glucagon-like peptide 1 decreases lipotoxicity in non-alcoholic steatohepatitis. J Hepatol. 2016;64:262–4.
178. Zhu R, Chen S. Proteomic analysis reveals semaglutide impacts lipogenic protein expression in epididymal adipose tissue of obese mice. Front Endocrinol (Lausanne). 2023;14:1095432. https://doi.org/10.3389/fendo.2023.1095432.
179. Launbo N, Zobel EH, von Scholten BJ, et al. Targeting epicardial adipose tissue with exercise, diet, bariatric surgery or pharmaceutical interventions: a systematic review and meta-analysis. Obes Rev. 2021;22(1):e13136. https://doi.org/10.1111/obr.13136.
180. Malavazos AE, Meregalli C, Sorrentino F, et al. Semaglutide therapy decreases epicardial fat inflammation and improves psoriasis severity in patients affected by abdominal obesity and type-2 diabetes. Endocrinol Diabetes Metab Case Rep. 2023;2023(3):23-0017. https://doi.org/10.1530/EDM-23-0017.
181. Masson W, Lavalle-Cobo A, Nogueira JP. Effect of sglt2-inhibitors on epicardial adipose tissue: a meta-analysis. Cells. 2021;10:2050. https://doi.org/10.3390/cells10082150.
182. Yaribeygi H, Maleki M, Butler AE, et al. Sodium-glucose co-transporter-2 inhibitors and epicardial adiposity. Eur J Pharm Sci. 2023;180:106322. https://doi.org/10.1016/j.ejps.2022.106322.

# Chapter 6
# Adipose Tissue Angiogenesis and Clinical Implications

**Branislava Ilincic, Mia Manojlovic, Iva Barjaktarovic, Edita Stokic, and Esma R. Isenovic**

## Abbreviations

| | |
|---|---|
| AARP | Anti-angiogenic recombinant protein |
| ACs | Adipocytes |
| AT | Adipose tissue |
| BAT | Brown adipose tissue |
| CSD1 | Stearoyl-co-enzyme A desaturase 1 |
| Dll4 | Delta-like 4 |
| DM | Diabetes mellitus |
| ECs | Endothelial cells |
| ED | Endothelial dysfunction |
| eNOS | Endothelial nitric oxide synthase |
| FABP | Fatty acid-binding protein |
| FFA | Free fatty acid |
| FGF21 | Fibroblast growth factor 21 |
| FOXO1 | Forkhead box protein O1 |

B. Ilincic · I. Barjaktarovic
Faculty of Medicine in Novi Sad, University of Novi Sad, Novi Sad, Serbia

Center of Laboratory Diagnostic, University Clinical Center of Vojvodina, Novi Sad, Serbia
e-mail: iva.barjaktarovic@mf.uns.ac.rs

M. Manojlovic · E. Stokic
Faculty of Medicine in Novi Sad, University of Novi Sad, Novi Sad, Serbia

Clinic for Endocrinology, Diabetes and Metabolic Disorders, University Clinical Center of Vojvodina, Novi Sad, Serbia
e-mail: mia.manojlovic@mf.uns.ac.rs; edita.stokic@mf.uns.ac.rs

E. R. Isenovic (✉)
Department of Radiobiology and Molecular Genetics, "VINČA" Institute of Nuclear Sciences—National Institute of the Republic of Serbia, University of Belgrade, Belgrade, Serbia

Stroke Monitoring and Diagnostic Division, AtheroPoint™, Roseville, CA, USA


D. Avtanski (ed.), *Adipose Tissue*, Contemporary Endocrinology,
https://doi.org/10.1007/978-3-031-72570-8_6

| | |
|---|---|
| GLUT1 | Glucose transporter 1 |
| GPIHB1 | Glycosylphosphatidylinositol-anchored high-density lipoprotein binding protein 1 |
| IGF-I | Insulin-like growth factor 1 |
| IR | Insulin resistance |
| LpL | Lipoprotein lipase |
| miRNA | microRNA |
| ncRNA | Non-coding RNA |
| NO | Nitric oxide |
| PDGF | Platelet-derived growth factor |
| PDGFRβ | Platelet-derived growth factor receptor-beta |
| SORT1 | Silent information regulator 1, *s.* Sirtuin 1 |
| T2DM | Type 2 diabetes mellitus |
| VEGF | Vascular endothelial growth factor |
| VLDL | Very low-density lipoprotein |
| VLDLR | Very low-density lipoprotein receptor |
| α-SMA | Alpha-smooth muscle actin |

## Introduction

### *Brief Overview of Adipose Tissue Angiogenesis*

Adipocytes (ACs) and endothelial cells (ECs) represent two closely interconnected cell lines, determining each other's functions through constant and intriguing interplay. As a cornerstone for these facts, we can consider the results of studies indicating the presence of precursor cells with dual differentiation capacity into both adipocytes and EC within the adipose tissue (AT) blood vessels [1, 2].

The vasculature of the AT can be considered an orchestrator of internal shifts. The initial signal for this process is believed to originate from both the ACs themselves during dynamic modulation, marked as tissue remodeling (hypertrophy and hyperplasia), and from the metabolic signals accompanying these alterations. New blood vessel formation represents the response of healthy AT to increased demand for oxygen and nutrients due to its expansion. This well-coordinated angiogenic response maintains adequate oxygenation and prevents hypoxia. However, in obesity, due to rapid tissue growth and an imbalance of anti-angiogenic and pro-angiogenic factors, insufficient vasculature becomes prominent, contributing to the onset of inflammation and hypoxia, ultimately leading to the formation of dysfunctional AT [3].

## *Importance of Understanding Adipose Tissue Vascularization*

Given these findings along with the shared pathophysiological grounds of broad-spectrum diseases involving dysfunctional AT, there is a distinct need for active investigation of its vascularization. This is aimed at identifying effective preventive strategies and therapeutic modalities [2, 4].

# Unique Vascularization of Adipose Tissue

## *Description of the Vascular Structure in Adipose Tissue*

The analogy of a multifaceted picture of AT can also be extended to the diverse subclasses of ECs, which vary in function and transcriptional profiles. The potential for constant communication between AT and ECs is based on the fact that every adipocyte in a state of physiological nourishment is supplied by at least one microvascular vessel. The unique vascularization of AT is reflected by the presence of non-permeable blood vessels, namely, continuous capillaries, whose structure is based on a continuous basement membrane and tight junctions between ECs. Thus, it is evident that molecular transport is a highly sophisticated process in this tissue, strictly controlled by transport systems, without possible paracellular leakage as a form of transport typical for ECs in other organs [2, 5].

## *Role of Vascularization in Adipose Tissue Function*

### Role of Vasculature in Targeted Nutrient Transport

The significant and distinctive role of AT ECs is evident in their regulation of the qualitative-quantitative influx of lipids, both during and between meals, into tissue depots [5].

#### Free Fatty Acid Uptake

A notable instance of paracrine communication between ECs and AT is exemplified by the uptake of triglyceride-rich lipoproteins from the bloodstream. This process is mediated by lipoprotein lipase (LpL), which is bound to the luminal endothelial surface via a protein serving its anchoring known as glycosylphosphatidylinositol-anchored high-density lipoprotein binding protein 1 (GPIHBP1), given that ECs do not synthesize this enzyme. The role of LpL is in forming transportable patterns of triglyceride-rich lipoproteins known as free fatty acids (FFAs) [5, 6].

### Trans-endothelial Vesicle-Mediated Transport

Recalling the fact that the AT endothelium is of a continuous type, the scientific debate regarding potential mechanisms for further transport of liberated FFA towards tissue depots has progressed from assumptions of ECs being filled with numerous pores to eventually identifying, via electron microscopy, a type of transport labelled as vesicular transcytosis. It has been determined that these cells possess many surface vesicles, varying in type, thereby maintaining the constancy of interstitial fluid and plasma transport. Caveolae are the most abundant vesicle type in the AT. Their formation is regulated by cavins 1 and 2 and caveolins 1 and 2 [5, 7]. They exhibit a wide range of roles, far beyond transport, as they are considered to participate in the process of angiogenesis, signal transduction, and nitric oxide (NO) production. Thus, their loss in large arteries is believed to protect against the development of atherosclerotic plaques owing to reduced lipid transcytosis. However, diminution of caveolae in adipocytes and microvasculature, leads to the onset of type 2 diabetes mellitus (T2DM), disrupting the capacity for lipid storage [5].

### Trans-endothelial Receptor-Mediated Transport

The role of individual proteins in receptor-mediated transendothelial transport remains unclear, but several ligand-binding proteins have been established. Within the domain of FFA transport, ECs expression of fatty acid-binding proteins (FABPs), scavenger receptor CD36, and fatty acid transport proteins (FATPs) 3 and 4 have been reported. It is also essential to recall the potential role of the very low-density lipoprotein (VLDL) receptor (VLDLR) in serving as a receptor for a few lipoproteins. Specifically, it is believed to promote both LpL-mediated release of FFA and vesicular transcytosis. It is speculated that ECs are expressed it in multiple tissues, including AT [5, 8, 9].

### Adipocytokines, FFAs, and Glucose Transport

The uniqueness of AT is also reflected in the fact that it not only takes up a large quantity of FFAs, but also releases them into the bloodstream. Conceptualizing AT as a sizeable endocrine gland is comparable to molecule release from other endocrine glands, which occurs via fenestrated capillaries. In contrast, AT is characterized by continuous capillaries; thus, metabolites are released from AT through direct release via trans-endothelial transport into the bloodstream or through the lymphatic system. An exciting interplay has been established between AT, ECs, and glucose. Specifically, AT takes up significant amounts of glucose, and ECs themselves, besides transporting glucose, utilize it as a metabolic substrate and fuel, given their low percentage of mitochondria. Glucose uptake by ECs is believed to occur predominantly via glucose transporter 1 (GLUT1) and is insulin-independent [5, 10].

### *The Contributions of Adipose Tissue Angiogenesis to Overall Vascular Health*

Angiogenesis is considered a crucial and central component of the processes underlying both AT development and its intricate connections with systemic metabolism, exerting influence on progenitor cell proliferation, tissue remodelling, regulation of adipocyte metabolism, and comprehensive adipocyte communication with the rest of the organism. Furthermore, focusing on pathological processes, it is evident that endothelial dysfunction (ED) is a hallmark of obesity. Recent perspectives emphasize the significance of the bidirectional relationship between these two pathological entities, as obesity was previously thought to lead to ED. In contrast, it is now understood that changes in ECs function can also positively influence adipocyte metabolism alterations [2, 4].

## Mechanisms of Regulation of Angiogenesis

### *Regulation of Angiogenesis Is in Adipose Tissue*

The multistep process of angiogenesis in AT encompasses the following phases: [1] ECs activation and proliferation in response to stimulating factors, [2] infiltration and invasion of ECs into the AT, and [3] formation of lumen and capillary networks within the tissue [4].

However, the role of the vasculature in AT is far more complex than the traditionally understood roles of the vascular system, and herein lies in the intriguing relationship between adipocytes and ECs. While the vasculature of AT does create a microenvironment that supports adipocyte differentiation, adipocyte precursors (preadipocytes) and endothelial precursor cells are believed to reside in close proximity within AT, facilitating their interactions during adipogenesis and angiogenesis. It is important to note the substantial resemblance between markers expressed by adipocytes and pericytes [2, 5]. However, although adipocytes and pericytes express similar markers, such as platelet-derived growth factor receptor-beta (PDGFR$\beta$) and alpha-smooth muscle actin ($\alpha$-SMA), they have distinct roles and express different sets of markers specific to their cell types. The initial basis for a deeper understanding of the significance of the angiogenic process in AT was the discovery that AT expansion is halted by its inhibition. However, subsequent scientific findings have confirmed that this conclusion could be more complex, presenting a new perspective that provides insight into the onset of metabolic perturbations due to downstream regulation of angiogenesis. The formation of new vasculature is conditioned by the secretion of factors that stimulate this process, designated as pro-angiogenic factors, predominantly from the vascular endothelial growth factor (VEGF) family. Among these key regulators, VEGF-A provides chemoattractant and mitogenic messages directed toward ECs, leading to their multiplication and phenotypic differentiation [4].

Intersignaling originating from the Notch signaling pathway through the delta-like 4 (DLL4) mediates communication between angiogenesis and VEGFs. However, the interaction between Notch signaling and VEGFs is complex and context-dependent, with Notch able to both positively and negatively regulate VEGF signaling, depending on the specific cell types and conditions involved [2, 4]. Activation of these factors results in body weight loss via heightened thermogenesis and energy expenditure as well as a reduction in hypoxia and related metabolic comorbidities associated with obesity [5]. Angiogenesis represents a fascinating entity, as evidenced by the observation that it can act as a protective and exacerbating factor. Specifically, animal models [11, 12] have shown that the effect depends on the time of application of the anti-angiogenic therapy. If administered to previously healthy subjects during weight gain, it leads to ectopic accumulation of AT by inhibiting its expansion, resulting in the onset of insulin resistance (IR). Conversely, when applied in cases of pre-existing obesity, this therapy can lead to the removal of pro-inflammatory factors and improvement in AT dysfunction [4, 5].

## *Role of HIF-1α in Angiogenesis Regulation*

Obesity is marked by a decrease in capillary density and thinning of capillary networks due to the rapid expansion of AT, resulting in a notable increase in the distance between adjacent microvasculature that leads to reduced tissue oxygenation, macrophage infiltration, and inflammation promotion, all characteristic features of tissue hypoxia [4, 5]. Hypoxia-inducible factors (HIFs) are activated during this state. HIF-1 comprises two subunits, HIF-1α and HIF-1β, and it serves as a functional transcription factor [4]. The role of HIF-1α in AT has not yet been fully elucidated, despite it being identified as a critical regulator of the hypoxia response, and its effects may also be specific to different tissue depots. The findings from obese animal model studies [13, 14] have contributed to the proposition of potential roles for HIF-1α. It is considered to be involved in the tissue response to hypoxia by interfering with tissue mitophagy, inflammation, and fibrosis processes, and its effects on angiogenesis are context-dependent and potentially tissue-specific [14–16]. However, its role is highly heterogeneous, as demonstrated by its varied effects on preadipocytes and adipocytes in terms of differentiation and subsequent function [2].

Additionally, there are challenges in extrapolating the results obtained from animal models considering the substantial differences between human and rodent biology [17]. Specifically, HIF-1α levels are increased in the AT of obese mice [18]. In contrast, in human models, increased HIF-1α levels have been observed only in a specific group of obese individuals with obesity hypoventilation syndrome, leading to the scientific hypothesis that different types of hypoxia may induce varying alterations in HIF-1α [19]. For instance, early and transient activation of HIF-1α occurs after exposure of C57Bl/6J mice to intermittent hypoxia, followed by further changes, including decreased levels of endothelial nitric oxide synthase (eNOS) and VEGF, capillary reduction, dominance of a pro-inflammatory macrophage phenotype, and an increase in insulin resistance. However, these changes were not observed under prolonged hypoxic condition [20]. Nevertheless, one potential indirect role of HIF-1 in improving vascular density is the induction of platelet-derived growth factor (PDGF) expression, which is believed to aid in forming capillary-like tubular networks [2, 19, 21].

## *Mechanisms of Adipose Tissue Adaptation to Changing Metabolic Needs through Angiogenic Processes*

The specific role of the unique AT vasculature can be discussed regarding two fundamental mechanisms of tissue expansion: hyperplasia and hypertrophy. The primary difference between these two processes is that hyperplasia is predominantly genetically determined and occurs physiologically during growth and development. In contrast, AT hypertrophy reflects the capacity for lipid storage during its lifespan. AT expansion is accompanied by dynamic perturbations at the stromal vascular fraction level. However, the fundamental difference between hyperplasia and hypertrophy is rooted in the temporal determinant of vascular development, which precedes and accompanies hyperplasia while lagging behind hypertrophy, thus representing dysfunction onset [1, 3].

### Hyperplasia of AT

AT development via hyperplasia is synchronized with microvasculature evolution. Although specific to the growth period, as previously mentioned, there is potential for this later AT expansion, albeit with individual variation. This mode of expansion was positively associated with metabolic health. These assumptions indicate the significance of angiogenesis in this process. The presumed mechanisms are very complex, including the involvement of different genes and signaling pathways, with a wide range of metabolic and growth factors, such as insulin-like growth factor 1 (IGF-I), which are crucial in endothelial cell proliferation. IGF-I is pivotal in conditions such as pregnancy and post-acute myocardial infarction, representing striking examples of hyperplastic AT expansion during adulthood [3, 22].

### Hypertrophy of AT

Hypertrophy, as a modality of AT expansion, is characteristic of a response to hypercaloric intake, representing the starting point for metabolic dysfunction, and is thus particularly distinctive for ACs within the abdominal depot. Specifically, hypertrophy is accompanied by disruption of the capillary architecture. Scientific premises supporting these findings include direct interference with signaling pathways during hypernutrition and insufficient angiogenesis, further leading to hypertrophy development, formation of a niche by hypertrophic adipocytes impermeable to vascular development, and, ultimately, combined dysfunction of adipocytes and the capillary network [3, 23].

## The Regulatory Role of MicroRNAs in Angiogenesis in AT

MicroRNAs (miRNAs) are a class of small (19–25 nucleotides long) non-coding RNAs (ncRNAs) that play a significant role in post-transcriptional regulation by binding mRNA and suppressing translation or causing their degradation, thus impacting the metabolism in multiple ways [24]. Approximately 2600 miRNAs are present in the human genome [25]. In AT, they are important regulators of AC differentiation, lipogenesis, and angiogenesis and are essential for developing obesity and T2DM [26].

According to the action they perform, miRNAs regulating AT can be categorized into two primary groups: miRNAs that impair angiogenesis (miR-34a, miR-124, miR-29, etc.) and miRNAs that can be affected by pro- or anti-angiogenic factors (miR-21, miR-210, miR-296, etc.) [27].

MiR-34a targets the *SIRT1* (silent information regulator 1, *s.* sirtuin 1) gene and upregulates *FOXO1* (forkhead box protein O1) and p53 (*TP53*) genes in endothelial progenitor cells, inhibiting angiogenesis. However, its downregulation promotes brown AT (BAT) formation by up-regulating SIRT1 and FGF21 (fibroblast growth factor 21) genes [27, 28]. The miR-29 family has been proven effective in inhibiting angiogenesis in vitro and is under-expressed in numerous malignant diseases. However, miR-29a overexpression in adipocytes is associated with IR; therefore, neither miR-34a nor miR-29a is a suitable therapeutic option [29]. miR-124-3p has known anti-angiogenic and anti-obesity potential; it increases the thermogenesis of BAT targeting the SCD1 (stearoyl-co-enzyme A desaturase 1) gene (*SCD*) [29]. miR-126b-5p inhibits tumor angiogenesis in various cancers but stimulates angiogenesis after injury and contributes to diet-induced adipogenesis by increasing lipid deposition and insulin resistance, similar to miR-221 and miR-222 [27, 30, 31].

miR-21 and miR-210 promote angiogenesis induced by hypoxia and seem to be associated with the exacerbation of IR in the pathophysiological progression from obesity to T2DM [27, 32]. miR-296 and miR-155 enhance angiogenesis in AT, cancer, and vascular diseases by up-regulating VEGF expression [27, 28]. miRNA let-7f improves hypoxia-induced angiogenesis, shows elevated expression in visceral

AT in obese patients, and the miRNA-26 family inhibits angiogenesis and adipogenesis by targeting SMAD1–Id1–p21$^{WAF/CIP1}$/p27 and Fbxl19 signaling, respectively [27, 33, 34]. Additionally, miR-132-3p in exosomes originating from BAT has been found to reduce the expression of lipogenic genes in the liver [35].

miRNAs play an essential role in angiogenesis, making them potential indicators for metabolic disorders and a powerful novel therapeutic strategy, considering that their overexpression can be regulated by antisense and anti-miRNA oligonucleotides and underexpression by synthetic miRNA (miRNA mimics) [36].

## Clinical Implications of Adipose Tissue Angiogenesis

Considering the clinical implications and AT angiogenesis-targeted approach, it is essential to note that specific interactions within this intriguing relationship remain unclear. This pertains primarily to the role of the vasculature during various metabolic conditions in maintaining AT homeostasis. Additionally, the role of hypoxia in stimulating angiogenesis during the spread of AT is still unclear. Despite the complexity of the relationship between these two cellular populations, recent discoveries in single-cell transcriptomics have revealed many different populations within the ACs themselves [3, 37, 38].

New findings beyond the simple inhibition of angiogenesis for obesity treatment highlight the complexities of therapies in this domain. These insights introduce a novel therapeutic concept based on the promotion of angiogenesis, applicable not only to obesity but also to diabetes mellitus (DM). This emerging approach challenges previous paradigms and opens new avenues for clinical intervention in metabolic diseases [38].

In the domain of paradoxical regulation of angiogenesis, we have examples of therapies leading to overall improvements in metabolic health and reductions in adiposity. For instance, the angiogenesis inhibitor TNP-470 application to leptin-deficient *ob*/*ob* mice resulted in these benefits [39]. However, whether this was solely due to angiogenesis inhibition remains debatable, given that the drug also suppresses food intake [3]. Other compounds tested as anti-angiogenic inhibitors, including ALS-L1023 [40], anti-angiogenic recombinant protein (AARP) (a.k.a. CTT peptide-endostatin mimic-kringle 5) [41], angiostatin, and VEGF inhibitors also result in weight loss, but the underlying mechanisms remain unclear [4]. Considering all this, the potential of angiogenesis modulation as a therapeutic tool in obesity treatment has become apparent and prudent [3, 4].

## Conclusion

When discussing the potential therapeutic impact of drugs targeting angiogenesis for obesity treatment, it is essential to consider combining these drugs with existing therapies, including those from the field of obesity management and anti-inflammatory agents. There is an urgent need for rapid scientific efforts in this area to achieve the desired clinical outcomes, which would benefit patients suffering from obesity, a disease of pandemic proportions. Furthermore, this research could unravel the intriguing and enigmatic communication between AC and ECs and the various cellular populations involved in this process.

**Funding** This work was supported by the Ministry of Education, Science and Technological Development of the Republic of Serbia (Contract No# 451–03-66/2024–03/200017).

**Conflicts of Interest** The authors declare no conflict of interest.

## References

1. Herold J, Kalucka J. Angiogenesis in adipose tissue: the interplay between adipose and endothelial cells. Front Physiol. 2020;11:624903.
2. AlZaim I, de Rooij L, Sheikh BN, Borgeson E, Kalucka J. The evolving functions of the vasculature in regulating adipose tissue biology in health and obesity. Nat Rev Endocrinol. 2023;19(12):691–707.
3. Corvera S, Solivan-Rivera J, Yang LZ. Angiogenesis in adipose tissue and obesity. Angiogenesis. 2022;25(4):439–53.
4. Nijhawans P, Behl T, Bhardwaj S. Angiogenesis in obesity. Biomed Pharmacother. 2020;126:110103.
5. Ioannidou A, Fisher RM, Hagberg CE. The multifaceted roles of the adipose tissue vasculature. Obes Rev. 2022;23(4):e13403.
6. Wu SA, Kersten S, Qi L. Lipoprotein lipase and its regulators: an unfolding story. Trends Endocrinol Metab. 2021;32(1):48–61.
7. Hansen CG, Shvets E, Howard G, Riento K, Nichols BJ. Deletion of cavin genes reveals tissue-specific mechanisms for morphogenesis of endothelial caveolae. Nat Commun. 2013;4:1831.
8. Elmasri H, Karaaslan C, Teper Y, Ghelfi E, Weng M, Ince TA, et al. Fatty acid binding protein 4 is a target of VEGF and a regulator of cell proliferation in endothelial cells. FASEB J. 2009;23(11):3865–73.
9. Goudriaan JR, Espirito Santo SM, Voshol PJ, Teusink B, van Dijk KW, van Vlijmen BJ, et al. The VLDL receptor plays a major role in chylomicron metabolism by enhancing LPL-mediated triglyceride hydrolysis. J Lipid Res. 2004;45(8):1475–81.
10. Varaliova Z, Vlasak R, Cizkova T, Gojda J, Potockova J, Siklova M, et al. Lymphatic drainage affects lipolytic activity of femoral adipose tissue in women. Int J Obes. 2020;44(9):1974–8.
11. Sun K, Wernstedt Asterholm I, Kusminski CM, Bueno AC, Wang ZV, Pollard JW, et al. Dichotomous effects of VEGF-A on adipose tissue dysfunction. Proc Natl Acad Sci USA. 2012;109(15):5874–9.
12. Xue Y, Xu X, Zhang XQ, Farokhzad OC, Langer R. Preventing diet-induced obesity in mice by adipose tissue transformation and angiogenesis using targeted nanoparticles. Proc Natl Acad Sci USA. 2016;113(20):5552–7.

13. Zhang X, Lam KSL, Ye H, Chung SK, Zhou M, Wang Y, Xu A. Adipose tissue-specific inhibition of hypoxia-inducible factor 1alpha induces obesity and glucose intolerance by impeding energy expenditure in mice. J Biol Chem. 2010;285(43):32869–77.
14. Shimizu I, Aprahamian T, Kikuchi R, Shimizu A, Papanicolaou KN, MacLauchlan S, et al. Vascular rarefaction mediates whitening of brown fat in obesity. J Clin Invest. 2014;124(5):2099–112.
15. Sun K, Halberg N, Khan M, Magalang UJ, Scherer PE. Selective inhibition of hypoxia-inducible factor 1alpha ameliorates adipose tissue dysfunction. Mol Cell Biol. 2013;33(5):904–17.
16. Halberg N, Khan T, Trujillo ME, Wernstedt-Asterholm I, Attie AD, Sherwani S, et al. Hypoxia-inducible factor 1alpha induces fibrosis and insulin resistance in white adipose tissue. Mol Cell Biol. 2009;29(16):4467–83.
17. Borgeson E, Boucher J, Hagberg CE. Of mice and men: pinpointing species differences in adipose tissue biology. Front Cell Dev Biol. 2022;10:1003118.
18. He Q, Gao Z, Yin J, Zhang J, Yun Z, Ye J. Regulation of HIF-1alpha activity in adipose tissue by obesity-associated factors: adipogenesis, insulin, and hypoxia. Am J Physiol Endocrinol Metab. 2011;300(5):E877–85.
19. Todorcevic M, Manuel AR, Austen L, Michailidou Z, Hazlehurst JM, Neville M, et al. Markers of adipose tissue hypoxia are elevated in subcutaneous adipose tissue of severely obese patients with obesity hypoventilation syndrome but not in the moderately obese. Int J Obes. 2021;45(7):1618–22.
20. Gozal D, Gileles-Hillel A, Cortese R, Li Y, Almendros I, Qiao Z, et al. Visceral white adipose tissue after chronic intermittent and sustained hypoxia in mice. Am J Respir Cell Mol Biol. 2017;56(4):477–87.
21. Kihira Y, Fujimura Y, Tomita S, Tamaki T, Sato E. Hypoxia-inducible factor-1alpha regulates Lipin1 differently in pre-adipocytes and mature adipocytes. Mol Med Rep. 2020;22(1):559–65.
22. Al-Samerria S, Radovick S. The role of insulin-like growth factor-1 (IGF-1) in the control of neuroendocrine regulation of growth. Cells. 2021;10(10):2664. https://doi.org/10.3390/cells10102664.
23. McLaughlin T, Craig C, Liu LF, Perelman D, Allister C, Spielman D, Cushman SW. Adipose cell size and regional fat deposition as predictors of metabolic response to overfeeding in insulin-resistant and insulin-sensitive humans. Diabetes. 2016;65(5):1245–54.
24. Oliveto S, Mancino M, Manfrini N, Biffo S. Role of microRNAs in translation regulation and cancer. World J Biol Chem. 2017;8(1):45–56.
25. Alles J, Fehlmann T, Fischer U, Backes C, Galata V, Minet M, et al. An estimate of the total number of true human miRNAs. Nucleic Acids Res. 2019;47(7):3353–64.
26. Gentile AM, Lhamyani S, Mengual-Mesa M, Garcia-Fuentes E, Bermudez-Silva FJ, Rojo-Martinez G, et al. MiR-221-3p/222-3p cluster expression in human adipose tissue is related to obesity and type 2 diabetes. Int J Mol Sci. 2023;24(24):17449.
27. Sun LL, Li WD, Lei FR, Li XQ. The regulatory role of microRNAs in angiogenesis-related diseases. J Cell Mol Med. 2018;22(10):4568–87.
28. Fu T, Seok S, Choi S, Huang Z, Suino-Powell K, Xu HE, et al. MicroRNA 34a inhibits beige and brown fat formation in obesity in part by suppressing adipocyte fibroblast growth factor 21 signaling and SIRT1 function. Mol Cell Biol. 2014;34(22):4130–42.
29. Li Q, Wang L, Liu H, Ren W, Zhang Z, Xia B. Roles of miR-124-3p/Scd1 in urolithin A-induced brown adipocyte differentiation and succinate-dependent regulation of mitochondrial complex II. Biochem Biophys Res Commun. 2022;606:174–81.
30. Shen L, He J, Zhao Y, Niu L, Chen L, Tang G, et al. MicroRNA-126b-5p exacerbates development of adipose tissue and diet-induced obesity. Int J Mol Sci. 2021;22(19):10261.
31. Chan GCK, Than WH, Kwan BCH, Lai KB, Chan RCK, Teoh JYC, et al. Adipose and plasma microRNAs miR-221 and 222 associate with obesity, insulin resistance, and new onset diabetes after peritoneal dialysis. Nutrients. 2022;14(22):4889.

32. Guglielmi V, D'Adamo M, Menghini R, Cardellini M, Gentileschi P, Federici M, Sbraccia P. MicroRNA 21 is up-regulated in adipose tissue of obese diabetic subjects. Nutr Healthy Aging. 2017;4(2):141–5.
33. Gentile AM, Lhamyani S, Coin-Araguez L, Clemente-Postigo M, Oliva Olivera W, Romero-Zerbo SY, et al. miR-20b, miR-296, and Let-7f expression in human adipose tissue is related to obesity and type 2 diabetes. Obesity (Silver Spring). 2019;27(2):245–54.
34. Acharya A, Berry DC, Zhang H, Jiang Y, Jones BT, Hammer RE, et al. miR-26 suppresses adipocyte progenitor differentiation and fat production by targeting Fbxl19. Genes Dev. 2019;33(19–20):1367–80.
35. Kariba Y, Yoshizawa T, Sato Y, Tsuyama T, Araki E, Yamagata K. Brown adipocyte-derived exosomal miR-132-3p suppress hepatic Srebf1 expression and thereby attenuate expression of lipogenic genes. Biochem Biophys Res Commun. 2020;530(3):500–7.
36. Pagoni M, Cava C, Sideris DC, Avgeris M, Zoumpourlis V, Michalopoulos I, Drakoulis N. miRNA-based technologies in cancer therapy. J Pers Med. 2023;13(11):1586.
37. Chaurasiya V, Nidhina Haridas PA, Olkkonen VM. Adipocyte-endothelial cell interplay in adipose tissue physiology. Biochem Pharmacol. 2024;222:116081.
38. Cao Y. Adipose tissue angiogenesis as a therapeutic target for obesity and metabolic diseases. Nat Rev Drug Discov. 2010;9(2):107–15.
39. Brakenhielm E, Cao R, Gao B, Angelin B, Cannon B, Parini P, Cao Y. Angiogenesis inhibitor, TNP-470, prevents diet-induced and genetic obesity in mice. Circ Res. 2004;94(12):1579–88.
40. Park BY, Lee H, Woo S, Yoon M, Kim J, Hong Y, et al. Reduction of adipose tissue mass by the angiogenesis inhibitor ALS-L1023 from Melissa officinalis. PLoS One. 2015;10(11):e0141612.
41. Wang H, Shi Y, Gu J. A multitarget angiogenesis inhibitor, CTT peptide-endostatin mimic-kringle 5, prevents diet-induced obesity. J Mol Med (Berl). 2020;98(12):1753–65.

# Chapter 7
# Technological Advances in Adipose Tissue Research and Clinical Practice

**Radoslav Stojchevski and Dimiter Avtanski**

## Abbreviations

| | |
|---|---|
| AI | Artificial intelligence |
| ASC-1 | Alanine serine cysteine transporter 1 |
| ASCs | Adipose-derived mesenchymal stem cells |
| ASPC | Adipose stem and progenitor cells |
| AT | Adipose tissue |
| BAT | Brown adipose tissue |
| BMI | Body mass index |
| CLI | Cerenkov luminescence imaging |
| CT | Computed tomography |
| CU | Contrast ultrasound |
| CVD | Cardiovascular diseases |
| DL | Deep learning |
| FACS | Fluorescent-activated cell sorting |
| IRT | Infrared thermography imaging |
| ML | Machine learning |
| MRI | Magnetic resonance imaging |
| MSCs | Mesenchymal stem cells |

R. Stojchevski (✉)
Friedman Diabetes Institute, Lenox Hill Hospital, New York, NY, USA

Donald and Barbara Zucker School of Medicine at Hofstra/Northwell, Hempstead, NY, USA

Feinstein Institutes for Medical Research, Manhasset, NY, USA
e-mail: rstojchevski@northwell.edu

D. Avtanski
Friedman Diabetes Institute, Northwell Health, New York, NY, USA

Feinstein Institutes for Medical Research, Manhasset, NY, USA

Donald and Barbara Zucker School of Medicine at Hofstra/Northwell, Hempstead, NY, USA
e-mail: davtanski@northwell.edu



D. Avtanski (ed.), *Adipose Tissue*, Contemporary Endocrinology,
https://doi.org/10.1007/978-3-031-72570-8_7

| | |
|---|---|
| NAFLD | Non-alcoholic fatty liver disease |
| NGS | Next-generation sequencing |
| NIRS | Near-infrared spectroscopy |
| NIRTRS | Near-infrared time-resolved spectroscopy |
| PAT2 | Proton-coupled amino acid transporter 2 |
| P2RX5 | Purinergic receptor P2X ligand-gated ion channel 5 |
| PET | Positron emission tomography |
| sc-seq | Single-cell sequencing technology |
| scATAC-seq | Single-cell assay for transposase-accessible chromatin using sequencing |
| scDNA-seq | Single-cell DNA sequencing |
| scEpi-seq | Single-cell epigenome sequencing |
| scRNA-seq | Single-cell RNA sequencing |
| SCP | Single-cell proteomics |
| snmC-seq | Single-nucleus methylome sequencing |
| snRNA-seq | Single-nucleus RNA sequencing |
| T2DM | Type 2 diabetes mellitus |
| WAT | White adipose tissue |

## Introduction

Adipose tissue (AT) is a complex and dynamic organ that plays crucial roles in energy homeostasis and metabolic health. It is composed of adipocytes and a stromal vascular fraction with cells containing a heterogeneous population, including pre-adipocytes, mesenchymal, progenitor, and immune cells such as T cells and macrophages [1, 2]. AT secretes adipokines, cytokines/chemokines, and vasoactive molecules with systemic and local effects. Dysregulation of these bioactive molecules due to obesity contributes to a cascade of metabolic disorders, including type 2 diabetes mellitus (T2DM), hypertension, cardiopathy, atherosclerosis, and cancer [3–6].

Recent technological advances have transformed our understanding of AT involvement in endocrine processes, metabolic regulation, immunity, and its interactions with other organs. Single-cell sequencing provides detailed information about AT cellular composition and how individual cells respond to various conditions, such as obesity, insulin resistance, or inflammatory signals [7]. Multi-omics technologies, including genomics, transcriptomics, epigenomics, and metabolomics, provide insights into the complexity of the molecular pathways and interactions that contribute to the overall function of AT [8]. *In vivo* imaging offers a real-time and non-invasive look into AT distribution, the mechanisms behind AT dysfunction, and the efficacy of potential treatments. With the combination of artificial intelligence (AI) and machine learning (ML) technologies, these techniques can revolutionize the future of AT research [9, 10].

AI, as a subfield of computer science, is anticipated to be a promising tool in future medicine and healthcare. The digitalization of healthcare combined with AI-driven devices and ML algorithms has transformed medical practice, improving clinical data management, disease diagnosis, risk assessment, disease prevention, and drug discovery. AI has shown remarkable success in dissecting the complexities associated with AT using automated information retrieval from large-scale magnetic resonance imaging (MRI), computed tomography (CT) scans, and multi-omics data. This system reduces human error, improves predictions, and enhances clinical outcomes for AT dysfunction [11, 12]. ML models trained on multi-omics data can identify biomarkers and novel molecular pathways associated with inflammation, immune response, adiposity, and cardiovascular diseases (CVD), which can predict risk assessment, patient outcomes, and treatment plans for metabolic diseases and CVD [13].

This chapter summarizes the transformative impact of modern technology in AT research and presents new insights into the importance of targeting AT in clinical practice.

## Single-Cell Sequencing in Adipose Tissue Research

Before the high-throughput sequencing era, researchers used fluorescent-activated cell sorting (FACS) to study AT heterogeneity. This method of identifying different cell types within AT is based on the presence or absence of specific cell surface markers. The sorted cells were then propagated and analyzed using downstream analyses. The cell surface markers used for white, brown, and beige adipocytes were alanine serine cysteine transporter 1 (ASC-1), proton-coupled amino acid transporter 2 (PAT2), and purinergic receptor P2X ligan-gated ion channel 5 (P2RX5), respectively [14]. Adipose stem and progenitor cells (ASPC) were sorted based on a specific combination of cell surface markers ($Lin^-/CD34^+/CD29^+/SVA1^+/CD24^+$) [15]. Lineage tracing via Cre-recombinase has been employed to identify adipocyte precursors and characterize mature adipocytes within murine white AT (WAT) [16]. However, the lack of single markers limits the use of FACS and lineage-based strategies for identifying APSC heterogeneity.

Microarray analysis was the first high-throughput genomic method used to identify differentially expressed genes in distinct samples, a technology that significantly advanced drug development and basic research [17, 18]. However, because it relies on preset probes for a specific set of genes, its capacity to detect all genes and/or non-coding RNAs is limited. Next-generation sequencing (NGS) technologies, which emerged within the last two decades, addressed this limitation. Early NGS primarily focused on bulk RNA sequencing of whole tissues or isolated cell populations and could not identify distinct cell subsets [19].

To overcome the limitations of bulk RNA sequencing, single-cell transcriptome sequencing (scRNA-seq) was developed as the first single-cell sequencing technology (sc-seq) [20–22]. It provides detailed information about the variations between different cell clusters and facilitates the study of various cell-to-cell interactions. scRNA-seq is an effective method for exploring the cellular composition and dynamics of the stromal cells in AT. It offers comprehensive insights into cellular identities, gene expression profiles, and changes in cellular states under both normal and pathological conditions. Even more, scRNA-seq can help characterize cellular responses by identifying specific pathways modulated in the process [23]. Zheng et al. [24] utilized scRNAseq to reveal the heterogeneity among mesenchymal stem cells (MSCs), immune cells, mesothelial cells, and endothelial cells in WAT, which are crucial in maintaining AT homeostasis and influencing disease development.

Although a promising technique, scRNA-seq has several limitations that must be considered. These include prolonged enzymatic treatment involved in the cell isolation protocol, which may modify the gene expression profile of sensitive cells. Additionally, difficulties arise in isolating cells on the same day as tissue extraction, particularly with large human and animal tissue samples. Moreover, mature adipocytes often prove too large for effective analysis using scRNA-seq. These major challenges with scRNA-seq can be effectively overcome by utilizing single-nuclei RNA sequencing (snRNA-seq) [25–27]. snRNA-seq can be performed on fresh or frozen tissue samples and is not dependent on the enzymatic digestion of tissue to obtain single viable cells [28–30]. Since nuclei are much smaller than whole cells, using snRNA-seq makes it easier to identify the nuclei of all cell types, including larger cells such as adipocytes and neurons.

Another single-cell technique, single-cell proteomics (SCP), has enabled a thorough and extensive examination of protein expression at an individual cell level, which can provide even more detailed knowledge into AT heterogeneity, functional states, and disease mechanisms. Various types of SCP, including button-up and top-down proteomics, allow analyses of protein interactions, signaling pathways, and post-translational modifications [31], thus potentially revolutionizing AT research and identifying novel biomarkers and therapeutic targets. Other techniques within sc-seq and omics analyses include single-cell genome sequencing (scDNA-seq), single-cell epigenome sequencing (scEpi-seq), single-cell sequencing assay for transposase-accessible chromatin (scATAC-seq)), and single-nucleus methylome sequencing (snmC-seq), among others [32]. With this new technology, it is possible to profile the AT at a single-cell level, which helps to better understand the cellular and molecular landscape of AT structure, heterogeneity, and function [33, 34].

## Multi-omics Analyses in Adipose Tissue Research

Recent technological developments in high-throughput sequencing and mass spectrometry enabled researchers to perform integrative multi-omics analyses, combining data from genomics, transcriptomics, proteomics, metabolomics, epigenomics, and microbiota studies.

Genomics has opened up a new range of opportunities for identifying genetic variants that influence AT development and function [35]. The corresponding gene expression profiles indicative of critical regulators and pathways for adipogenesis, metabolism, and inflammatory response are identified by transcriptomics [36]. Proteomics is a high-throughput molecular tool for describing protein expression patterns, thus identifying the major functional proteins that may be involved in various metabolic processes [37]. Metabolomics provides the platform for analyzing a wide array of small molecules involved in adipocyte metabolic pathways and lipid metabolism [38]. Epigenomics focuses on DNA methylation and histone modifications and their effect on gene expression and AT phenotype [39, 40]. Finally, microbiota analyses explore the interactions between the gut microbiome and AT that have an impact on metabolic health and inflammation [41, 42]. Multi-omics analyses allow scientists to analyze complex biological data and find novel correlations between different molecular layers within the AT. These new perspectives can lead to a better understanding of the AT function and its role in metabolic health and diseases.

Through the multi-omics approach by proteomic/transcriptomic analysis of visceral and subcutaneous pig AT, researchers revealed several depot-specific dysregulations of proteins involved in glucose and lipid metabolism as a response to energy stress, such as insulin deficiency or hyperglycemia [43]. Another study has combined transcriptomics, non-coding RNAomics (miRNAomics), and epigenomics to investigate the aging-related changes in the subcutaneous AT composition, structure, and function of female individuals [44]. In addition, Yang et al. [45] carried out transcriptomic/metabolomic analyses and revealed distinct patterns of differentially expressed genes and metabolites between adipose-derived mesenchymal stem cells (ASCs) from various sites, including subcutaneous, perirenal, and epididymal AT. These multisite ASCs exhibit diversity in gene expression and metabolite profiles, which may have potential relevance in associated diseases. Moreover, Si et al. [46] provided a detailed molecular profile of how air pollution affects AT, lipid metabolism, and mitochondrial function in mice by conducting metabolomic, lipidomic, and transcriptomic analyses.

In-depth analysis of multi-omics data sets is essential for advancing research on AT. It is an effective approach for uncovering novel therapeutic targets for obesity, T2DM, and other metabolic diseases.

## *In Vivo* Imaging and Visualization of Adipose Tissue

*In vivo* imaging involves several advanced techniques that allow a non-invasive observation of AT in living organisms for research or diagnostic purposes. Key *in vivo* imaging methods include positron emission tomography (PET), magnetic resonance imaging (MRI), computed tomography (CT), ultrasound imaging, and near-infrared spectroscopy (NIRS), among others. They provide valuable information about the distribution, dynamics, function, and metabolic activity of AT.

One particularly effective approach for imaging AT involves a combination of PET and CT. While PET depends on the emission of positrons from isotope-labeled molecules, CT utilizes X-rays to create images based on differences between tissue densities. This combination of techniques allows the capture of distinct metabolic changes within the body. As AT is less dense, it has specific X-ray attenuation properties, allowing for visualization using CT scans [47]. At the same time, PET imaging highlights areas of high metabolic activity using radiotracers, such as 2-deoxy-2-[18F]-fluoro-D-glucose (18F-FDG) [48]. Using the PET/CT technique, researchers have identified brown AT (BAT) depots in the body and established correlations between their presence and variables including age, BMI, and glucose levels [49–51]. However, both PET and CT techniques are radiation-based, limiting their use. To address this concern, alternative radiation-free methods, such as MRI, are gaining attention. MRI relies on differences in the resonance frequencies of water and fat protons in a magnetic field, which allows for a cost-effective capturing of images of soft tissues at any anatomical angle. MRI is an effective method for visually distinguishing between WAT and BAT based on the difference in the water and fat content of these tissues [52].

Apart from the techniques described above, other imaging techniques are also efficient for exploring AT. These techniques include infrared thermography imaging (IRT), near-infrared spectroscopy (NIRS), near-infrared time-resolved spectroscopy (NIRTRS), contrast ultrasound (CU), Cerenkov luminescence imaging (CLI), and ultrasonography. IRT employs infrared-sensitive cameras to evaluate temperature distribution patterns and detect BAT by monitoring skin temperature, mainly in the interscapular region [51, 53]. NIRS relies on the varying absorption of oxy- and deoxyhemoglobin and offers a method for assessing tissue oxygen perfusion in highly vascularized BAT [54]. NIRTRS enables the analysis of BAT activity by measuring hemoglobin concentration and mitochondrial density. Similarly, CU can detect BAT activation by detecting its heat patterns and blood flow. In conjunction with standard optical imaging instruments, CLI is a quicker and less expensive replacement for PET imaging [55].

Raman spectroscopy is a powerful and non-invasive tool for exploring AT anatomy and physiology. Its method is based on measuring the inelastic scattering of light by molecules with a laser source providing the necessary excitation light. Compared to the other techniques discussed above, Raman spectroscopy has the least interference from water and has proven to be very efficient in tracking the distribution and composition of lipid droplets within cells [56, 57]. This technique can successfully visualize subcutaneous AT; however, it might be challenging for deeper AT depots such as perivascular or visceral AT [58].

In recent years, the outdated view of AT as a mere passive energy storage and insulator has already been reassessed. Modern physiology recognizes AT as a crucial endocrine organ with a significant role in actively regulating metabolic processes. This new understanding has profound implications in clinical imaging. Existing techniques must be re-evaluated to meet current demands, and new methods need to be developed to provide even more detailed insights into AT's role at various levels of granularity [59]. Moreover, high throughput techniques and *in vivo* imaging generate a vast amount of data that requires accurate interpretation. The incorporation of advanced computational approaches such as AI and ML aids researchers in efficiently processing and analyzing multi-omics and imagining data. In the following section, we will explore the advantages of AI and ML in AT research.

## Artificial Intelligence and Machine Learning in Adipose Tissue Research

AI involves the development of computer systems capable of performing tasks that typically require human intelligence by utilizing algorithms, data, and computing power. ML is a subset of AI that can analyze large-scale data and extract features to improve over time as it is exposed to more data. ML can be further categorized as either supervised or unsupervised. As the name suggests, in supervised ML, the algorithm trains itself when exposed to input data with designated output labels. In unsupervised ML, the algorithm trains itself without any labels entered into the algorithm and finds hidden features in the data. Another subset of AI, deep learning (DL), involves artificial neural network algorithms that can handle structured data without human intervention and learn and predict high-level abstractions in the data. DL models can be 1D or 2D, as applied in image analyses [11, 12].

AI and ML are powerful tools for studying the physiology and molecular etiology of AT. They have proven extremely useful in analyzing high-throughput data from genomic, transcriptomic, proteomic, epigenomic, and metabolomic studies. AI and ML can help identify subtle changes in gene or protein expression and identify indistinct patterns and correlations that may evade human analyses [13]. Investigating multi-omics data through AI models leads to discovery of novel biomarkers and developing targeted therapeutics by identifying the molecular pathways involved in metabolic diseases, T2DM, and CVD [60].

AI has rapidly integrated into medicine and healthcare and is widely used in maintaining patient records, disease management, facilitating drug discovery, and improving medical diagnostic devices. AI has accelerated disease diagnosis by automating data-intensive stages, thereby improving the accuracy of diagnoses. Recent studies have demonstrated that unsupervised ML can identify subtypes in patients with persistent atrial fibrillation based on adipocyte markers [61, 62]. Another study by Castañé et al. [63] highlighted the use of ML to identify specific lipids that serve as predictive biomarkers for non-fatty acid liver disease (NAFLD). Incorporating AI/ML algorithms and methodologies with diagnostic tools, such as MRI and CT scans, has revolutionized the field of healthcare technology, leading to accurate and precise diagnostic, therapeutic, and decision-making processes. Body composition imaging is a breakthrough concept in quantifying AT segmentation. Manual segmentation analyses are tedious and time-consuming, whereas AI has automated this process with precision and accuracy [64]. In addition to their role in data analyses, disease diagnosis, and risk prediction, AI and ML are highly effective in personalizing healthcare and therapy planning. Personalized medicine empowers healthcare providers to tailor treatment approaches, recommend accurate medications and dosages, and improve diagnostic accuracy across all diseases, including AT-related disorders [65].

Recent technological advancements have improved the real-time monitoring of body parameters using advanced sensors available in wearable forms, such as smartwatches. They can easily record personalized data in real-time, including BMI, oxygen level, heart rate, and fat proposition. In this regard, AI helps clinicians deliver personalized lifestyle interventions such as behavioral, physical, or nutritional modifications to prevent AT-related diseases [66, 67]. Thus, AI has immense potential to revolutionize personalized healthcare for individuals with AT-related conditions, including obesity, using advanced data-driven computational techniques and predictive analytics.

Although AI has successfully advanced the fields of medicine and healthcare, it is still associated with several significant challenges. The biggest challenges are the lack of high-quality data, risks of bias in data interpretation, and ethical and regulatory issues associated with healthcare [68]. As the field evolves, AI and other advancements in AT research are expected to increase their contribution to improving patient outcomes, diagnostic accuracy, treatment efficacy, and overall understanding of metabolic diseases and related conditions.

## Clinical Trials on Adipose Tissue Reduction

The AT has a major role in various biological processes, including energy storage, hormone secretion, and metabolic regulation. However, excessive accumulation of AT, especially visceral AT, is linked with multiple metabolic disorders such as obesity, T2DM, CVD, and cancer [4, 69]. The growing prevalence of these conditions highlights the urgent need for effective strategies and therapies to reduce excessive AT and its negative impact on health that focus on weight loss and prevention of weight regain. To achieve this, both lifestyle and pharmacotherapeutic interventions are being evaluated in ongoing clinical trials, which are summarized in Table 7.1.

## Conclusion

Recent technological advancements have revealed the dynamic nature of AT and highlighted their important role in elucidating the complex cellular and molecular mechanisms that support its function. Cutting-edge technologies such as single-cell sequencing and multi-omics analyses are crucial for understanding the role of AT in metabolic diseases such as obesity and T2DM. Applying innovative methods in AT research provides deeper understanding and potential identification of novel diagnostic biomarkers and therapeutic targets.

Integrating various imaging techniques has significantly advanced our knowledge on AT far beyond its traditional role as a passive energy reservoir. These techniques are essential for developing a broad perspective of AT's anatomy and physiology, revealing its active role in metabolic regulation.

**Table 7.1** List of completed (phases 3 and 4) and ongoing (recruiting in phases 1, 2, 3, and 4) clinical trials related to AT reduction

| ClinicalTrials.gov ID | Official title | Condition | Intervention/treatment | Phase | Status |
|---|---|---|---|---|---|
| NCT03241563 | A Randomized, Double-blind Comparison Trial of Subcutaneous Sodium Deoxycholate Injections With or Without Triamcinolone for Reduction of Submental Fat | Fat reduction | Drugs: Sodium Deoxycholate, Triamcinolone | 4 | Completed |
| NCT04897867 | CoolSculpting® Elite: Multi-Country Study to Evaluate Patient Satisfaction for Non-Invasive Fat Reduction in Abdomen, Flanks, Upper Arms, Inner Thighs, Outer Thighs and Submental Area | Non-invasive fat reduction | Device: CoolSculpting® Elite System with CoolSculpting® Elite applicators | 4 | Completed |
| NCT03510598 | Fat Reduction in the Submental Area –a Sequential Treatment Approach With CoolMini and KYBELLA® | Body fat disorder | Device: The ZELTIQ CoolSculpting System<br>Drug: Kybella 20 mg in 2 mL injection | 4 | Completed |

(continued)

**Table 7.1** (continued)

| ClinicalTrials.gov ID | Official title | Condition | Intervention/treatment | Phase | Status |
|---|---|---|---|---|---|
| NCT02163902 | A Multicenter, Double-blind, Nontreatment, Long-term Follow up Study of Subjects Who Completed ATX-101 (Deoxycholic Acid Injection) Clinical Trials ATX-101-11-22 or ATX-101-11-23 for the Reduction of Localized Subcutaneous Fat in the Submental Area | Submental fat: Healthy | Drug: ATX-101 | 3 | Completed |
| NCT02123134 | A Multicenter, Double-blind, Placebo-controlled Safety Study of ATX-101 (Deoxycholic Acid) Injection for the Reduction of Localized Subcutaneous Fat in the Submental Area in Subjects 65 to 75 Years of Age | Moderate to severe convexity of submental fat: safety, efficacy | Drug: ATX-101 | 3 | Completed |
| NCT02035267 | A Multicenter, Double-blind, Placebo-controlled Safety Study of ATX-101 (Deoxycholic Acid) Injection for the Reduction of Localized Subcutaneous Fat in the Submental Area in Subjects With Clinician-Reported Submental Fat Rating Scale (CR-SMFRS) Grade 1 or CR-SMFRS Grade 4 | Moderate to severe convexity of submental fat: safety, efficacy | Drug: Deoxycholic acid | 3 | Completed |

(continued)

**Table 7.1** (continued)

| ClinicalTrials.gov ID | Official title | Condition | Intervention/treatment | Phase | Status |
|---|---|---|---|---|---|
| NCT02010359 | Omega-3 PUFA Suppress Adipochemokines | Obesity | Drug: Lovaza | 4 | Completed |
| NCT03849729 | Effectiveness and Tolerability of Phentermine in the Reduction of Intrahepatic Fat Infiltration, Adipose Tissue and Postoperative Complications in Patients Under Bariatric Surgery | Obesity, morbid obesity, bariatric surgery, non-alcoholic fatty liver, weight loss | Drug: Phentermine | 4 | Completed |
| NCT00299325 | A Randomized, Double Blind, Two-arm Placebo Controlled, 12-Month Study of the Effects of Rimonabant 20 mg Once Daily on the Amount and the Activity of Visceral Fat in Abdominally Obese Patients With Metabolic Syndrome | Metabolic syndrome | Drugs: Rimonabant, Placebo (for Rimonabant) Other: Mild hypocaloric diet | 3 | Completed |
| NCT00147264 | A Randomized, Double Blind, 2X2 Factorial Design Study to Evaluate the Effects of Telmisartan vs. Placebo, and of a Low-Glycemic Diet vs. Control Diet, in Reducing Intra-Myocellular Lipids In Individuals With Abdominal Obesity | Metabolic syndrome X | Drug: Telmisartan (Micardis®) vs. Placebo Behavioral: Low-Glycemic Index Diet vs. Control Diet | 3 | Completed |

(continued)

**Table 7.1** (continued)

| ClinicalTrials.gov ID | Official title | Condition | Intervention/treatment | Phase | Status |
|---|---|---|---|---|---|
| NCT01305577 | Phase 3, Multicenter, Randomized, Double-blind, Placebo-controlled Study of ATX-101 (Sodium Deoxycholate Injection) Versus Placebo for the Reduction of Localized Subcutaneous Fat in the Submental Area | Moderate or severe submental fullness | Drug: Deoxycholic acid injection | 3 | Completed |
| NCT01294644 | Phase 3, Multicenter, Randomized, Double-blind, Placebo-controlled Study of ATX-101 (Sodium Deoxycholate Injection) Versus Placebo for the Reduction of Localized Subcutaneous Fat in the Submental Area | Moderate or severe submental fullness | Drug: Deoxycholic acid injection | 3 | Completed |
| NCT01426373 | Multicenter, Open-label Study of ATX-101 (Sodium Deoxycholate Injection) for the Reduction of Localized Subcutaneous Fat in the Submental Area | Moderate or severe submental fullness | Drug: Deoxycholic acid injection | 3 | Completed |
| NCT02397499 | A Multi-Center, Randomized, Double-Blinded, Placebo-Controlled Study Evaluating the Safety and Efficacy of LIPO-202 for the Reduction of Central Abdominal Bulging Due to Subcutaneous Fat in Non-Obese Subjects | Central abdominal bulging | Drug: LIPO-202 | 3 | Completed |

(continued)

**Table 7.1** (continued)

| ClinicalTrials.gov ID | Official title | Condition | Intervention/treatment | Phase | Status |
|---|---|---|---|---|---|
| NCT02398188 | A Multi-Center, Randomized, Double-Blinded, Placebo-Controlled Study Evaluating the Safety and Efficacy of LIPO-202 for the Reduction of Central Abdominal Bulging Due to Subcutaneous Fat in Non-Obese Subjects | Central abdominal bulging | Drug: LIPO-202 | 3 | Completed |
| NCT00353054 | A Double-Blind, Randomized, Placebo Controlled, Parallel Group Study to Evaluate the Effect of Calcium/Vitamin D Supplementation (Caltrate 600+ D) on Body Weight and Fat Loss in Overweight and Obese Individuals on a Weight-Reducing Program | Obesity, overweight | Drug: Caltrate® 600 + D<br>Behavioral: Weight loss intervention | 4 | Completed |
| NCT01546142 | Multicenter, Randomized, Double-Blind, Placebo-Controlled, Phase 3 Study of ATX-101 (Sodium Deoxycholate Injection) Versus Placebo for the Reduction of Localized Subcutaneous Fat in the Submental Area | Moderate or severe submental fullness | Drug: Deoxycholic acid injection | 3 | Completed |

(continued)

**Table 7.1** (continued)

| ClinicalTrials. gov ID | Official title | Condition | Intervention/ treatment | Phase | Status |
|---|---|---|---|---|---|
| NCT01542034 | Multicenter, Randomized, Double-Blind, Placebo-Controlled, Phase 3 Study of ATX-101 (Sodium Deoxycholate Injection) Versus Placebo for the Reduction of Localized Subcutaneous Fat in the Submental Area | Moderate or severe submental fullness | Drug: Deoxycholic acid injection | 3 | Completed |
| NCT01423617 | Double-blind, Randomized, Placebo-controlled, Bicentric Study to Evaluate the Safety and Efficacy of Zenoctil in Reducing Body Weight of Overweight Subjects | Overweight | Dietary Supplement: Zenoctil | 3 | Completed |
| NCT04142749 | A Multi-center, Randomized, Double-blind, Placebo-controlled, Parallel, Phase III Clinical Trial to Evaluate the Efficacy and Safety of Oltipraz | Non-alcoholic fatty liver disease | Drug: Oltipraz | 3 | Completed |
| NCT03361176 | A Randomized, Double-blind, Dose-ranging Trial of Subcutaneous Sodium Deoxycholate Injections With or Without Low Dose Triamcinolone and Low Dose Lidocaine for Reduction of Submental Fat With Reduction of Pain and Swelling | Adiposity | Drug: Injectable sodium deoxycholate Combination Product: Injectable sodium deoxycholate with Triamcinolone acetate | 4 | Completed |

(continued)

**Table 7.1** (continued)

| ClinicalTrials. gov ID | Official title | Condition | Intervention/ treatment | Phase | Status |
|---|---|---|---|---|---|
| NCT01407393 | Double-blind, Randomized, Placebo-controlled, Bicentric Clinical Investigation to Evaluate Safety and Efficacy of Glucosanol in Reducing Body Weight in Overweight and Obese Subjects | Overweight and obesity | Device: Glucosanol | 4 | Completed |
| NCT02068339 | A Multicenter, Randomized, Double-blind, Placebo-controlled, Parallel Phase 3 Clinical Trial to Evaluate the Efficacy and Safety of Oltipraz for Liver Fat Reduction in Patients With Non-Alcoholic Fatty Liver Disease Except for Liver Cirrhosis | Non-alcholic fatty liver disease | Drugs: Oltipraz (90 mg and 120 mg) | 3 | Completed |
| NCT02007434 | A Single Center, Double-blind, Parallel-group, Two Factor Patient-Experience Management Study of ATX-101 (Deoxycholic Acid Injection) for the Reduction of Localized Subcutaneous Fat in the Submental Area | Moderate or severe submental fullness | Drugs: Deoxycholic Acid Injection, Lidocaine/ Epinephrine, Loratadine, Ibuprofen<br>Other: Cold Compression chin strap | 3 | Completed |

(continued)

**Table 7.1** (continued)

| ClinicalTrials.gov ID | Official title | Condition | Intervention/treatment | Phase | Status |
|---|---|---|---|---|---|
| NCT01233349 | Double-blind, Randomized, Placebo-controlled, Bicentric Clinical Investigation to Evaluate Safety and Efficacy of LitramineTM in Reducing Body Weight in Overweight and Obese Subjects | Overweight, obesity, weight loss | Device: Litramine | 3 | Completed |
| NCT03877406 | Decrease in Body Fat and Increase in Ketone Body After 6-month Treatment of Empagliflozin in Korean Patients With Type 2 Diabetes | Diabetes Mellitus Fat; Intolerance, Pancreas | Drug: Empagliflozin (10 mg), Hypoglycemics Oral | 4 | Recruiting |
| NCT06256445 | A Multicenter, Randomized, Double-blind, Placebo-controlled, Phase III Clinical Study to Evaluate Efficacy and Safety of AYP-101 for the Reduction of Submental Fat in Adults | Healthy | Drug: Polyene phosphatidylcholine | 3 | Recruiting |
| NCT06005441 | A Phase 2b, Randomized, Placebo-Controlled Factorial Study to Evaluate the Efficacy, Safety and Tolerability of CBL-514 Injection Compared With CBL-A1 and CBL-A2 for Reducing Abdominal Subcutaneous Fat | Subcutaneous fat | Drugs: CBL-514 injection, CBL-A1 Injection, CBL-A2 Injection | 2 | Recruiting |

(continued)

**Table 7.1** (continued)

| ClinicalTrials.gov ID | Official title | Condition | Intervention/treatment | Phase | Status |
|---|---|---|---|---|---|
| NCT04392206 | A Phase I, Open Label Study of Allogeneic Adipose Derived Mesenchymal Stem Cells in Reducing Anastomotic Stenosis in Primary Arteriovenous Anastomoses | End-stage renal disease | Drug: Adipose Derived Mesenchymal Stem Cells | 1 | Recruiting |

The transformative impact of AI, particularly through ML and DL, is evident in medicine and healthcare. AI improves patient outcomes, diagnostic accuracy, treatment efficacy, and the overall understanding of metabolic diseases by analyzing large-scale, high-throughput data from various omics studies.

Finally, the importance of ongoing clinical trials on AT reduction lies in their potential to advance the medical knowledge and treatment options for obesity and related conditions. These trials evaluate the effectiveness of new interventions and provide valuable data for improving existing therapies, thus enhancing the quality of life and addressing the global health challenge of obesity.

## References

1. Gustafson B. Adipose tissue, inflammation and atherosclerosis. J Atheroscler Thromb. 2010;17(4):332–41. https://doi.org/10.5551/jat.3939.
2. Corrêa LH, Heyn GS, Magalhaes KG. The impact of the adipose organ plasticity on inflammation and cancer progression. Cells. 2019;8(7):662. https://doi.org/10.3390/cells8070662.
3. Luo L, Liu M. Adipose tissue in control of metabolism. J Endocrinol. 2016;231(3):R77–99. https://doi.org/10.1530/JOE-16-0211.
4. Chait A, den Hartigh LJ. Adipose tissue distribution, inflammation and its metabolic consequences, including diabetes and cardiovascular disease. Front Cardiovasc Med. 2020;7:22. https://doi.org/10.3389/fcvm.2020.00022.
5. Lengyel E, Makowski L, DiGiovanni J, Kolonin MG. Cancer as a matter of fat: the cross-talk between adipose tissue and tumors. Trends Cancer. 2018;4(5):374–84. https://doi.org/10.1016/j.trecan.2018.03.004.
6. Must A. The disease burden associated with overweight and obesity. JAMA. 1999;282(16):1523. https://doi.org/10.1001/jama.282.16.1523.
7. Maniyadath B, Zhang Q, Gupta RK, Mandrup S. Adipose tissue at single-cell resolution. Cell Metab. 2023;35(3):386–413. https://doi.org/10.1016/j.cmet.2023.02.002.
8. Paczkowska-Abdulsalam M, Kretowski A. Obesity, metabolic health and omics: current status and future directions. World J Diabetes. 2021;12(4):420–36. https://doi.org/10.4239/wjd.v12.i4.420.
9. Avtanski D, Hadzi-Petrushev N, Josifovska S, Mladenov M, Reddy V. Emerging technologies in adipose tissue research. Adipocytes. 2023;12(1):2248673. https://doi.org/10.1080/21623945.2023.2248673.

10. Latteri S, Sofia M, Puleo S, Di Vincenzo A, Cinti S, Castorina S. Mechanisms linking bariatric surgery to adipose tissue, glucose metabolism, fatty liver disease and gut microbiota. Langenbeck's Arch Surg. 2023;408(1):101. https://doi.org/10.1007/s00423-023-02821-8.
11. Greco F, Mallio CA. Artificial intelligence and abdominal adipose tissue analysis: a literature review. Quant Imaging Med Surg. 2021;11(10):4461–74. https://doi.org/10.21037/qims-21-370.
12. Umapathy VR, et al. Perspective of artificial intelligence in disease diagnosis: a review of current and future endeavours in the medical field. Cureus. 2023;15(9):e45684. https://doi.org/10.7759/cureus.45684.
13. Lee Y-C, et al. Using machine learning to predict obesity based on genome-wide and epigenome-wide gene–gene and gene–diet interactions. Front Genet. 2022;12:783845. https://doi.org/10.3389/fgene.2021.783845.
14. Ussar S, et al. ASC-1, PAT2, and P2RX5 are cell surface markers for white, beige, and brown adipocytes. Sci Transl Med. 2014;6(247):247ra103. https://doi.org/10.1126/scitranslmed.3008490.
15. Rodeheffer MS, Birsoy K, Friedman JM. Identification of white adipocyte progenitor cells in vivo. Cell. 2008;135(2):240–9. https://doi.org/10.1016/j.cell.2008.09.036.
16. Berry R, Rodeheffer MS. Characterization of the adipocyte cellular lineage in vivo. Nat Cell Biol. 2013;15(3):302–8. https://doi.org/10.1038/ncb2696.
17. Trevino V, Falciani F, Barrera-Saldaña HA. DNA microarrays: a powerful genomic tool for biomedical and clinical research. Mol Med. 2007;13(9–10):527–41. https://doi.org/10.2119/2006-00107.Trevino.
18. Fernandes TG, Diogo MM, Clark DS, Dordick JS, Cabral JMS. High-throughput cellular microarray platforms: applications in drug discovery, toxicology and stem cell research. Trends Biotechnol. 2009;27(6):342–9. https://doi.org/10.1016/j.tibtech.2009.02.009.
19. Li X, Wang C-Y. From bulk, single-cell to spatial RNA sequencing. Int J Oral Sci. 2021;13(1):36. https://doi.org/10.1038/s41368-021-00146-0.
20. van Dijk EL, Auger H, Jaszczyszyn Y, Thermes C. Ten years of next-generation sequencing technology. Trends Genet. 2014;30(9):418–26. https://doi.org/10.1016/j.tig.2014.07.001.
21. Metzker ML. Sequencing technologies—the next generation. Nat Rev Genet. 2010;11(1):31–46. https://doi.org/10.1038/nrg2626.
22. Wen L, et al. Single-cell technologies: from research to application. Innovation (Camb). 2022;3(6):100342. https://doi.org/10.1016/j.xinn.2022.100342.
23. Liu X, et al. Single-cell RNA sequencing of subcutaneous adipose tissues identifies therapeutic targets for cancer-associated lymphedema. Cell Discov. 2022;8(1):58. https://doi.org/10.1038/s41421-022-00402-5.
24. Zheng GXY, et al. Massively parallel digital transcriptional profiling of single cells. Nat Commun. 2017;8(1):14049. https://doi.org/10.1038/ncomms14049.
25. Emont MP, et al. A single-cell atlas of human and mouse white adipose tissue. Nature. 2022;603(7903):926–33. https://doi.org/10.1038/s41586-022-04518-2.
26. Sun W, et al. snRNA-seq reveals a subpopulation of adipocytes that regulates thermogenesis. Nature. 2020;587(7832):98–102. https://doi.org/10.1038/s41586-020-2856-x.
27. Sárvári AK, et al. Plasticity of epididymal adipose tissue in response to diet-induced obesity at single-nucleus resolution. Cell Metab. 2021;33(2):437–453.e5. https://doi.org/10.1016/j.cmet.2020.12.004.
28. Van Hauwaert EL, et al. Isolation of nuclei from mouse white adipose tissues for single-nucleus genomics. STAR Protoc. 2021;2(3):100612. https://doi.org/10.1016/j.xpro.2021.100612.
29. Denisenko E, et al. Systematic assessment of tissue dissociation and storage biases in single-cell and single-nucleus RNA-seq workflows. Genome Biol. 2020;21(1):130. https://doi.org/10.1186/s13059-020-02048-6.
30. van den Brink SC, et al. Single-cell sequencing reveals dissociation-induced gene expression in tissue subpopulations. Nat Methods. 2017;14(10):935–6. https://doi.org/10.1038/nmeth.4437.

31. Botelho D, Wall MJ, Vieira DB, Fitzsimmons S, Liu F, Doucette A. Top-down and bottom-up proteomics of SDS-containing solutions following mass-based separation. J Proteome Res. 2010;9(6):2863–70. https://doi.org/10.1021/pr900949p.
32. Kashima Y, Sakamoto Y, Kaneko K, Seki M, Suzuki Y, Suzuki A. Single-cell sequencing techniques from individual to multiomics analyses. Exp Mol Med. 2020;52(9):1419–27. https://doi.org/10.1038/s12276-020-00499-2.
33. Tang F, et al. mRNA-Seq whole-transcriptome analysis of a single cell. Nat Methods. 2009;6(5):377–82. https://doi.org/10.1038/nmeth.1315.
34. Hwang B, Lee JH, Bang D. Single-cell RNA sequencing technologies and bioinformatics pipelines. Exp Mol Med. 2018;50(8):1–14. https://doi.org/10.1038/s12276-018-0071-8.
35. Deutsch A, Feng D, Pessin JE, Shinoda K. The impact of single-cell genomics on adipose tissue research. Int J Mol Sci. 2020;21(13):4773. https://doi.org/10.3390/ijms21134773.
36. Kostecka A, et al. Adipose-derived mesenchymal stromal cells in clinical trials: insights from single-cell studies. Life Sci. 2024;351:122761. https://doi.org/10.1016/j.lfs.2024.122761.
37. Li Y, et al. Proteomics reveals different pathological processes of adipose tissue, liver, and skeletal muscle under insulin resistance. J Cell Physiol. 2020;235(10):6441–61. https://doi.org/10.1002/jcp.29658.
38. Kučera J, Spáčil Z, Friedecký D, Novák J, Pekař M, Bienertová-Vašků J. Human white adipose tissue metabolome: current perspective. Obesity. 2018;26(12):1870–8. https://doi.org/10.1002/oby.22336.
39. Ling C, Rönn T. Epigenetics in human obesity and type 2 diabetes. Cell Metab. 2019;29(5):1028–44. https://doi.org/10.1016/j.cmet.2019.03.009.
40. Stojchevski R, Velichkovikj S, Arsov T. Genetic and epigenetic basis of obesity-induced inflammation and diabetes. In: Obesity, diabetes and inflammation; 2023. p. 129–46. https://doi.org/10.1007/978-3-031-39721-9_6.
41. Cani PD, Van Hul M. Gut microbiota in overweight and obesity: crosstalk with adipose tissue. Nat Rev Gastroenterol Hepatol. 2023;21(3):164–83. https://doi.org/10.1038/s41575-023-00867-z.
42. Avtanski D, Reddy V, Stojchevski R, Hadzi-Petrushev N, Mladenov M. The microbiome in the obesity-breast cancer Axis: diagnostic and therapeutic potential. Pathogens. 2023;12(12):1402. https://doi.org/10.3390/pathogens12121402.
43. Flenkenthaler F, et al. Differential effects of insulin-deficient diabetes mellitus on visceral vs. subcutaneous adipose tissue—multi-omics insights from the Munich MIDY Pig Model. Front Med. 2021;8:751277. https://doi.org/10.3389/fmed.2021.751277.
44. Li Z, et al. New insights into aging-associated characteristics of female subcutaneous adipose tissue through integrative analysis of multi-omics data. Bioengineered. 2022;13(2):2044–57. https://doi.org/10.1080/21655979.2021.2020467.
45. Yang C, et al. Multi-omics analysis to examine gene expression and metabolites from multisite adipose-derived mesenchymal stem cells. Front Genet. 2021;12:627347. https://doi.org/10.3389/fgene.2021.627347.
46. Si H, et al. Multi-omics reveals hypertrophy of adipose tissue and lipid metabolism disorder via mitochondria in young mice under real-ambient exposure to air pollution. Front Pharmacol. 2023;14:1122615. https://doi.org/10.3389/fphar.2023.1122615.
47. Frankl J, Sherwood A, Clegg DJ, Scherer PE, Öz OK. Imaging metabolically active fat: a literature review and mechanistic insights. Int J Mol Sci. 2019;20(21):5509. https://doi.org/10.3390/ijms20215509.
48. Marzola P, Boschi F, Moneta F, Sbarbati A, Zancanaro C. Preclinical in vivo imaging for fat tissue identification, quantification, and functional characterization. Front Pharmacol. 2016;7:336. https://doi.org/10.3389/fphar.2016.00336.
49. Borga M, Virtanen KA, Romu T, Leinhard OD, Persson A, Nuutila P, Enerbäck S. Brown adipose tissue in humans: detection and functional analysis using PET (positron emission tomography), MRI (magnetic resonance imaging), and DECT (dual energy computed tomography). Methods Enzymol. 2014;537:141–59. https://doi.org/10.1016/B978-0-12-411619-1.00008-2.

50. Reber J, et al. Non-invasive measurement of brown fat metabolism based on optoacoustic imaging of hemoglobin gradients. Cell Metab. 2018;27(3):689–701.e4. https://doi.org/10.1016/j.cmet.2018.02.002.
51. Chondronikola M, Beeman SC, Wahl RL. Non-invasive methods for the assessment of brown adipose tissue in humans. J Physiol. 2018;596(3):363–78. https://doi.org/10.1113/JP274255.
52. Borga M. MRI adipose tissue and muscle composition analysis—a review of automation techniques. Br J Radiol. 2018;91(1089):20180252. https://doi.org/10.1259/bjr.20180252.
53. Brasil S, Renck AC, de Meneck F, Brioschi ML, Costa EF, Teixeira MJ. A systematic review on the role of infrared thermography in the Brown adipose tissue assessment. Rev Endocr Metab Disord. 2020;21(1):37–44. https://doi.org/10.1007/s11154-020-09539-8.
54. Hamaoka T, et al. Near-infrared time-resolved spectroscopy for assessing Brown adipose tissue density in humans: a review. Front Endocrinol (Lausanne). 2020;11:261. https://doi.org/10.3389/fendo.2020.00261.
55. Zhang X, Kuo C, Moore A, Ran C. In vivo optical imaging of interscapular brown adipose tissue with 18F-FDG via cerenkov luminescence imaging. PLoS One. 2013;8(4):e62007. https://doi.org/10.1371/journal.pone.0062007.
56. Butler HJ, et al. Using Raman spectroscopy to characterize biological materials. Nat Protoc. 2016;11(4):664–87. https://doi.org/10.1038/nprot.2016.036.
57. Mitchell A, Ashton L, Yang XB, Goodacre R, Smith A, Kirkham J. Detection of early stage changes associated with adipogenesis using Raman spectroscopy under aseptic conditions. Cytom Part A. 2015;87(11):1012–9. https://doi.org/10.1002/cyto.a.22777.
58. Czamara K, Majka Z, Stanek E, Hachlica N, Kaczor A. Raman studies of the adipose tissue: current state-of-art and future perspectives in diagnostics. Prog Lipid Res. 2022;87:101183. https://doi.org/10.1016/j.plipres.2022.101183.
59. Nikiforaki K, Marias K. MRI methods to visualize and quantify adipose tissue in health and disease. Biomedicines. 2023;11(12):3179. https://doi.org/10.3390/biomedicines11123179.
60. Henninger J, Eliasson B, Smith U, Rawshani A. Identification of markers that distinguish adipose tissue and glucose and insulin metabolism using a multi-modal machine learning approach. Sci Rep. 2021;11(1):17050. https://doi.org/10.1038/s41598-021-95688-y.
61. Baykaner T, Narayan S. Machine learning of adipose tissue in atrial fibrillation. Hear Rhythm. 2022;19(12):2042–3. https://doi.org/10.1016/j.hrthm.2022.08.027.
62. Cui C, et al. Unsupervised machine learning reveals epicardial adipose tissue subtypes with distinct atrial fibrosis profiles in patients with persistent atrial fibrillation: a prospective 2-center cohort study. Hear Rhythm. 2022;19(12):2033–41. https://doi.org/10.1016/j.hrthm.2022.07.030.
63. Castañé H, et al. Coupling machine learning and lipidomics as a tool to investigate metabolic dysfunction-associated fatty liver disease. A general overview. Biomol Ther. 2021;11(3):473. https://doi.org/10.3390/biom11030473.
64. Greco F, Salgado R, Van Hecke W, Del Buono R, Parizel PM, Mallio CA. Epicardial and pericardial fat analysis on CT images and artificial intelligence: a literature review. Quant Imaging Med Surg. 2022;12(3):2075–89. https://doi.org/10.21037/qims-21-945.
65. Johnson KB, et al. Precision medicine, AI, and the future of personalized health care. Clin Transl Sci. 2021;14(1):86–93. https://doi.org/10.1111/cts.12884.
66. Wang W, Cheng J, Song W, Shen Y. The effectiveness of wearable devices as physical activity interventions for preventing and treating obesity in children and adolescents: systematic review and meta-analysis. JMIR Mhealth Uhealth. 2022;10(4):e32435. https://doi.org/10.2196/32435.
67. Liu Y, et al. Skin-interfaced deep-tissue sensing patch via microneedle waveguides. Adv Mater Technol. 2022;7(9):2200468. https://doi.org/10.1002/admt.202200468.
68. Jiang L, et al. Opportunities and challenges of artificial intelligence in the medical field: current application, emerging problems, and problem-solving strategies. J Int Med Res. 2021;49(3):030006052110001. https://doi.org/10.1177/03000605211000157.
69. Wajchenberg BL. Subcutaneous and visceral adipose tissue: their relation to the metabolic syndrome. Endocr Rev. 2000;21(6):697–738. https://doi.org/10.1210/edrv.21.6.0415.

# Chapter 8
# Targeting Adipose Tissue: Therapeutic Strategies and Future Directions

**Bingyang Dai, Lizhen Zheng, and Ling Qin**

Bingyang Dai, Lizhen Zheng and Ling Qin contributed equally to this work.

B. Dai
Musculoskeletal Research Laboratory of Department of Orthopaedics & Traumatology and Innovative Orthopaedic Biomaterial & Drug Translational Research Laboratory, Li Ka Shing Institute of Health Sciences, The Chinese University of Hong Kong, Hong Kong, China

Areas of Excellence Centre for Musculoskeletal Degeneration and Regeneration, Department of Orthopaedics and Traumatology, Faculty of Medicine, The Chinese University of Hong Kong, Hong Kong, China

Department of Biomedical Engineering, Faculty of Engineering, The Hong Kong Polytechnic University, Hong Kong, China

The Hong Kong Polytechnic University Shenzhen Research Institute, Shenzhen, China
e-mail: bingydai@polyu.edu.hk

L. Zheng
Musculoskeletal Research Laboratory of Department of Orthopaedics & Traumatology and Innovative Orthopaedic Biomaterial & Drug Translational Research Laboratory, Li Ka Shing Institute of Health Sciences, The Chinese University of Hong Kong, Hong Kong, China

Areas of Excellence Centre for Musculoskeletal Degeneration and Regeneration, Department of Orthopaedics and Traumatology, Faculty of Medicine, The Chinese University of Hong Kong, Hong Kong, China

Centre for Regenerative Medicine and Health, Hong Kong Institute of Science and Innovation, Chinese Academy of Sciences, Hong Kong, China
e-mail: zhenglz@cuhk.edu.hk

L. Qin (✉)
Musculoskeletal Research Laboratory of Department of Orthopaedics & Traumatology and Innovative Orthopaedic Biomaterial & Drug Translational Research Laboratory, Li Ka Shing Institute of Health Sciences, The Chinese University of Hong Kong, Hong Kong, China

Areas of Excellence Centre for Musculoskeletal Degeneration and Regeneration, Department of Orthopaedics and Traumatology, Faculty of Medicine, The Chinese University of Hong Kong, Hong Kong, China
e-mail: lingqin@cuhk.edu.hk


D. Avtanski (ed.), *Adipose Tissue*, Contemporary Endocrinology,
https://doi.org/10.1007/978-3-031-72570-8_8

## Abbreviations

| | |
|---|---|
| AAV | Adeno-associated virus |
| Ad | Adenovirus |
| ATM | Adipose tissue-derived macrophage |
| AT-MSC | Adipose tissue-derived mesenchymal stem cells |
| BAT | Brown adipose tissue |
| BMI | Body mass index |
| cfDNA | Cell-free DNA |
| CNS | Central nervous system |
| CPT1 | Carnitine palmitoyltransferase 1 |
| CVD | Cardiovascular disease |
| DA | Dopamine |
| DRG | Dorsal root ganglion |
| ECM | Extracellular matrix |
| EMA | European Medicines Agency |
| ER | Extended-release |
| FDA | Food and Drug Administration |
| FFA | Free fatty acid |
| GIPR | Gastric inhibitory polypeptide receptor |
| GWAS | Genome-wide association studies |
| HFD | High-fat diet |
| KSR2 | Kinase suppressor of Ras 2 |
| MAOI | Monoamine oxidase inhibitor |
| MC4R | Melanocortin 4 receptor |
| NE | Norepinephrine |
| NK | Natural killer (cell) |
| PDGF | Platelet-derived growth factor |
| OPA1 | Optic atrophy 1 |
| PET | Positron-emission tomography |
| PM | Precision medicine |
| PRIME | Priority Medicines |
| SAT | Subcutaneous adipose tissue |
| SNS | Sympathetic nervous system |
| SR | Sustained-release |
| T2DM | Type 2 diabetes mellitus |
| TNFα | Tumor necrosis factor-alpha |
| UCP1 | Uncoupling protein-1 |
| VEGF | Vascular endothelial growth factor |
| WAT | White adipose tissue |

## Introduction

Self-awareness of the dangers of being overweight and maintaining a healthy lifestyle are the ideal strategies for combating obesity. Lifestyle intervention, pharmacotherapy, and bariatric surgery are the primary treatments of obesity [1]. Nevertheless, comprehensive therapies should be aligned with severe obesity coexisting with chronic diseases such as lipodystrophies, metabolic syndrome, insulin resistance, type 2 diabetes mellitus (T2DM), and cardiovascular disease (CVD). Based on the recent advances in understanding the pathogenesis of obesity-induced chronic diseases, recent research focuses on the intercellular networks of adipose tissue linking the disorders of immunometabolism with energy storage. These findings are essential for the development of new therapeutic strategies targeting adipose tissue.

There is still unmet need for the effective treatment of these obesity-related metabolic disorders. Therefore, in-depth and comprehensive analysis of the underlying mechanisms is crucial for designing innovative approaches. In this chapter, we summarize the gaps in the current understanding of adipose tissue pathophysiology and provide insights into the potential therapies for metabolic disorders that target adipose tissue.

## Current Therapeutic Strategies

Severe overweight is associated with chronic metabolic diseases. Approaches to weight control that are synchronously beneficial for compromising chronic metabolic diseases are needed. The main existing therapeutic strategies and clinical practice guidelines for managing adipose tissue-related conditions are lifestyle interventions, pharmacological treatments, and bariatric surgery [2].

### *Lifestyle Interventions*

Lifestyle interventions as the preferred obesity management include dietary and activity modifications with minimal side effects and socio-economic burden [3]. Low-calorie diets and increased daily activities are tailored to individuals with obesity [4]. High-intensity lifestyle interventions result in a mean weight loss of 5–8%, and approximately 60% of participants have more than 5% weight loss compared to their initial weight within 1 year [2]. High-intensity lifestyle interventions demonstrated similar results compared to joint intervention with pharmacotherapy and low-to-moderate-intensity lifestyle [2]. Histological analysis has shown improved mitochondrial biogenesis, decreased adipocyte size and lipid content, and low-grade inflammation in white adipose tissue (WAT) under dietary and exercise interventions [5]. In addition, lifestyle interventions revive metabolic homeostasis and reduce the incidence of T2DM by 58% in obesity, which overrides the 31% reduction with metformin treatment [6]. Nevertheless, it is a challenge to persistently adhere to lifestyle interventions without achieving the desired effect of reducing body mass.

## *Pharmacological Treatments*

Pharmacotherapy is indicated adducting with lifestyle interventions when an individual has a body mass index (BMI) $\geq$ 30 kg/m$^2$ or BMI $\geq$ 27 kg/m$^2$ + at least one comorbid condition, such as diabetes mellitus, hypertension, hyperlipidemia, or sleep apnea [4].

The Food and Drug Administration (FDA) criteria for new drug approvals should strictly follow the guidelines: placebo-adjusted weight loss >5% at 1 year or >35% of patients achieving weight loss >5% (which must be at least twice that induced by placebo), additionally showing evidence of metabolic improvements including blood pressure, lipid content, and glycemia [7]. Currently, the major FDA-approved anti-obesity medications include orlistat, lorcaserin, liraglutide (the only injectable formulation), phentermine/topiramate extended-release (ER), and naltrexone sustained-release (SR)/bupropion SR [7] (Table 8.1). Orlistat, a pancreatic and gastric lipase inhibitor, results in fat malabsorption and a net energy absorption disorder [8]. Lorcaserin, a 5-HT$_{2c}$ (5-hydroxytryptamine receptor 2C) agonist, induces satiety to dampen food intake [9]. Liraglutide is a GLP1 receptor agonist that delays gastric emptying and inhibits food intake [10]. Phentermine, the most widely prescribed anti-obesity medication, was approved by the FDA in 1959 for ages >16 years [11]. Phentermine is a sympathomimetic amine that reduces appetite and food consumption. However, the prescription of phentermine is limited for short-term use ($\leq$90 days) owing to its potential adverse effects [11]. Either phentermine/topiramate ER, a combination of sympathomimetic amine and ER antiepileptic drug, or naltrexone SR/bupropion SR, a combination opioid antagonist and aminoketone antidepressant, acts on the central nervous system (CNS) to cumulatively suppress food intake [12]. Clinical trials for three monotherapies (orlistat, lorcaserin, and liraglutide) and two combined medications (phentermine-topiramate and naltrexone-bupropion) showed approximately 5–10% weight loss compared to the initial body weight within 1 year trials [2]. However, every anti-obesity medication has contraindications and common adverse effects, which are concerns for physicians recommending pharmacotherapy, especially for the long-term use of weight-loss medications.

**Table 8.1** Current anti-obesity medications

| Drug (Trade name) | Mechanisms of action | Administration | Contraindications | Adverse effects |
|---|---|---|---|---|
| Orlistat (Xenical, Alli) | Inhibits gastric and pancreatic lipases to reduce intestinal fat digestion | 120 mg orally before meals, three times daily | Malabsorption syndrome, cholestasis, pregnancy | Flatulence, bloating, diarrhea |
| Lorcaserin (Belviq) | Activates hypothalamic neurons through 5-hydroxytryptamine receptor 2C to promote satiety | 10 mg orally twice daily or 20 mg once daily | Pregnancy | Headache, dizziness, fatigue, dry mouth, constipation, upper respiratory tract-like symptoms; hypoglycemia in patients with diabetes |
| Liraglutide (Saxenda) | GLP-1 receptor agonist. Reduces appetite | Starting dose 0.6 mg, increased weekly by 0.6 mg to a maximum of 3.0 mg subcutaneously once daily | Family or personal history of medullary thyroid carcinoma or multiple endocrine neoplasia syndrome type 2, pregnancy | Nausea, hypoglycemia, diarrhea, constipation, vomiting, headache, decreased appetite, dyspepsia, abdominal pain, fatigue, dizziness, increased lipase levels |
| Phentermine–topiramate (Qsymia) | Sympathomimetic amine, anorectic and ER antiepileptic drug. Suppresses food intake and increases energy expenditure | Starting dose 3.75 mg/23 mg for 2 weeks, recommended dose 7.5 mg/46 mg, maximum dose 15 mg/92 mg | Glaucoma, hyperthyroidism or monoamine oxidase inhibitors (MAOIs) used within 14 days, hypersensitivity to sympathomimetic amines, pregnancy | Insomnia, dry mouth, constipation, paresthesia, dizziness, dysgeusia |
| Naltrexone–bupropion (Contrave, Mysimba) | Opioid receptor antagonist, dopamine (DA) and norepinephrine (NE) reuptake inhibitor. Increases energy expenditure and suppresses food intake | One tablet (8 mg Naltrexone/90 mg of bupropion) daily for 1 week, dose increased by one tablet per day weekly, maximum 2 tablets twice a day at week 4 | Uncontrolled hypertension, seizure disorders, anorexia nervosa or bulimia, chronic opioid use, MAOIs used within 14 days, alcohol or drug withdrawal, pregnancy | Nausea, constipation, headache, vomiting, dizziness, dry mouth, diarrhea |

### *Bariatric Surgery*

Bariatric surgery is more effective than lifestyle and pharmacological interventions in weight loss for long-term survival [13, 14]. It is considered for individuals with BMI ≥ 40 kg/m$^2$ or BMI ≥ 35 kg/m$^2$ + risk factors when other weight-loss attempts have failed [4]. For the Asian population, the BMI thresholds for bariatric surgery should be adjusted to BMI ≥ 27.5 kg/m$^2$ [15]. There are three standard surgical techniques: restrictive procedures for food limitation, such as gastric bypass sleeve gastrectomy; operations limiting the absorption of food calories (such as biliopancreatic diversion); and combined operations (such as biliopancreatic diversion with duodenal or distal gastric bypass) [16]. The laparoscopic technique is regarded as the first choice for bariatric surgery because of its safety and lower risks of complications [17]. The mechanisms are ill-defined, except for restriction of food intake, malabsorption of food calories, enhanced metabolic rate, increased satiety, heightened activities of other gastrointestinal hormones, elevated circulating bile acids, and tuned gut microbiome activities [18]. Weight loss can reach approximately 25% of the initial body weight within one-year post-surgery. Although bariatric surgery improves the quality of life [19], there are still several limitations, including high costs, risks of short/long-term complications, and weight regain in approximately 5–20% of patients [2].

## Adipose Tissue as a Potential Therapeutic Target

Adipose tissue-secreted adipokines are crucial regulators of numerous physiological processes, including energy storage and release, lipid metabolism, glucose homeostasis, and thermogenesis. Recent research findings suggest potential targets within adipose tissue as therapeutic interventions for metabolic homeostasis, insulin resistance, and many other related diseases.

### *Targeting Thermogenic Adipose Tissue*

Brown adipose tissue (BAT) is characterized by high mitochondrial content and a relatively smaller capacity of fat storage compared with WAT, which has low mitochondrial content and unilocular lipid droplets [20]. BAT is responsible for non-shivering thermogenesis and energy expenditure [21]. An investigation using positron-emission tomography (PET) observed cold activation with substantial amounts of metabolic activity in BAT of healthy adults [22]. Beige adipocytes, which belong to WAT, have a thermogenic attribute like that in BAT. Although beige adipocytes contain small lipid droplets, the number of mitochondria increases after cold exposure or exercise stimulation. Activation of adipocyte thermogenesis can promote energy balance, improve insulin sensitivity, and reduce the risk of obesity-related diseases [23]. Uncoupling protein-1 (UCP1) plays a core role in the thermogenic function of brown/beige adipocytes and directly separates mitochondrial combustion energy to produce heat [23]. BAT activation and WAT browning can be induced by cold exposure and pharmacological activation using $\beta_3$AR agonists [24]. Of late, TGR5 agonists and FGF21 replacement therapy may be the nascent potential drugs targeting thermogenic activation in obese individuals [25]. Motivated by these findings, several strategies based on diet-induced BAT thermogenesis have been developed to enhance sympathetic nervous system (SNS) activity. Effective ingredients, such as capsaicin, capsinoids, tea caffeine, and catechins, in these foods induce thermogenesis that suppresses obesity [25].

## *Targeting Macrophages in Adipose Tissue*

Adipose tissue-derived macrophages (ATMs) accumulate in the extracellular matrix (ECM) and actively express proinflammatory cytokines in obese individuals [26]. ATMs maintain adipose tissue homeostasis in the steady lean state, engulf dead cells, release lysosomal enzymes, and take up free fatty acids (FFAs) [27]. Adipocyte apoptosis recruits activated anti-inflammatory M2 macrophages to the adipose tissue [28]. Chronic nutrient overload in adipocytes causes excessive oxidative stress and hypoxic conditions in adipose tissue, triggering the accumulation of ATMs and an unbalanced phenotypic switch between proinflammatory M1 and anti-inflammatory M2. M2 ATMs produce the anti-inflammatory cytokine IL-10, which preserves insulin sensitivity by directly counteracting TNF [29]. Conversely, M1 ATMs produce inflammatory cytokines, including tumor necrosis factor-alpha (TNF$\alpha$) [30]. Tipping the balance from M2 ATMs towards M1 ATMs in obesity determines the processes of several human pathologies, such as T2DM and CVD. In addition, ATMs-induced chronic low-grade inflammation can mediate tumor initiation, progression, and metastasis [31, 32]. The depletion of ATMs in obese tumor-bearing mice induces tumor necrosis and arrests tumor angiogenesis [33]. ATMs can produce a platelet-derived growth factor (PDGF) and vascular endothelial growth factor (VEGF)-family ortholog (Pvf3) to regulate energy storage in adipose tissue, and blockade of this growth factor ortholog can redirect excess lipids to thermogenesis in BAT [34]. Therefore, ATMs may be druggable targets for the clinical prevention and treatment of obesity-induced metabolic disorders.

## *Targeting Fibrosis in Adipose Tissue*

Fibrosis usually occurs in chronic inflammation during a pathological repair process. WAT undergoes expansion with hyperplasia (formation of new adipocytes) and hypertrophy (enlargement of adipocytes), with activated inflammatory conditions that initiate the fibrotic repair process during obesity [35]. Fibrosis is the cumulative production of stiff ECM, where dynamic remodeling is lost due to physical constraints on expansion [36]. Inhibited expression of Collagen VI, a marker of fibrosis in adipose tissue, releases the limitation of expansion in adipocytes for improved homeostasis in obese mice [37]. Although bariatric surgery could effectively reduce fat mass in obesity, WAT fibrosis cannot be reversed [38]. The investigation demonstrated that fibrosis is negatively associated with the success of bariatric surgery [39]. Eplerenone, a mineralocorticoid receptor antagonist, suppresses fibrosis in WAT. The combination of eplerenone and bariatric surgery could be a potential therapeutic approach to prevent fibrosis in obese individuals [40, 41].

### *Targeting Visceral and Ectopic Fat*

The distribution and accumulation of ectopic adipose tissue are determinants of metabolism in lean and obese individuals [42]. Subcutaneous adipose tissue (SAT) dysfunction, such as chronic low-level inflammation and insulin resistance, limits its expandability [43]. Released FFAs lead to ectopic adipose tissue deposition, which in turn aggravates the progression of inflammation and insulin resistance and increases systemic FFA content [43]. Hepatic, epicardial, perivascular, and visceral adipose tissues contribute to atherosclerosis and cardiometabolic disease [44–46]. Re-distribution of adipose tissue by increasing the tolerability of SAT and ectopic adipose tissue browning could be an effective approach to ameliorate metabolic diseases [47, 48].

## Novel Therapeutic Approaches

Advancements in adipose tissue research have revealed the cellular and molecular pathophysiology of obesity and the underlying mechanisms of its comorbidities. Motivated by these findings, novel therapeutic strategies, including gene therapy, cell-based therapies, and precision medicine, are emerging to address the global obesity epidemic.

### *Gene Therapy*

The approaches to gene therapies include replacing (a disease gene with a healthy gene), inactivating (a disease gene), and introducing a gene [49]. Given that the gene delivered by the adeno-associated virus (AAV) does not integrate into the patient genome with respect to low immunogenicity, stability of the viral capsid, and long-term effects in vivo, AAV is the most popular approach in clinical trials [50]. Mice with a mutation in leptin (*ob* gene) exhibit obesity and severe diabetes [51]. Subsequently, introducing the leptin gene carried by adenovirus (Ad) rescued obesity and diabetes in *ob/ob* mice [52]. AAV8, a serotype of AAVs, has been demonstrated to effectively mediate the transduction in WAT and BAT in mice, indicating that AAV8 may be an ideal tool for further studies of adipose pathophysiology and the development of therapeutic strategies [53]. For instance, the overexpression of Akt2 using AAV8 in BAT of mice achieved thermogenesis through the mTORC2-Akt2 pathway [54]. Rec2, an engineered serotype of recombinant AAV (rAAV), effectively transfects BAT by oral administration, providing a new avenue to genetically manipulate BAT function [55]. In addition, a brain-adipose circuit controls the energy expenditure in adipocytes by using AAV [56]. Adenovirus-mediated human kallistatin injection in gonadal WAT improves adipose tissue remodeling and function in high-fat diet (HFD)-fed mice [57].

## *Cell-Based Therapies*

Cell-based therapy is defined as injecting, grafting, or implanting live cells into patients to treat or cure diseases [58]. Cell-based therapy has been successfully applied to the transplantation of allogeneic human hematopoietic stem cells for bone marrow regeneration [58]. Preclinical investigations have validated that mesenchymal stromal cell (MSC)-based therapies mitigate obesity and metabolic syndromes induced by a high-fat diet in mice [59]. Dysfunctional adipogenic precursors can be reprogrammed and induced to differentiate into beige adipocytes for use in potential cell-based therapies [60]. Using CRISPR technology to overexpress UCP1 in human white adipocytes can give rise to brown-like adipocytes that prevent diet-induced obesity and ameliorate metabolic syndrome in mice. These CRISPR-engineered human brown-like adipocytes have the potential for clinical applications in obesity and the associated metabolic syndrome [61]. Ex vivo gene therapy approaches have facilitated research on cell-based therapies and achievements in preclinical studies. Implantation of carnitine palmitoyltransferase 1 (CPT1) (a mitochondrial enzyme) in adipocytes derived from adipose tissue-derived mesenchymal stem cells (AT-MSC) improves weight loss and glucose intolerance in obese mice [62]. Natural killer (NK) cells are currently considered a new cell source for cell-based therapy of obesity-associated cancer in clinical applications [63]. With the expansion of technologies and the understanding of pathological mechanisms, we look forward to novel therapeutic strategies for clinical applications.

### *Precision Medicine*

Precision medicine (PM) is a revolution relative to traditional medicine, including diverse aspects from diagnosis to treatment. Advances in genome sequencing technology, molecular classification of clinical markers for diagnosis and treatment, drug target design, clinical cohort studies, and biomedical big data analysis algorithms support the development of PM. For instance, under clinical assessment and molecular profiling of cancer, predicted cancer genomics can guide specific therapies for a patient [64]. The well-established and widely used sequencing plasma circulating cell-free DNA (cfDNA) technique in prenatal diagnosis has enormous potential for multiple cancer screening [65]. To date, a WAT-on-chip system provides human-based, autologous, and immunocompetent information to recapitulate tissue heterogeneity, and this system can become a tool for PM application in metabolic diseases [66]. In addition, a multicenter collaboration must define and characterize methodologies for diagnosing asymptomatic adults with immunometabolic risk [67].

## Gaps in Understanding and Future Research Directions

### *The Gaps in our Current Understanding of Adipose Tissue Physiology and Its Clinical Implications*

Basic research on adipose tissue regarding metabolic disorders in obese individuals is compelling since the obtained scientific knowledge can be applied to only a small proportion of obese patients [68]. First, most investigations are preclinical studies taken on rodents, in which the conclusions acquired need further translation into humans [69]. Second, in addition to the adipose tissue itself, metabolic interactions among multiple systems in the body should be dissected. For instance, it has been demonstrated that adipose tissue communicates with the central and peripheral nervous systems to maintain whole-body energy homeostasis [70]. Additionally, when nutrient overload activates multi-systemic inflammatory responses in the liver, musculoskeletal tissue, pancreas, and hypothalamus, individuals are subject to insulin resistance and glucose intolerance [30, 71]. However, clinical therapies mainly focus on addressing the persistent complications of obesity rather than directly targeting adipose tissue itself. Finally, current technologies for analyzing adipose metabolism, especially for metabolic mining in humans, are relatively limited. Muti-technology combinations and technological innovation are particularly necessary.

## Ongoing Research in Uncovering Adipose Tissue's Complexities

We have a limited understanding of the fate of adipocytes and of how adipogenesis is governed by both internal and external signals. In addition, the adipose tissue is simultaneously an independent entity, while also contributing to a cascading multi-systemic metabolic disorder. Hence, exploration of obesity should not only encompass drawing a blueprint of whole-body fat metabolism but also regarding adipose tissue as a metabolic crossroad linking back and forth to other organs.

## Neuronal Regulations

The CNS bidirectionally communicates with adipose tissue by collecting peripheral nerve information and responding to sympathetic nerve outflow [72, 73] (Fig. 8.1). More specifically, the CNS neurochemically modulates BAT sympathetic nerve activity on thermoregulation and metabolic alterations in response to cutaneous and deep body thermoreceptor signals [75]. Dorsal root ganglia (DRG) sensory neurons arrest local sympathetic function, linking the role of vagal baroreceptors in modulating blood pressure [70]. Sensory nerves also serve as essential modulators for adipose tissue

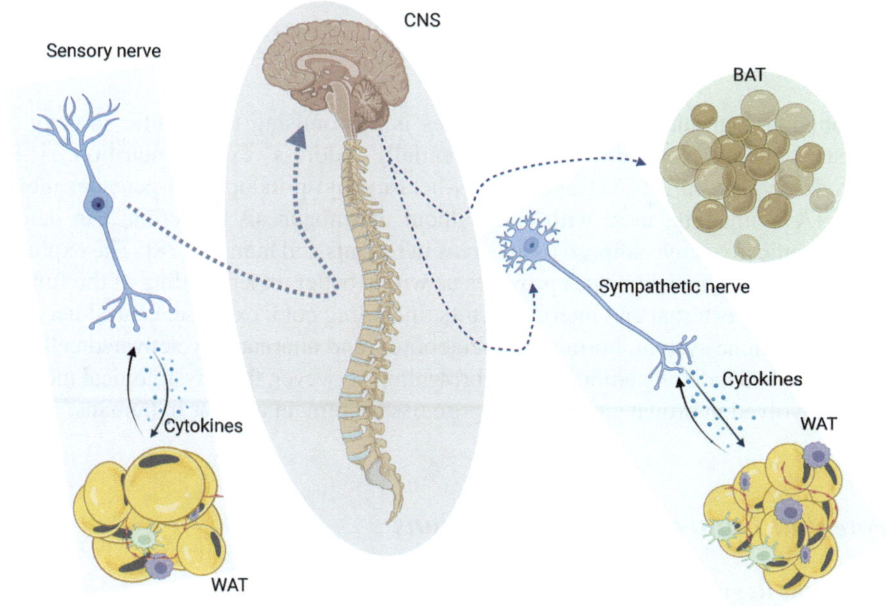


**Fig. 8.1** Adipose tissue microenvironment in regulating systemic metabolism through signaling to local nerve fibers. *BAT* brown adipose tissue, *WAT* white adipose tissue, *CNS* central nervous system. (Modified after ref. [74] and created with BioRender.com)

activities and overall metabolic homeostasis [70]. Infiltrated endothelial and immune cells in adipose tissue finely tune the processes of neurotransmitter tone and blood flow with nerve fibers that favor adipose tissue browning, adipocyte activities, and energy expenditure [74]. However, using current research and measurement techniques, it is challenging to discern sequential responses in either nerve or adipose tissue building.

## *Immunometabolism*

The joint exploration of inflammation and metabolism has led to the concept of immunometabolism, a field in which many questions remain unanswered. Activated or recruited immune cells in adipose tissues secrete a variety of cytokines that contribute to local and systemic metabolic aberrations [30]. The interconnected, often overlapping intracellular signals in inflammation pose a significant challenge to research. Both pro- and anti-inflammatory cytokines originating from the innate and adaptive immune systems contribute to unhealthy WAT conditions, including fibrosis and metabolic disorders [76, 77]. Hence, in-depth analysis of immunoregulation in obese patients must extend beyond individual cytokine interactions to encompass a broader system-level perspective [76].

## *Adipose Tissue Browning*

Targeting brown and beige adipose tissues is a promising therapeutic strategy to increase energy expenditure and potentially address excess nutrition. [$^{18}$F] fluorodeoxyglucose-PET/CT and [$^{123/125}$I]-beta-methyl-p-iodophenyl-pentadecanoic-SPECT/CT imaging, used with and without thermogenesis induction, can detect thermogenically active adipose tissue areas in rodents and humans [78]. The exploration of thermogenic adipocytes provides us with a better understanding of the fundamental role of external and internal factors, including cold, exercise, central nervous system, immune system, hormones, metabolites, and alternatively activated cells, in regulating and inducing adipose tissue browning. However, the physiological mechanisms involved in brown and beige adipose tissues remain unclear in humans.

## *Potential Future Research Directions*

### Neuronal Regulations

The brain picks up peripheral information via electrical signals or adipose tissue-secreted cytokines and subsequently returns integrated signals to adipocytes through the noradrenergic sympathetic pathway [70]. To date, substantial unmet needs still exist to achieve clinical treatment. First, advanced technologies and methodological techniques are urgently required observing the interaction among adipose tissue, sensory nervous systems, and sympathetic nervous systems, as well as to dissect synaptic junctions and electrochemical structures formed in transparent processed adipose tissue and nerves. Second, several key aspects, such as the time and type of innervation, axonal subtypes, branching patterns, and neuropeptide release responses to circumstances, remain unclear. The organ-targeted single-cell RNA-seq approach can identify endogenous signals for finding the putative receptors [79]. Long-term calcium imaging techniques can reveal projection-specific neuronal or DRG activity in vivo to confirm the signals [70].

### Immunoregulation

Transient inflammation is essential for tissue remodeling [80]. The time at which the anti-inflammatory status switches to a chronic proinflammatory status is crucial for preventing metabolic disorders in an individual. Persistent inflammation in adipose tissue, particularly visceral adipose deposits, is a determinant of the development of chronic diseases. It explains why the cumulative abdominal adipose tissue is correlated with a high risk of death [81]. The limited success of interventions focused on a single cytokine or receptor in patients with obesity highlights the need for new therapeutic approaches. Combining therapies targeting multiple immune cell types or upstream signaling pathways could potentially provide greater efficacy [30].

### Adipose Tissue Browning

Human active adipose tissue, comprising brown and beige adipose tissues, can potentially fuel energy expenditure [82]. One promising strategy is the direct injection of mitochondria into adipose tissue, leading to browning via the enhanced expression of optic atrophy 1 (OPA1)/UCP1 [83]. Pharmacological approaches can also activate the CNS to stimulate BAT and burn excess calories [84]. Moreover, genetic engineering techniques can increase the expression of β3-adrenergic receptors and transcriptional co-regulators (such as Prdm16 or FOXC2) in WAT to promote browning [85, 86]. These strategies demonstrate the potential of inducing thermogenic function in human brown/beige adipose tissue, leading to increased energy expenditure.

### Anti-fibrosis

Fibrosis is a hallmark of ECM dysregulation in dysfunctional WAT [87]. Although no specific anti-fibrosis drug exists for adipose tissue, several potential pro-fibrotic targets have been identified, such as anti-endotrophin and anti-PDGFRα [88]. To date, the most effective anti-fibrotic interventions are dietary modulation, exercise, anti-inflammation, and anti-obesity-associated T2DM [88]. Given the difficulty of explicitly targeting fibrosis, a comprehensive understanding of the mechanisms driving adipose tissue fibrosis is essential for preventing metabolic disorders.

## Personalized Medicine and Tailored Interventions

A one-drug-fits-all model yields unsatisfactory results in treating obesity due to patient heterogeneity. Although common symptoms exist, obese individuals may also experience variations in their symptoms. Additionally, conditions, such as Cushing syndrome, can contribute to obesity and its complications [89]. This finding emphasizes that patients presenting with similar clinical symptoms may have vastly different underlying causes. Hence, PM can provide customization of healthcare under precise analysis that guides the prescription of treatment, with medical decisions, treatments, practices, or products being tailored to the patient's needs [90].

Clinical decision-making is driven by a checklist and personal experience. The diagnostic "measure" is important. PM can encompass genetic, genomic, geographical, environmental, radiological imaging, and clinical symptom information to accurately predict a person's treatment [91, 92].

A patient's molecular and cellular analysis or genetic content can guide the choice of treatment. For example, $^{99m}$Tc-methoxyisobutylisonitrile, used clinically in cardiac perfusion and parathyroid imaging, can localize to the supraclavicular adipose tissue because of its relatively high blood flow and mitochondrial density [93]. [$^{123}$I]-Metaiodobenzylguanidine, a radiolabeled norepinephrine analog and a marker of sympathetic innervation, can simultaneously visualize metabolic activity and innervation in BAT [94]. $^{18F}$-fluoro-thiaheptadecanoic acid, a fatty acid tracer, has been used to quantify BAT oxidative metabolism and glucose and non-esterified fatty acid turnover [95]. Hence, the combined techniques of mitochondrial implantation and contrast-enhanced imaging can mediate adipose tissue browning. The systemic identification of active adipose deposits using this functional analysis clearly guides medical and surgical interventions.

Severe and persistent obesity often has underlying genetic components. Genome-wide association studies (GWAS) have identified numerous loci linked to BMI, including genes influencing waist and hip circumferences [96]. One notable example is the kinase suppressor of Ras 2 (KSR2), which is implicated in the early onset of severe obesity accompanied by food intake disorders and metabolic dysregulation [97]. Additionally, mutations in melanocortin 4 receptor (MC4R) and gastric inhibitory polypeptide receptor (GIPR) have been associated with extreme obesity. Notably, these variants exhibited a considerably stronger effect on BMI than the common variants [98]. MC4R deficiency, a monogenic syndrome, can now be precisely treated with a medication-granted Priority Medicines (PRIME) designation by the European Medicines Agency (EMA) [69]. Furthermore, studies have indicated that enhancer activity (H3K4me1and H3K27ac) in adipose tissue may regulate gene expression related to metabolism and insulin resistance [96]. Combining genetic screening with BMI analysis revealed the enrichment of neuronal genes in adipose tissue that influenced energy expenditure. This offers a promising avenue for the future targeting of certain central obesity and associated metabolic disorders via genetic re-editing.

Compared to molecular or genetic targets, the environmental drivers of obesity pose a significant challenge for PM. However, clustering techniques can help tackle this complexity by analyzing geographical, environmental, and clinical data algorithms. Integrating these clusters with healthcare AI can pinpoint the "right patient" for targeted interventions. Clustering provides a starting point for treatment, but the ultimate goal is to develop effective real-word applications for these treatments. However, the gap between preclinical findings and patient outcomes remains a significant obstacle. First, a deeper understanding of adipose tissue physiology and its clinical implications is needed. Then, scientists and clinicians must collaborate with engineers to translate preclinical findings into well-defined codes for use in AI systems. This integration will enable the dissemination of knowledge and facilitate PM strategies.

## Conclusions

As part of the global challenge of blunting the chronic disease epidemic, multifaceted efforts must be made to prevent and treat obesity. The promotion of healthy living in modern environments is a crucial goal. Pharmacotherapy combined with accessible lifestyle programs remains the mainstay of treatment for those already facing obesity and its associated chronic conditions. The current treatments for obesity-related conditions entail a growing global, multidisciplinary network of certified medical professionals and the exploration of innovative therapeutic strategies. We have emphasized the ongoing evolution of adipose tissue research and its potential to transform healthcare. We hope our efforts can marginally hasten a healthier state of living for everyone.

**Acknowledgement** This work was supported by the Areas of Excellence Scheme (AoE/M-402/20) provided by the Research Grant Council of Hong Kong and Health@InnoHK Program launched by Innovation Technology Commission of the Hong Kong SAR, China.

**Competing InterestThe authors declare that they have no competing interests.**

## References

1. Heymsfield SB, Wadden TA. Mechanisms, pathophysiology, and management of obesity. N Engl J Med. 2017b;376(3):254–66. https://doi.org/10.1056/NEJMra1514009.
2. Heymsfield SB, Wadden TA. Mechanisms, pathophysiology, and management of obesity. N Engl J Med. 2017a;376(15):1492. https://doi.org/10.1056/NEJMc1701944.

3. American College of Cardiology/American Heart Association Task Force on Practice Guidelines, Obesity Expert Panel. Expert panel report: guidelines (2013) for the management of overweight and obesity in adults. Obesity (Silver Spring, Md). 2014;22:S41–S410.
4. Tsigos C, Hainer V, Basdevant A, Finer N, Fried M, Mathus-Vliegen E, Micic D, Maislos M, Roman G, Schutz Y, Toplak H, Zahorska-Markiewicz B, Obesity Management Task Force of the European Association for the Study of Obesity. Management of obesity in adults: European clinical practice guidelines. Obes Facts. 2008;1(2):106–16. https://doi.org/10.1159/000126822.
5. Linden MA, Pincu Y, Martin SA, Woods JA, Baynard T. Moderate exercise training provides modest protection against adipose tissue inflammatory gene expression in response to high-fat feeding. Physiol Rep. 2014;2(7):e12071.
6. Knowler WC, Barrett-Connor E, Fowler SE, Hamman RF, Lachin JM, Walker EA, Nathan DM, Diabetes Prevention Program Research Group. Reduction in the incidence of type 2 diabetes with lifestyle intervention or metformin. N Engl J Med. 2002;346(6):393–403.
7. Srivastava G, Apovian CM. Current pharmacotherapy for obesity. Nat Rev Endocrinol. 2018;14(1):12–24.
8. Hvizdos KM, Markham A. Orlistat: a review of its use in the management of obesity. Drugs. 1999;58(4):743–60. https://doi.org/10.2165/00003495-199958040-00015.
9. Lam DD, Garfield AS, Marston OJ, Shaw J, Heisler LK. Brain serotonin system in the coordination of food intake and body weight. Pharmacol Biochem Behav. 2010;97(1):84–91. https://doi.org/10.1016/j.pbb.2010.09.003.
10. van Can J, Sloth B, Jensen CB, Flint A, Blaak EE, Saris WH. Effects of the once-daily GLP-1 analog liraglutide on gastric emptying, glycemic parameters, appetite and energy metabolism in obese, non-diabetic adults. Int J Obes. 2014;38(6):784–93. https://doi.org/10.1038/ijo.2013.162.
11. Apovian CM, Aronne LJ, Bessesen DH, McDonnell ME, Murad MH, Pagotto U, et al. Pharmacological management of obesity: an Endocrine Society clinical practice guideline. J Clin Endocrinol Metabol. 2015;100(2):342–62.
12. Kelly MJ, Loose MD, Ronnekleiv OK. Opioids hyperpolarize beta-endorphin neurons via mu-receptor activation of a potassium conductance. Neuroendocrinology. 1990;52(3):268–75. https://doi.org/10.1159/000125597.
13. Sjöström L, Lindroos A-K, Peltonen M, Torgerson J, Bouchard C, Carlsson B, et al. Lifestyle, diabetes, and cardiovascular risk factors 10 years after bariatric surgery. N Engl J Med. 2004;351(26):2683–93.
14. Syn NL, Cummings DE, Wang LZ, Lin DJ, Zhao JJ, Loh M, et al. Association of metabolic–bariatric surgery with long-term survival in adults with and without diabetes: a one-stage meta-analysis of matched cohort and prospective controlled studies with 174 772 participants. Lancet. 2021;397(10287):1830–41.
15. Eisenberg D, Shikora SA, Aarts E, Aminian A, Angrisani L, Cohen RV, de Luca M, Faria SL, Goodpaster KPS, Haddad A, Himpens JM, Kow L, Kurian M, Loi K, Mahawar K, Nimeri A, O'Kane M, Papasavas PK, Ponce J, Pratt JSA, Rogers AM, Steele KE, Suter M, Kothari SN. 2022 American Society of Metabolic and Bariatric Surgery (ASMBS) and International Federation for the Surgery of Obesity and Metabolic Disorders (IFSO) indications for metabolic and bariatric surgery. Obes Surg. 2023;33(1):3–14. https://doi.org/10.1007/s11695-022-06332-1.
16. Rogers AM. Current state of bariatric surgery: procedures, data, and patient management. Tech Vasc Interv Radiol. 2020;23(1):100654.
17. Reoch J, Mottillo S, Shimony A, Filion KB, Christou NV, Joseph L, et al. Safety of laparoscopic vs open bariatric surgery: a systematic review and meta-analysis. Arch Surg. 2011;146(11):1314–22.
18. Albaugh VL, Abumrad NN. Surgical treatment of obesity. F1000Res. 2018;7:F1000 Faculty Rev-617. https://doi.org/10.12688/f1000research.13515.1.
19. Dumon KR, Murayama KM. Bariatric surgery outcomes. Surg Clin North AM. 2011;91(6):1313–38.

20. Shinde AB, Song A, Wang QA. Brown adipose tissue heterogeneity, energy metabolism, and beyond. Front Endocrinol (Lausanne). 2021;12:651763. https://doi.org/10.3389/fendo.2021.651763.
21. Chen Y, Hu Q, Wang C, Wang T. The crosstalk between BAT thermogenesis and skeletal muscle dysfunction. Front Physiol. 2023;14:1132830. https://doi.org/10.3389/fphys.2023.1132830.
22. Virtanen KA, Lidell ME, Orava J, Heglind M, Westergren R, Niemi T, et al. Functional brown adipose tissue in healthy adults. N Engl J Med. 2009;360(15):1518–25.
23. Porter C. Quantification of UCP1 function in human brown adipose tissue. Adipocytes. 2017;6(2):167–74.
24. Blondin DP, Nielsen S, Kuipers EN, Severinsen MC, Jensen VH, Miard S, et al. Human brown adipocyte thermogenesis is driven by β2-AR stimulation. Cell Metab. 2020;32(2):287–300. e287
25. Saito M, Matsushita M, Yoneshiro T, Okamatsu-Ogura Y. Brown adipose tissue, diet-induced thermogenesis, and thermogenic food ingredients: from mice to men. Front Endocrinol. 2020;11:222.
26. Weisberg SP, McCann D, Desai M, Rosenbaum M, Leibel RL, Ferrante AW. Obesity is associated with macrophage accumulation in adipose tissue. J Clin Invest. 2003;112(12):1796–808.
27. Catrysse L, van Loo G. Adipose tissue macrophages and their polarization in health and obesity. Cell Immunol. 2018;330:114–9.
28. Fischer-Posovszky P, Wang QA, Asterholm IW, Rutkowski JM, Scherer PE. Targeted deletion of adipocytes by apoptosis leads to adipose tissue recruitment of alternatively activated M2 macrophages. Endocrinology. 2011;152(8):3074–81.
29. Lumeng CN, Bodzin JL, Saltiel AR. Obesity induces a phenotypic switch in adipose tissue macrophage polarization. J Clin Invest. 2007;117(1):175–84.
30. Osborn O, Olefsky JM. The cellular and signaling networks linking the immune system and metabolism in disease. Nat Med. 2012;18(3):363–74.
31. Li B, Sun S, Li J-J, Yuan J-P, Sun S-R, Wu Q. Adipose tissue macrophages: implications for obesity-associated cancer. Mil Med Res. 2023;10(1):1–15.
32. Quail DF, Dannenberg AJ. The obese adipose tissue microenvironment in cancer development and progression. Nat Rev Endocrinol. 2019;15(3):139–54.
33. Hillers-Ziemer LE, McMahon RQ, Hietpas M, Paderta G, LeBeau J, McCready J, Arendt LM. Obesity promotes cooperation of cancer stem-like cells and macrophages to enhance mammary tumor angiogenesis. Cancers. 2020;12(2):502.
34. Cox N, Crozet L, Holtman IR, Loyher P-L, Lazarov T, White JB, et al. Diet-regulated production of PDGFcc by macrophages controls energy storage. Science. 2021;373(6550):eabe9383.
35. Verboven K, Wouters K, Gaens K, Hansen D, Bijnen M, Wetzels S, et al. Abdominal subcutaneous and visceral adipocyte size, lipolysis and inflammation relate to insulin resistance in male obese humans. Sci Rep. 2018;8(1):4677.
36. Khan T, Muise ES, Iyengar P, Wang ZV, Chandalia M, Abate N, et al. Metabolic dysregulation and adipose tissue fibrosis: role of collagen VI. Mol Cell Biol. 2009;29(6):1575–91.
37. Williams L, Layton T, Yang N, Feldmann M, Nanchahal J. Collagen VI as a driver and disease biomarker in human fibrosis. FEBS J. 2022;289(13):3603–29.
38. Liu Y, Aron-Wisnewsky J, Marcelin G, Genser L, Le Naour G, Torcivia A, et al. Accumulation and changes in composition of collagens in subcutaneous adipose tissue after bariatric surgery. J Clin Endocrinol. 2016;101(1):293–304.
39. Bel Lassen P, Charlotte F, Liu Y, Bedossa P, Le Naour G, Tordjman J, et al. The FAT score, a fibrosis score of adipose tissue: predicting weight-loss outcome after gastric bypass. J Clin Endocrinol Metabol. 2017;102(7):2443–53.
40. Johansen ML, Ibarrola J, Fernández-Celis A, Schou M, Sonne MP, Refsgaard Holm M, et al. The mineralocorticoid receptor antagonist eplerenone suppresses interstitial fibrosis in subcutaneous adipose tissue in patients with type 2 diabetes. Diabetes. 2021;70(1):196–203.
41. Marcelin G, Gautier EL, Clément K. Adipose tissue fibrosis in obesity: etiology and challenges. Annu Rev Physiol. 2022;84:135–55.

42. Després J-P. Body fat distribution and risk of cardiovascular disease: an update. Circulation. 2012;126(10):1301–13.
43. Britton KA, Fox CS. Ectopic fat depots and cardiovascular disease. Circulation. 2011;124(24):e837–41.
44. Kishida K, Funahashi T, Matsuzawa Y, Shimomura I. Visceral adiposity as a target for the management of the metabolic syndrome. Ann Med. 2012;44(3):233–41.
45. Neeland IJ, Ross R, Després J-P, Matsuzawa Y, Yamashita S, Shai I, et al. Visceral and ectopic fat, atherosclerosis, and cardiometabolic disease: a position statement. Lancet Diabetes Endocrinol. 2019;7(9):715–25.
46. Rafeh R, Viveiros A, Oudit GY, El-Yazbi AF. Targeting perivascular and epicardial adipose tissue inflammation: therapeutic opportunities for cardiovascular disease. Clin Sci. 2020;134(7):827–51.
47. Ghaben AL, Scherer PE. Adipogenesis and metabolic health. Nat Rev Mol Cell Biol. 2019;20(4):242–58.
48. Karastergiou K, Fried SK. Multiple adipose depots increase cardiovascular risk via local and systemic effects. Curr Atheroscler Rep. 2013;15:1–11.
49. Ginn SL, Amaya AK, Alexander IE, Edelstein M, Abedi MR. Gene therapy clinical trials worldwide to 2017: an update. J Gene Med. 2018;20(5):e3015.
50. Mendell JR, Al-Zaidy SA, Rodino-Klapac LR, Goodspeed K, Gray SJ, Kay CN, et al. Current clinical applications of in vivo gene therapy with AAVs. Mol Ther. 2021;29(2):464–88.
51. Zhang Y, Proenca R, Maffei M, Barone M, Leopold L, Friedman JM. Positional cloning of the mouse obese gene and its human homologue. Nature. 1994;372(6505):425–32.
52. Muzzin P, Eisensmith RC, Copeland KC, Woo SL. Correction of obesity and diabetes in genetically obese mice by leptin gene therapy. Proc Natl Acad Sci. 1996;93(25):14804–8.
53. Jimenez V, Muñoz S, Casana E, Mallol C, Elias I, Jambrina C, et al. In vivo adeno-associated viral vector–mediated genetic engineering of white and brown adipose tissue in adult mice. Diabetes. 2013;62(12):4012–22.
54. Albert V, Svensson K, Shimobayashi M, Colombi M, Muñoz S, Jimenez V, et al. mTORC 2 sustains thermogenesis via Akt-induced glucose uptake and glycolysis in brown adipose tissue. EMBO Mol Med. 2016;8(3):232–46.
55. Huang W, McMurphy T, Liu X, Wang C, Cao L. Genetic manipulation of brown fat via oral administration of an engineered recombinant adeno-associated viral serotype vector. Mol Ther. 2016;24(6):1062–9.
56. Cardoso F, Klein Wolterink RG, Godinho-Silva C, Domingues RG, Ribeiro H, Da Silva JA, et al. Neuro-mesenchymal units control ILC2 and obesity via a brain–adipose circuit. Nature. 2021;597(7876):410–4.
57. Wu Y, Jiang Y, Guo JQ, Yang ZW, Carvalho A, Qian LL, Ji JJ, Ji ZJ, Ma GS, Yao YY. Visceral adipose tissue-directed human kallistatin gene therapy improves adipose tissue remodeling and metabolic health in obese mice. Cell Signal. 2023;106:110637.
58. Heathman TR, Nienow AW, McCall MJ, Coopman K, Kara B, Hewitt CJ. The translation of cell-based therapies: clinical landscape and manufacturing challenges. Regen Med. 2015;10(1):49–64.
59. Lee C-W, Hsiao W-T, Lee OK-S. Mesenchymal stromal cell-based therapies reduce obesity and metabolic syndromes induced by a high-fat diet. Transl Res. 2017;182(61–74):e68.
60. Su S, Guntur AR, Nguyen DC, Fakory SS, Doucette CC, Leech C, et al. A renewable source of human beige adipocytes for development of therapies to treat metabolic syndrome. Cell Rep. 2018;25(11):3215–28. e3219
61. Wang C-H, Lundh M, Fu A, Kriszt R, Huang TL, Lynes MD, et al. CRISPR-engineered human brown-like adipocytes prevent diet-induced obesity and ameliorate metabolic syndrome in mice. Sci Transl Med. 2020;12(558):eaaz8664.
62. Soler-Vázquez MC, del Mar Romero M, Todorcevic M, Delgado K, Calatayud C, Benitez-Amaro A, et al. Implantation of CPT1AM-expressing adipocytes reduces obesity and glucose intolerance in mice. Metab Eng. 2023;77:256–72.

63. Mylod E, Lysaght J, Conroy MJ. Natural killer cell therapy: a new frontier for obesity-associated cancer. Cancer Lett. 2022;535:215620.
64. Pincu Y, Yoel U, Haim Y, Makarenkov N, Maixner N, Shaco-Levy R, et al. Assessing obesity-related adipose tissue disease (OrAD) to improve precision medicine for patients living with obesity. Front Endocrinol. 2022;13:860799.
65. Liu MC, Oxnard G, Klein E, Swanton C, Seiden M, Liu MC, et al. Sensitive and specific multi-cancer detection and localization using methylation signatures in cell-free DNA. Ann Oncol. 2020;31(6):745–59.
66. Rogal J, Roosz J, Teufel C, Cipriano M, Xu R, Eisler W, et al. Autologous human immunocompetent white adipose tissue-on-chip. Adv Sci (Weinh). 2022;9(18):e2104451.
67. Rodriguez-Ayala E, Gallegos-Cabrales EC, Gonzalez-Lopez L, Laviada-Molina HA, Salinas-Osornio RA, Nava-Gonzalez EJ, et al. Towards precision medicine: defining and characterizing adipose tissue dysfunction to identify early immunometabolic risk in symptom-free adults from the GEMM family study. Adipocytes. 2020;9(1):153–69.
68. Lee SJ, Shin SW. Mechanisms, pathophysiology, and management of obesity. N Engl J Med. 2017;376(15):1491–2. https://doi.org/10.1056/NEJMc1701944.
69. Angelidi AM, Belanger MJ, Kokkinos A, Koliaki CC, Mantzoros CS. Novel noninvasive approaches to the treatment of obesity: from pharmacotherapy to gene therapy. Endocr Rev. 2022;43(3):507–57. https://doi.org/10.1210/endrev/bnab034.
70. Wang Y, Leung VH, Zhang Y, Nudell VS, Loud M, Servin-Vences MR, et al. The role of somatosensory innervation of adipose tissues. Nature. 2022;609(7927):569–74. https://doi.org/10.1038/s41586-022-05137-7.
71. Dai B, Xu J, Li X, Huang L, Hopkins C, Wang H, et al. Macrophages in epididymal adipose tissue secrete osteopontin to regulate bone homeostasis. Nat Commun. 2022;13(1):427. https://doi.org/10.1038/s41467-021-27683-w.
72. Mishra G, Townsend KL. The metabolic and functional roles of sensory nerves in adipose tissues. Nat Metab. 2023;5(9):1461–74. https://doi.org/10.1038/s42255-023-00868-x.
73. Puente-Ruiz SC, Jais A. Reciprocal signaling between adipose tissue depots and the central nervous system. Front Cell Dev Biol. 2022;10:979251. https://doi.org/10.3389/fcell.2022.979251.
74. Guilherme A, Henriques F, Bedard AH, Czech MP. Molecular pathways linking adipose innervation to insulin action in obesity and diabetes mellitus. Nat Rev Endocrinol. 2019;15(4):207–25. https://doi.org/10.1038/s41574-019-0165-y.
75. Morrison SF, Madden CJ. Central nervous system regulation of brown adipose tissue. Compr Physiol. 2014;4(4):1677–713. https://doi.org/10.1002/cphy.c140013.
76. Mathis D. Immunological goings-on in visceral adipose tissue. Cell Metab. 2013;17(6):851–9. https://doi.org/10.1016/j.cmet.2013.05.008.
77. Sun K, Tordjman J, Clément K, Scherer PE. Fibrosis and adipose tissue dysfunction. Cell Metab. 2013;18(4):470–7. https://doi.org/10.1016/j.cmet.2013.06.016.
78. Zhang F, Hao G, Shao M, Nham K, An Y, Wang Q, et al. An adipose tissue atlas: an image-guided identification of human-like BAT and beige depots in rodents. Cell Metab. 2018;27(1):252–262.e253. https://doi.org/10.1016/j.cmet.2017.12.004.
79. Gawel DR, Serra-Musach J, Lilja S, Aagesen J, Arenas A, Asking B, et al. A validated single-cell-based strategy to identify diagnostic and therapeutic targets in complex diseases. Genome Med. 2019;11(1):47. https://doi.org/10.1186/s13073-019-0657-3.
80. Karin M, Clevers H. Reparative inflammation takes charge of tissue regeneration. Nature. 2016;529(7586):307–15. https://doi.org/10.1038/nature17039.
81. Pischon T, Boeing H, Hoffmann K, Bergmann M, Schulze MB, Overvad K, et al. General and abdominal adiposity and risk of death in Europe. N Engl J Med. 2008;359(20):2105–20. https://doi.org/10.1056/NEJMoa0801891.
82. Pan R, Chen Y. Latest advancements on combating obesity by targeting human brown/beige adipose tissues. Front Endocrinol (Lausanne). 2022;13:884944. https://doi.org/10.3389/fendo.2022.884944.

83. Zeng J, Liu J, Ni H, Zhang L, Wang J, Li Y, et al. Mitochondrial transplantation reduces lower limb ischemia-reperfusion injury by increasing skeletal muscle energy and adipocyte browning. Mol Ther Methods Clin Dev. 2023;31:101152. https://doi.org/10.1016/j.omtm.2023.101152.
84. Sell H, Berger JP, Samson P, Castriota G, Lalonde J, Deshaies Y, Richard D. Peroxisome proliferator-activated receptor gamma agonism increases the capacity for sympathetically mediated thermogenesis in lean and ob/ob mice. Endocrinology. 2004;145(8):3925–34. https://doi.org/10.1210/en.2004-0321.
85. Cederberg A, Grønning LM, Ahrén B, Taskén K, Carlsson P, Enerbäck S. FOXC2 is a winged helix gene that counteracts obesity, hypertriglyceridemia, and diet-induced insulin resistance. Cell. 2001;106(5):563–73. https://doi.org/10.1016/s0092-8674(01)00474-3.
86. Lee YH, Petkova AP, Mottillo EP, Granneman JG. In vivo identification of bipotential adipocyte progenitors recruited by β3-adrenoceptor activation and high-fat feeding. Cell Metab. 2012;15(4):480–91. https://doi.org/10.1016/j.cmet.2012.03.009.
87. Ricard-Blum S. The collagen family. Cold Spring Harb Perspect Biol. 2011;3(1):a004978. https://doi.org/10.1101/cshperspect.a004978.
88. Kusminski CM, Bickel PE, Scherer PE. Targeting adipose tissue in the treatment of obesity-associated diabetes. Nat Rev Drug Discov. 2016;15(9):639–60. https://doi.org/10.1038/nrd.2016.75.
89. Pivonello R, Isidori AM, De Martino MC, Newell-Price J, Biller BM, Colao A. Complications of Cushing's syndrome: state of the art. Lancet Diabetes Endocrinol. 2016;4(7):611–29. https://doi.org/10.1016/s2213-8587(16)00086-3.
90. Deswal S, Bulusu KC, Agapow P-M, Khan FM. Chapter 8—Precision medicine. In: Ashenden SK, editor. The era of artificial intelligence, machine learning, and data science in the pharmaceutical industry. Cambridge, MA: Academic Press; 2021. p. 139–57.
91. Belgrave D, Henderson J, Simpson A, Buchan I, Bishop C, Custovic A. Disaggregating asthma: big investigation versus big data. J Allergy Clin Immunol. 2017;139(2):400–7. https://doi.org/10.1016/j.jaci.2016.11.003.
92. Pembrey L, Barreto ML, Douwes J, Cooper P, Henderson J, Mpairwe H, et al. Understanding asthma phenotypes: the World Asthma Phenotypes (WASP) international collaboration. ERJ Open Res. 2018;4(3):00013-02018. https://doi.org/10.1183/23120541.00013-2018.
93. Goetze S, Lavely WC, Ziessman HA, Wahl RL. Visualization of brown adipose tissue with 99mTc-methoxyisobutylisonitrile on SPECT/CT. J Nucl Med. 2008;49(5):752–6. https://doi.org/10.2967/jnumed.107.048074.
94. Admiraal WM, Holleman F, Bahler L, Soeters MR, Hoekstra JB, Verberne HJ. Combining 123I-metaiodobenzylguanidine SPECT/CT and 18F-FDG PET/CT for the assessment of brown adipose tissue activity in humans during cold exposure. J Nucl Med. 2013;54(2):208–12. https://doi.org/10.2967/jnumed.112.111849.
95. Ouellet V, Labbé SM, Blondin DP, Phoenix S, Guérin B, Haman F, et al. Brown adipose tissue oxidative metabolism contributes to energy expenditure during acute cold exposure in humans. J Clin Invest. 2012;122(2):545–52. https://doi.org/10.1172/jci60433.
96. Shungin D, Winkler TW, Croteau-Chonka DC, Ferreira T, Locke AE, Mägi R, et al. New genetic loci link adipose and insulin biology to body fat distribution. Nature. 2015;518(7538):187–96. https://doi.org/10.1038/nature14132.
97. Pearce LR, Atanassova N, Banton MC, Bottomley B, van der Klaauw AA, Revelli JP, et al. KSR2 mutations are associated with obesity, insulin resistance, and impaired cellular fuel oxidation. Cell. 2013;155(4):765–77. https://doi.org/10.1016/j.cell.2013.09.058.
98. Turcot V, Lu Y, Highland HM, Schurmann C, Justice AE, Fine RS, et al. Protein-altering variants associated with body mass index implicate pathways that control energy intake and expenditure in obesity. Nat Genet. 2018;50(1):26–41. https://doi.org/10.1038/s41588-017-0011-x.

# Index





D. Avtanski (ed.), *Adipose Tissue*, Contemporary Endocrinology,
https://doi.org/10.1007/978-3-031-72570-8

GPSR Compliance

*The European Union's (EU) General Product Safety Regulation (GPSR) is a set of rules that requires consumer products to be safe and our obligations to ensure this.*

*If you have any concerns about our products, you can contact us on ProductSafety@springernature.com*

In case Publisher is established outside the EU, the EU authorized representative is:

Springer Nature Customer Service Center GmbH
Europaplatz 3
69115 Heidelberg, Germany

**Batch number: 10739801**

Printed by Printforce, the Netherlands